Lance

1st with MMDS-ITFS-MDS Oper

Don't take a chance, choose the only design already proven in field tests to offer full range MMDS/ITFS performance

Models to suit every requirement.

Microwave 3 Ft., 4 Ft. or 6 Ft. Dish Parabolics – Microwave Section Parabolics - Corner Reflector "The Angle" for MDS 2150-2162 MHz or NEW MMDS/ITFS 2500-2690 MHz

Plus – A quality line of masts, jiffy and roof mounts, related hardware, video control centers, switches, hookup cables and other TV & Video Accessories.

Los Angeles and Chicago facilities for fast delivery.
Call Joe Barris, Vice-President, 818-367-1811

Modern Design, **M**ore Features
- MDS to ITFS/MMDS Multi Channel
- Horizontal or Vertical Polarization
- Dual or Multi Channel Polarization – 45° mounting
- Multi Channel Ready
- Lowest Wind Loading
- Electronically Welded
- Pressure Tested Dipole

Designed for Performance
- Models Offer 12 dbi to 32 dbi
- Metro to Far Fringe Reception
- All models include RG8 Cable with N Connector
- Compatible with all down converters

Selection – More Models to Choose From
- 3' and 4' Grid Type – Dish Parabolics –
  Models 36, 48 –
  Specify 2150 MHz or ITFS/MMDS Frequencies
- Closed Loop Grid Type – Section Parabolics –
  Models 18, 21, 24 –
  Specify 2150 MHz or ITFS/MMDS Frequencies
- Corner Reflector – "The Angle"
  Models 12 –
  Specify 2150 MHz or ITFS/MMDS Frequencies

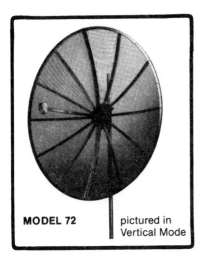

MODEL 72 — pictured in Vertical Mode

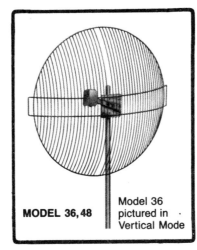

MODEL 36, 48 — Model 36 pictured in Vertical Mode

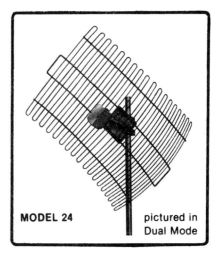

MODEL 24 — pictured in Dual Mode

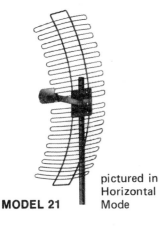

MODEL 21 — pictured in Horizontal Mode

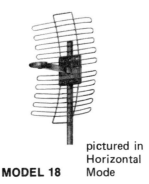

MODEL 18 — pictured in Horizontal Mode

"THE ANGLE" MODEL 12 — pictured in Vertical Mode

Des. Pats. 269009, 268343   Lic. under U.S. Pat. 4295143   Other Pats. Pending

Lance Industries   13001 Bradley Avenue, Sylmar, California 91342 (818) 367-1811 FAX (818) 362-3594
4440 North Clark Street, Chicago, Illinois 60640 (312) 275-7000 FAX (312) 275-1259

How to get and keep your business on autopilot.

It really is easier than you think.

WSN Subscriber Marketing Services works to not only get your business off the ground, but to keep it on course and flying high. Producing subscriber driven support, custom-fit to your needs through aggressive, market-smart materials and management saves you time, cost and tail-spins. Providing easy access to all available programming, all day, everyday keeps your on-time performance consistent. With WSN, subscriber revenues take flight.

WORLD SATELLITE NETWORK
WSN
The Subscriber Network.

1·800·367·3193

We give great service, blah.

It's subscribers you want, not lipservice.

"Service" means little, unless it's backed up with commitment and action. Having easy access to all available programming, all day, everyday is a given. Programming doesn't sell itself. WSN Subscriber Marketing Services provides subscriber driven support, custom-fit to your needs through aggressive, market smart materials and management. Bottom line: increased subscribers — improved revenue. Immediately.

WORLD SATELLITE NETWORK
WSN
The Subscriber Network.

1·800·367·3193

WIRELESS CABLE and SMATV

Frank Baylin and Steve Berkoff

Published by Baylin Publications

(This book replaces an earlier work entitled *Satellite, Off-Air and SMATV*)

DEDICATION

We dedicate this book to the positive effects that wireless cable and SMATV systems potentially can have on worldwide communication and education.

ACKNOWLEDGMENTS

Many people provided valuable help in producing this book. Barbara Payne of the National Satellite Programming Network authored the bulk of the chapter on programming and the satellite programmers. Robert Corazinni, Attorney at Law, contributed the entire chapter "A Legal Perspective on Wireless Cable." Steve Johnson of American Wireless System provided a thorough review of all the chapters in Section II of this book. Jim MacNaughton, Attorney at Law, contributed a valuable section on SMATV regulation in the United States. Jim Powell of WTI contributed extensively to the section on wireless installation. Tim Meintz contributed to an earlier version of the book entitled *Satellite, Off-Air and SMATV*, a precursor to this text. Our thanks to them all.

For speaking engagements or technical consulting services contact:

Frank Baylin
1905 Mariposa
Boulder, CO 80302
Tel: (303) 449-4551
FAX: (303) 939-8720

Steve Berkoff
Wireless Technologies
7607 Eastmark Drive, Suite 250
College Station, TX 77840
Telephone: 409-696-8840
FAX: 409-696-2143

First Edition – August 1992

This book is a completely revised version of *Satellite, Off-Air and SMATV– The Private Cable, Multi-Unit Handbook*, first published in March, 1987. It supplements and replaces this original work.

ISBN: 0-917893-17-4

Copyright 1992 by Baylin Publications

All rights reserved. Reproduction or publication of the content in any manner, without express written permission of the publisher, is prohibited. No liability is assumed with respect to the use of this information herein.

SHORT TABLE OF CONTENTS

SECTION I. SMATV

Chapter 1.	Overview and History of Private Cable TV	3
Chapter 2.	Economics, Sales Contracts and Regulations	11
Chapter 3.	Programming	21
Chapter 4.	The Site Survey and Planning	27
Chapter 5.	Headend System Design	53
Chapter 6.	Distribution System Design	117
Chapter 7.	Project Bidding	153
Chapter 8.	Construction and Installation	157
Chapter 9.	System Operations	193
Chapter 10.	Troubleshooting and Test Instruments	201

SECTION II. WIRELESS CABLE

Chapter 11.	Wireless Cable System Overview	223
Chapter 12.	Wireless Cable Components	237
Chapter 13.	Wireless Signal Coverage	269
Chapter 14.	Receive-Site Installation	279
Chapter 15.	System Planning and Operation	297
Chapter 16.	A Legal Perspective on Wireless Cable in the United States	317
Chapter 17.	The Wireless Cable Deal in the U.S.	329

SECTION III. PRIVATE CABLE SECURITY

Chapter 18.	Private Cable Security	337

APPENDICES

Appendix A.	Additional Examples of SMATV Headends	347
Appendix B.	Equations and Technical Details	353
Appendix C.	Programmers in North America	359
Appendix D.	Satellite TV Guides and Publications	361
Appendix E.	Off-Air and Cable TV Channels	363
Appendix F.	Glossary	365
Appendix G.	Manufacturers	375

TABLE of CONTENTS

SECTION I. BASICS and SMATV SYSTEMS

CHAPTER 1. OVERVIEW and HISTORY of PRIVATE CABLE TV 3
   A. BRIEF HISTORY of SATELLITE TELEVISION 3
   B. THE DEVELOPMENT of COMMUNITY ANTENNA TV SYSTEMS 5
   C. SMATV versus FRANCHISED CABLE TV SYSTEMS 6
   D. THE DEVELOPMENT of SMATV and WIRELESS CABLE TV 7
      Wireless Cable 8
   E. THE ADVANTAGES of SMATV 8
      The Market for SMATV and Wireless Cable 9

CHAPTER 2. ECONOMICS, SALES CONTRACTS and REGULATIONS 11
   A. ECONOMIC OPTIONS for INSTALLERS, OPERATORS and OWNERS 11
   B. SALES CONTRACTS 12
      The Owner/Operator Contract 12
      The Direct Sales Contract 14
      Legal Requirement of a Sales Contract 14
      The Warranty 15
      A Sample Contract 16
   E. REGULATIONS 19
      The United States 19
      Canada and European Nations 20

CHAPTER 3. PROGRAMMING 21
   A. AVAILABLE SATELLITE PROGRAMMING 21
      Technical Factors Relating to $2°$ Spacing 21
   B. THE MECHANICS of PURCHASING PROGRAMMING 24
      Co-operative Service Organizations 24
      Pay-per-view Services 25
   C. CURRENT TRENDS and ISSUES in the U.S. 25

CHAPTER 4. THE SITE SURVEY and PLANNING 27
   A. THE SITE SURVEY PROCESS 27
   B. BACKGROUND TECHNICAL INFORMATION 28
      Basic Electrical Theory 28
         Ohm's Law and Resistance 28
         Impedance and Alternating Currents 28
      The Decibel Notation 28
         Example I 30
         Example II 30
      Signal-to-Noise Ratio and Picture Quality 31
         Television Picture Quality and S/N 31
      Noise 32
      Amplifier Noise 32
      Calculating Television S/N Ratio 33
   C. THE SATELLITE PROGRAMMING SURVEY 34
   D. THE TVRO SITE SURVEY 34
      Ensuring a Clear View of the Arc 34
   E. AVOIDING TERRESTRIAL INTERFERENCE 35
      Sources of TI 35
         The Ingress Interference Band 36
         The Antenna Interference Band 37
            In-Band TI 37
            Out-of-band TI 38
      Effects of TI on Satellite TV 38
         Ingress Interference 38
         Out-of-band TI 38
         In-band TI 39
      Using a Portable Satellite System to Detect TI 43
   F. THE OFF-AIR SITE SURVEY 44
      Measuring Off-Air Signal Levels 45
      Estimating Off-Air Signals 46
      Standard NTSC Channel Frequency Assignments 47
   G. OFF-AIR INTERFERENCE and RECEPTION PROBLEMS 47
      The Bottom Line 48
   H. HEADEND SURVEY 49

I. DISTRIBUTION SYSTEM SURVEY	49
Home-Run vs. Loop-Run Systems	49
Trunk and Branch Wiring	50
Checking System Condition	50
Surveying for New Construction	51
Final Evaluation	52
CHAPTER 5. HEADEND SYSTEM DESIGN	53
A. THE BASICS of HEADEND DESIGN	53
Balancing Video Signals	54
B. OFF-AIR ANTENNAS	56
Antenna Designs	56
Yagi Antennas	57
Log Periodic Antennas	57
Aperture Antennas	58
Antenna Performance Parameters	59
Gain	59
Bandwidth	59
Directivity and Front-to-Back Ratio	60
Impedance Matching and Ghosts	61
Antenna Configurations	61
Purchasing an Off-Air Antenna	62
Obtaining Required Signal Levels	63
Increasing Antenna Height	63
Stacking Antennas	63
C. OFF-AIR ELECTRONIC COMPONENTS	65
Preamplifiers	65
Gain and Noise Figure	66
Bandwidth	66
Input and Output Specs and Distortion	66
Bandpass Filters and Traps	67
Performance Characteristics	68
Channel Separators	69
Off-Air Processors	71
Strip Amplifiers and On-Channel Processors	71
Demodulation/Re-Modulation Systems	72
Heterodyne Signal Processors	73
UHF-to-VHF Conversion	75
Prohibited Conversions	76
Incidental and Self-Conversions	76
Mixing and Combining Signals	77
Basic Requirements for Signal Mixing	77
Channel Separators/Mixers	78
Active Mixers	79
Simple Examples of Mixing Techniques	79
Launch Amplifiers	80
D. SOLVING OFF-AIR SIGNAL LEVEL PROBLEMS	81
Fading and Weak Off-Air Signals	81
AGC Design Considerations	81
Overloading	82
Ghosting and Stacking Antennas	82
Processing Strong Signals	85
E. SATELLITE TV RECEPTION COMPONENTS	86
Satellite Antennas	86
Feedhorns and Multi-Satellite Reception	87
Low Noise Amplifiers	90
The Downconversion Step	91
Satellite Receivers	93
Modulators	95
Descramblers	96
Component Reliability	97
Combining and Balancing the Signal	97
F. DESIGNING THE SATELLITE HEADEND	101
Calculating Receiver Input Levels	101
G. A STRATEGY FOR COMBATING TI	102
Charting TI Problem Sites	102
Methods to Screen Out Interfering Signals	103
Combating TI With Filters	104
Out-of-Band TI	104
In-Band TI	104
Selection of RF and Microwave Filters	105

H. EXAMPLES OF HEADEND DESIGNS 107
    Example I – Typical Off-Air Processing Requirements 107
        On-Channel Processor 107
        Heterodyne Channel Processor 107
        Notch Filter 107
    Example II – A Basic Headend 107
    Example III – More Elaborate Headend Design Issues 109
    Example IV - Basic One Channel Satellite System 111
    Example V - 4-Channel, Single Polarity TVRO Headend 112
    Example VI - 7-Channel, Dual Polarity Satellite Headend 112
    Example VII - Basic 8-Channel Satellite Headend 112
    Example VIII - 11-Channel, C/Ku-band Satellite Headend 112

CHAPTER 6. DISTRIBUTION SYSTEM DESIGN 117
  A. THE BASIC COMPONENTS of DISTRIBUTION SYSTEMS 117
    Cables and Connectors 117
        Types of Coax 117
        Cable Performance 119
            Installation Details 121
        Water Damage to Coaxial Cables 121
    Splitters 121
    Distribution Amplifiers 122
        Gain, Tilt, Bandwidth and Noise Figure 122
        Signal Distortions in Amplifiers 124
        Cross Modulation 124
        Minimizing Signal Distortions 125
    Attenuation Pads 125
    Terminators 126
    Taps 126
        Resistive/Backmatched Taps and Directional Couplers 127
        Installing Taps and Wall Terminals 127
        Special Design Requirements for Taps 128
    Matching Transformers 128
    Converters and Decoders 128
  B. CABLE TV DESIGN ISSUES RELATING to SMATV SYSTEMS 129
    Trunk and Feed Distribution Systems 129
    System Design Methods 130
    Distribution versus Trunk Amps 130
    System Powering 131
    CATV Quality 131
  C. TELEVISION INPUT SIGNAL REQUIREMENTS 131
  D. MIDBAND and HIGHER FREQUENCY DISTRIBUTION SYSTEMS 132
    Mid, Super and Hyperband Distribution 132
    UHF Distribution 132
    Satellite IF Distribution 136
  E. SAMPLE SYSTEM DESIGNS and CALCULATIONS 137
    Example I – The Basic System 137
    Example II – Calculating Tap Values 138
    Example III – Distribution Amplifiers and Gain 139
    Example IV – 32-Unit High-Rise Apartment Building 139
    Example V – Distribution Amplifiers and Tilt 141
    Example VI – Noise in Cascaded Line Amps - Part I 141
    Example VII - Distortion in Cascaded Distribution Amplifiers 142
    Example VIII – Distortion in Cascaded Line Amps - Part II 142
  F. 18 GHz MICROWAVE LINE-OF-SIGHT TRANSMISSION SYSTEMS 144
    Applications 144
    History 145
    Licensing in the United States 145
        CARS Band Licenses 145
        18 GHz Licenses 145
    Technical Operation 146
    Technical Design Consideration 147
    The Financial Advantage 150
  G. LASER LINKS 150

CHAPTER 7. PROJECT BIDDING 153
  A. BID PREPARATION 153
  B. THE BID DOCUMENT 155
  C. NEGOTIATIONS 156

CHAPTER 8. CONSTRUCTION and INSTALLATION 157
A. THE PERMIT PROCESS 157
B. SUBCONTRACTING the CONSTRUCTION 159
- Finding a Subcontractor 159
- Subcontractor Costs and Fees 160
- Legal Considerations 161
- Subcontractor Management 163
C. IN-HOUSE CONSTRUCTION 163
- The Basic Approach 163
- Pre-Installation Testing 163
- Headend Construction 164
- Satellite Antenna Installation 166
- Off-Air Antenna Installation 167
  - Grounding and Surge Protection 169
- Distribution System Construction 169
  - Above-Ground Signal Distribution 170
  - Underground Signal Distribution 170
- System Activation 172
  - System Performance Levels and Maintenance 174
- Other Requirements for In-House Construction 175
  - Construction Planning 175
  - The Final Decision 175
D. LOCAL ORIGINATION and ADDITIONAL SMATV SERVICES 183
- One-Way Services 183
  - Character Generators 183
  - Video Cameras 184
  - Video Cassette Recorders 184
- Two-Way Services 184
  - Design Considerations 186
E. STEREO AUDIO SERVICES 188
- Satellite Audio Reception 188
  - Multiplex Stereo 189
  - Warner Amex Stereo 189
  - Discrete Stereo 189
  - Processed Narrow Deviation Stereo 189
- Off-Air Stereo 190
  - Receiving MTS Stereo 190
- Stereo Processing and Distribution 191

CHAPTER 9. SYSTEM OPERATIONS 193
A. THE CUSTOMER RELATIONS OFFICE 194
B. COMPUTER SYSTEMS 195
C. THE TECHNICAL SUPPORT GROUP 196
- Staffing 196
D. THE OVERALL MANAGEMENT STRUCTURE 197
E. MARKETING THE PROGRAMMING 198

CHAPTER 10. TROUBLESHOOTING and TEST INSTRUMENTS 201
A. TROUBLESHOOTING 201
- The Basic Approach 201
- Test Instruments 202
B. A CATALOG of SYSTEM PROBLEMS 203
- Inadequate S/N Ratio 203
- High Audio-to-Video Carrier Ratio 204
- Hum Modulation 204
- Signal Egress and Ingress 204
- Interference from Video and FM Broadcasts 205
- Ghosting and Ringing 206
  - Ghosting 206
  - Ringing 208
- Sync Compression 208
- Other Sources of Interference 208
  - Electrical Interference 208
  - Ignition Noise 209
C. THE FIELD STRENGTH METER 209
- System Performance Tests 211
  - Signal Levels 211
  - Signal-to-Noise Ratio 211
  - Hum 212
  - Signal Leakage and Return Loss 212
  - Return Loss 213
    - Testing Passive System Components 214
- Splitters 214
- Taps and Directional Couplers 214

- D. THE SPECTRUM ANALYZER..215
  - Measuring TI Levels...216
- E. POINT-BY-POINT DIAGNOSTICS..217
  - The Customer Drop...218
  - Troubleshooting the Headend...219
    - Satellite Signals...219
    - Off-Air Signals...219
    - The Combiner...219
  - Equipment Repair and Periodic Maintenance..220

CHAPTER 11. WIRELESS CABLE SYSTEM OVERVIEW...223
- A. TECHNICAL OVERVIEW..224
- B. HISTORY..226
  - International Wireless Development..227
  - Technical History...228
- C. FREQUENCY ALLOCATIONS, CHANNEL AVAILABILITY and REGULATIONS..........................229
  - The Optimal Frequency Range?...229
    - Transmission..229
    - Reception...230
  - Frequency Availability and the Licensing Procedure in the U. S..230
    - Frequencies...230
    - The License Application Process...234
  - Comparison with Cable and Conventional TV..235
    - Equipment and Cost Comparison..235

SECTION II. WIRELESS CABLE /MMDS SYSTEMS

CHAPTER 12. WIRELESS CABLE COMPONENTS..237
- A. THE TRANSMITTER FACILITY..237
  - Transmission Antennas..237
  - The Transmitter Site..241
    - Geographic Considerations and Location...241
    - Space and Facilities Requirements..241
- B. SIGNAL PROCESSING and TRANSMISSION...241
  - Overview..241
  - Microwave Transmitters Technical Details..245
    - Diplexing Method..245
    - Control of Output Power Levels..246
    - Linearity...246
    - Frequency Stability..246
    - Feedback and Crosstalk...246
  - Addressing and Encoding..246
    - The Comband Encoding System – Details...249

  - Evaluating and Maintaining Headend System Performance..250
  - Transmission Formats and Signal Security..250
    - Broadcast Formats...250
    - Signal Security...250
- C. OPERATIONS FACILITIES..250
  - Business Office Site..253
    - Space and Facilities Requirements..253
    - Billing Computer Interface...254
    - Addressing System Features..255
    - Communication to Transmit Site...255
- D. RECEIVE SITES...255
  - Receive Antenna..255
  - Block Downconverter..256
    - Noise Figure...259
    - Gain..259
    - Dynamic Range and Signal Distortions...260
    - Interfering Signals..260
    - Image Rejection...261
    - The Effect of Oscillator Performance...261
    - Return Loss..261
    - Environmental Effects on Performance...262
    - Downconverter Designs..262

- Set-Top Converter 262
  - Pay-Per-View Capability 263
  - Technical Demands in a Wireless versus a Wired Cable Environment 264
  - Converter Design Considerations 264
    - RF Input Requirements 264
    - Scrambling System Requirements 265
    - Addressing System Requirements 265
- E. SINGLE and MULTI-UNIT RECEIVE-SITE CONFIGURATIONS 265
  - Single Family Residences 265
  - Multiple Dwelling Residences 265
- F. VIEWER QUALITY ASSESSMENT and S/N RATIO 268

CHAPTER 13. WIRELESS SIGNAL COVERAGE 269
- A. SIGNAL COVERAGE and LIMITATIONS 269
- B. COVERAGE PREDICTIONS 272
  - Service Area Coverage 272
  - General Coverage Calculations 272
- C. EXTENDED RECEPTION TECHNIQUES 275
  - Blocked Areas 275
  - Shadow Areas 275
  - Localized Blocked and Shadow Areas 275
    - Repeaters or Beambenders© 275
    - Receiving Antenna Enhancements 277
    - Local Cable Distribution Systems 277
    - Higher Gain Receive Antennas 277

CHAPTER 14. RECEIVE-SITE INSTALLATION 279
- A. INSTALLATION PLANNING and SITE SURVEY 279
  - Arrival at the Customer Site 279
  - Site Survey 279
    - Customer Approval 280
- B. MAST, ANTENNA and DOWNCONVERTER INSTALLATION 281
  - Guy-Wires 283
  - Antenna Polarity Orientation 284
  - Antenna Tuning 285
  - Off-Air Antennas 285
- C. GROUNDING and WEATHERPROOFING 285
  - Grounding 285
  - Weatherproofing 286
- D. CABLE and CONVERTER INSTALLATION 287
  - Fitting and Connectors 288
  - Drilling the Cable Entry 288
  - Under Carpet Installations 289
  - Connection to the Television Set 289
  - Telephone Connections 290
- E. CUSTOMER EDUCATION 291
  - Additional Outlets 291
  - Clean-Up 291
  - Customer Sign-Off 291
- F. SERVICE PROCEDURES and SAFETY 292
  - Common Problems 292
    - Poor Fittings 292
    - Operator Error 292
    - Blown Power Supplies 292
    - Spun Antennas 293
    - Broken Masts 293
    - Damaged Cable Drops 293
    - Blown Downconverters 293
    - Water in the Antenna Dipole 293
    - Signal Blockage 293
- G. SAFETY 294
  - Roof-Top Safety 294
  - Electrical Safety 294
  - Eye Safety 295
  - Night Operations 295

CHAPTER 15. SYSTEM PLANNING and OPERATION 297
A. THE BUSINESS PLAN 297
Market Study 297
Costs and Income 297
Supporting Documentation 298
B. SYSTEM OPERATIONS PLANNING 298
The Operations Plan 311
Construction/Pre-Marketing Phase 311
Subscriber Installations 311
Customer Service 311
Billing 312
Organization and Personnel 313
Management Personnel 313
System General Manager 313
Marketing Director 313
Director of Finance 314
Director of Engineering 314
Administration 314
Executive Secretary 314
Receptionist/Typist 314
Customer Service Representatives 314
Computer Operator/Billing Clerk 314
Programming Director 314
Personnel Director 314
Technical/Engineering 314
Dispatchers 314
Field Service Technicians 314
Warehouse Supervisor 315
Sales and Marketing 315
Sales Manager 315
Outside Sales Counselors 315
Telephone Sales Representatives 315
Construction Schedule 315
Permits and Licensing 315
Capital Expenditure Schedule 315
Programming 315
Availability 315
Packages and Rates 315

CHAPTER 16. A LEGAL PERSPECTIVE on WIRELESS CABLE in the UNITED STATES 317
A. AN HISTORICAL PERSPECTIVE 318
Allocation of the Electromagnetic Spectrum 318
Education Television 319
Reallocation 319
The Reallocation Plan 319
How the Rules Evolved 320
Lottery 320
Leasing 321
Common Carrier Status 321
Non-Interference 322
Resale of MDS Licenses 322
Recent Changes in the Rules 322
Filing Procedures 323
B. THE APPLICATION PROCESS FOR AN MDS OPERATION 324
Eligibility to Apply 324
The Market 324
Pending Applications 324
The Application 324
Interference Analysis 324
Settlement Groups 325
Selection Process 325
Station Operation 325
C. RECENT TRENDS 326

CHAPTER 17. The WIRELESS DEAL in the U.S. 329
A. THE WIRELESS DEAL 329
The Deal from both Buyer's and Operator's Perspective 330
In Conclusion 332
B. THE WIRELESS CABLE ASSOCIATION INTERNATIONAL, INC. 332

SECTION III. PRIVATE CABLE SECURITY SYSTEMS

CHAPTER 18. PRIVATE CABLE SECURITY 337
   A. Wireless Cable Security Systems 338
   B. SMATV Security Systems 340
      Passive Traps 340
      Active Traps 341
      Off-Premise Addressable Security 341

APPENDICES

APPENDIX A. ADDITIONAL EXAMPLES of DISTRIBUTION SYSTEMS 347

APPENDIX B. EQUATIONS and TECHNICAL DETAILS 353
   A. Satellite and Cable TV Equations 353
      Wavelength to Frequency Conversion 353
         Wavelength 353
         Ohm's Law 353
         Power 353
         Inductive and Capacitive Reactance 353
         Resonant Frequency 354
      Link Equations 354
      Antenna Gain 354
      Loss of Gain with Surface Irregularities 355
      Antenna Beamwidth 355
      Noise Temperature and Figure 355
      The Effect of Bandwidth on System Noise Power 356
      Declination Angle 356
      Azimuth and Elevation Angles 356
      Voltage Standing Wave Ratio 356
      Antenna Parabolic Geometry 357
      Coaxial Cable Attenuation 357
   B. Wireless Cable System Parameters 357
      Performance 357
      Conversion of FSM Voltage to Field Intensity Readings 358

APPENDIX C. PROGRAMMERS 359

APPENDIX D. SATELLITE TV GUIDES and PUBLICATIONS 361

APPENDIX E. OFF-AIR and CABLE TV CHANNELS 363

APPENDIX F. GLOSSARY 365

APPENDIX G. MANUFACTURERS 375
   MANUFACTURERS of SMATV COMPONENTS 375
      North America 375
   MANUFACTURERS of WIRELESS COMPONENTS 377

Forward

Congratulations to the authors of and contributors to this much needed and excellent publication! Because of their foresight and professional quality standards, the information contained in this new book is both accurate, concise and timely.

The wireless industry, while still in its infancy, has evolved into its present form as an emerging, viable competitor to cable television. This manual will not only aid novices in understanding and comprehending this new and explosive industry, but will also stand alone as a guide and reference for those already well experienced in wireless cable. Through exhaustive hours of research combined with invaluable hands-on experience, the authors have created a work that artfully guides the reader through all facets of this industry.

This book explores all aspects of and conflicts within wireless cable from technical to legal to contractual. A reader in possession of the knowledge contained in this book will be able to ask the right questions as he searches the for necessary contacts to create a viable and successful career in wireless cable.

Steve Johnson, American Wireless Systems
Phoenix, Arizona
July 1992

SECTION I

BASICS and SMATV SYSTEMS

OVERVIEW & HISTORY

CHAPTER 1. OVERVIEW and HISTORY of PRIVATE CABLE TV

A. BRIEF HISTORY of SATELLITE TELEVISION

In a 1945 article entitled "Extra Terrestrial Relays" that appeared in Wireless World, Arthur C. Clarke first described a network of satellite repeating stations that would form a complete worldwide communications system. These electronic "mountain tops" were designed to receive and transmit signals from any location or nation on earth to any other. His vision has come to pass. Today geostationary communication satellites hover over the equator at an altitude of 22,247 miles and provide us with a fixed point in space from which to broadcast uplinked signals. This belt which encircles the earth is named the "Clarke Belt" in honor of the man who first envisioned it.

Satellite communications at first involved the use of massive, expensive and often unreliable equipment even though the pioneer research had begun long before the first satellites were launched. Communication using microwaves in space was first accomplished in 1948 when the U.S. Army Signal Corps relayed radar signals to the moon and bounced them back to earth. This research group proved that relatively low powers could be used to transmit and detect signals over extremely long distances. Then in 1954, the U.S. Navy successfully reflected the first voice messages from earth to the moon and back again. These signals took a few seconds to travel this complete circuit.

It was twelve years after Clarke's bold prediction before mankind developed the capability to launch spacecraft. The world witnessed the birth of the space race on October 4, 1957, when the Soviet Union succeeded in orbiting a basketball-sized satellite named Sputnik. This tiny object was the focus of international attention as it passed overhead once every ninety minutes. The United States promptly entered the space race with the launch of its first satellite, Explorer 1, on January 31, 1958.

The pioneer satellites collected an enormous amount of information about space and the technology needed for its exploration. Following the establishment of the National Aeronautics and Space Administration (NASA) in 1958 SCORE was launched into orbit. It received messages at 150 MHz, taped them and then re-transmitted the information back to earth at a frequency of 122 MHz. On December 19, 1958, President Eisenhower broadcasted the very first satellite message, a Christmas greeting. In 1960, the U.S. launched Echo 1, a passive satellite that functioned simply as an orbiting reflector. It was a 100-foot-high balloon made of Mylar just 5/10,000 of an inch thick. The signal reflected back to earth was only 10^{12}, a millionth of a millionth, of the original 10,000 watt uplinked transmission. This experiment demonstrated that it was possible to relay two-way telephone conversations via satellite across the entire continent even when the downlinked signals were of extremely low powers. Subsequently, RCA's Relay, launched in 1961, provided improved broadcasting of trans-Atlantic telephone, telegraph and television messages.

OVERVIEW & HISTORY

Figure 1-1. Satcom Satellite. *The large wing-like structures on this satellite are solar cells which provide on-board power. The antennas receive and transmit television broadcast relayed via microwaves. (Courtesy of RCA American Communications)*

In 1961 the Hughes Aircraft Corporation built Syncom I, an experimental, active satellite that was to be launched into a geosynchronous orbit. Previous satellites had followed the standard elliptical orbits. Such space vehicles were therefore much more difficult to track and were visible from any one location on earth for only limited periods of time each day. Telstar, designed and built by AT&T, linked Europe and North America via television on July 10, 1962. It was the first satellite capable of broadcasting television signals. This vehicle was considered to be "active" because it carried "transponders" which detected, amplified and re-transmitted signals to earth. Telstar II and Syncom II, also geosynchronous spacecraft, were launched in mid-1963. Soon thereafter, Syncom III, relayed the 1964 Tokyo Olympic Games to the United States (see Figure 1-1).

In 1965 Intelsat I, also known as Early Bird because of its position in the far eastern sky, was launched for Intelsat, the International Satellite Telecommunications Consortium (see Figure 1-2). In 1973, Canada also began taking an active role in satellite communications when NASA orbited the first in its Anik series of birds. Anik I was the pioneer satellite to serve the U.S. domestic market. The Westar I and II vehicles, launched the following year, formed the basis of the first 12-transponder commercial satellite network. RCA soon launched the first 24-transponder satellite, Satcom F-1. Today this trend continues with the development of high-powered Ku-band and futuristic, experimental satellites.

Figure 1-2. Intelsat VI. *This communication satellite stands nearly 39 feet high with its antennas unfolded and aft solar panel extended. (Courtesy Hughes Aircraft Company)*

B. THE DEVELOPMENT of COMMUNITY ANTENNA TV SYSTEMS

Community antenna television, also known as CATV or cable TV, was the forerunner of smaller SMATV or private cable systems. The concepts underlying both technologies are quite straightforward. Programming relayed via satellite is combined with signals received from off-air sources and is distributed throughout a community via coaxial cables.

The first cable TV service was installed by Ed Parsons in Astoria, Oregon in 1948. Using a few antennas and primitive amplifiers, he collected distant signals for distribution to his friends and neighbors in town. Pennsylvania was another area of early cable TV growth with enterprising television retailers setting up systems to bolster their sales of television sets. Others followed suit in areas where reception of off-air broadcasts was difficult with conventional, smaller-scale antennas. High-gain, off-air antennas were combined with amplifiers and other electronic devices to make programming from nearby metropolitan areas available to cabled customers.

The earlier systems had access to only a limited number of off-air channels due to the limited signal processing capability of the equipment in use at that time. In these pioneer days, non-adjacent VHF channels 2, 4 and 6 were all that could be transmitted on MATV systems. High signal losses due to excessive cable attenuation, typically over 10 dB per 100 feet, and poor amplifier performance limited the number of channels that could be distributed as well as the system range. Amplifiers had to be installed every few hundred feet. The development of the 12 channel transistorized broadband amplifier that contributed relatively low amounts of signal distortion gave the cable operator a chance to insert more channels into the line-up while extending the service range. The development of standard distribution and wiring methods also came slowly. These varied widely among the various entrepreneurs' systems.

The early days of the cable industry were characteristic of any new, promising industry. Stimulating, creative ideas excited many but few clearly understood how to accomplish their technical and management objectives.

With the launch of the Westar and Satcom satellite networks, a new source of programming became available. Cable operators no longer had to rely only on local programming to satisfy the expanding tastes and requirements of their customers. Until the mid-1970s, CATV companies had essentially built and operated large MATV (master antenna TV) systems. In 1975, an east coast-based operation, the Home Box Office (HBO), started delivering programming via satellite to its 57,000 subscribers. On September 30th of that year it staged its first satellite cable event with "The Thriller In Manila" title fight between Mohammed Ali and Joe Frazier. Following HBO's lead, a young entrepreneur named Ted Turner established the first "superstation" when he uplinked the WTBS Atlanta television signal in 1976. In a relatively short time, satellite programming became the basic staple of cable systems everywhere.

As more programming became available to CATV system operators, channel capacity began to increase. This was speeded by the development of the push-pull amplifier that eliminated midband distortion created by the use of single-ended amps. The previously unusable mid, super and hyperbands also became available to cable operators and the rush was on for more channels. Subscribers were offered 24, 36 and even 60 or more channels. At the same time in the United States, the Federal Communications Commission, the FCC, which regulated all forms of satellite TV activities, reduced the required minimum diameter of TVRO (television receive-only) antennas from 30 to 15 feet. This decision speeded the decline in the costs of earth stations and subsequently brought this technology within the economic reach of numerous small cable television operators.

As the number of cable systems increased, the more aggressive, astute owners began acquiring numerous smaller systems and started forming the large cable conglomerates. Unlike the regulations affecting conventional television broadcast stations, the FCC in the United States does not limit the number of cable TV systems that an organization can own. As a consequence, the multi-system operator, the MSO, was born. Among the most prominent MSOs today are Telecommunications, Inc. (TCI), United Cable, and Rogers Cable.

Cable TV has come a long way since the early 1950s. Some modern systems are capable of delivering over a hundred channels of programming to their customers while the average number of channels delivered is now about sixty. Computer controllers now allow an operator to alter a subscriber's service and provide impulse pay-per-view events at the touch of a button. In addition to regular programming, two-way cable provides security services and computer terminal interfacing to subscribers. In many cases, fiber optic networks are now being used for trunk distribution of signals. The result is an increase in system efficiency coupled with a reduction in installation problems. The vast amounts of bandwidth offered by modern cable systems gives them the technical capability to compete with telephone lines as common carrier delivery systems. Clearly, cable TV technologies are continuing the process of developing new and innovative solutions to design problems. Fortunately, most of these developments are readily adaptable to SMATV applications.

C. SMATV versus FRANCHISED CABLE TV SYSTEMS

Franchised CATV and private SMATV industries have common roots, employ similar technologies and clearly share many other common features. However, two fundamental differences exist. SMATV cable distribution systems are substantially smaller in size and the overall business activity is subjected to fewer government regulations than the cable TV industry.

Franchised cable TV systems require a substantially larger up-front investment than private systems. The costs associated with wiring up an entire community can be tremendous. As a consequence, a large subscriber base is necessary to create a profit stream that can justify the capital outlay needed to construct the average franchised system. In contrast, a well designed, installed and managed private cable system can serve as few as 100 subscribers and still generate healthy profits. This is especially true when an SMATV system is installed during the construction phase of a multi-family dwelling because expensive aerial and underground work can be minimized.

Government regulations also play an important part in franchise cable operations. Before a system can begin to operate it must be granted a franchise agreement by the local municipal government. This agreement is often reached only after months or even years of competitive bidding and negotiations. The franchise document details the area and method of system operation and the rates an operator can charge. Most franchise agreements require that the cable system share some of its revenues with the parent city. This franchise may be subject to periodic renewal based on past performance. In addition to local regulations, the franchise operator must also contend with extensive requirements relating to system operation and performance. These are mandated by the FCC in the United States, the CRTC and DOC in Canada and the PTTs in European countries. In contrast, private cable systems generally do not have to conform to government regulations as extensive as those required of the CATV operator because SMATV systems are much smaller in scope and construction occurs on private property. No public roads need to be crossed by cables. Installation can be particularly simple when an SMATV system is being added onto an existing multi-family dwelling with an in-place master antenna system. In general, an operator must comply with local construction ordinances and, in most municipalities, with zoning and permit requirements. However, whatever the existing regulations on SMATV systems, operating a system in a professional manner can go a long way towards avoiding unneeded scrutiny.

SMATV operators should be aware of the issue of forced access in dealing with franchised operators. In some cases a franchised CATV operator has insisted that tenants have the right to choose between cable and an existing SMATV system. The cable company has then actually built their system over the SMATV installation. This forced access can potentially be disastrous because of the revenues lost.

Private cable operators have legally battled for the right of the property owner to make the decision between the two alternatives, especially after an SMATV installation has been completed and is in operation.

In summary, there are three major difference between a private and franchised cable system, programming cost, installation cost and government regulations. The fact that a private cable system involves less red tape and has lower construction costs makes it an appealing and ideal venture for the smaller-scale entrepreneur wishing to enter the cable TV market.

D. THE DEVELOPMENT of SMATV and WIRELESS CABLE TV

When broadcast television first appeared in the late 1940s, the demand for television receivers was phenomenal. Along with this demand came the need for antennas to receive the signals. In cities everywhere a vast aluminum forest began to rise. Nowhere was this concentration of hardware more prevalent than on the roofs of apartment buildings. These antennas were not only unsightly, but actually caused reception problems when these devices were spaced too closely together. As a result, a method was developed that allowed a single antenna to simultaneously serve many apartments. This concept became known as master antenna television or MATV.

The pioneer MATV systems did nothing more than receive and distribute local off-air programming. Nevertheless, some early operators in problem areas such as mobile home parks where installing multiple antennas was difficult, found that they could charge customers for improved television service. These systems provided the basis for the private cable industry.

As franchised cable became increasingly commonplace in North America, MATV operators began to feel pressure to upgrade the services they were offering. Although MATV systems had been on equal grounds with cable TV in distributing off-air television signals, as CATV developed and acquired new forms of programming many MATV systems were either abandoned or converted to cable. Fortunately, during this period, MDS (multipoint distribution service) operators had started using microwave and UHF signals to beam a single premium channel, such as Premier, ON-TV and Select TV, into non-cabled homes in major metropolitan areas. Local MATV operators found that they could easily intercept and integrate these signals into their existing systems in order to create an additional revenue stream. This development was the beginning of modern MATV and became the first serious competition to the franchised cable industry.

In 1977, when the FCC eliminated the licensing of earth station terminals for non-commercial, home use, satellite antennas began a rather rapid fall in price. As a consequence of this development, MATV operators found that they could economically have access to the same satellite programming as the franchised cable outfits. When satellite received programming was linked into their MATV systems, they created the first truly private cable systems.

Out of necessity, these pioneer SMATV operators pirated the signals they received. The program suppliers, in part because of their affiliation with franchise cable, would not sell programming to the new competitor on the block, the private cable industry. After all, most satellite programmers had grown hand-in-hand with the CATV industry. However, over a period of time as more SMATV systems were installed and as penalties for signal theft became stiffer, agreements were reached for suppliers to offer their services to SMATV operators. These accords were often reached only following protracted legal battles and negotiations. Today, after years of painstaking work, almost every service available to franchise operators is also available to the SMATV industry even though programming rates may be higher and deposits may be necessary.

SMATV has now become a full fledged industry with its own trade association, programming cooperatives and specialized equipment manufactur-

OVERVIEW & HISTORY

ers who produce electronic components that rival the performance and features of more expensive CATV grade equipment. Modern technology borrowed from CATV operations gives the SMATV operator features such as fully addressable control systems and pay-per-view programming. Today, SMATV cable systems can rival franchise cable networks in available features offered as well as in technical performance and can also provide subscribers with services custom tailored to their needs.

Wireless Cable

Wireless cable television or MMDS, which has evolved from a North American educational service known as Instructional Television Fixed Service (ITFS), is today another alternative to franchised cable delivery of audio-video entertainment. This new technology is a particularly attractive mate for SMATV systems because it simplifies and reduces the costs of necessary headends. Up to 33 video channels can be relayed to subscribers. In addition, this technology has inherent advantages in third world countries because it does not require installation of costly cable networks that require constant monitoring and periodic maintenance by trained technicians.

The history, evolution and impact of wireless cable television is explored in detail in Chapter 11.

E. THE ADVANTAGES of SMATV

SMATV systems offer their customers many advantages, some of which are rather unique to this technology. First and foremost, these systems, like cable TV networks, have access to a great diversity of programming including satellite broadcasts, off-air signals and taped entertainment. In addition, these audio and video signals are under the control of a local headend and therefore have the potential of delivering superb technical quality. Properly designed and managed SMATV systems also have the advantage of being relatively small-scale so that the response in the event of any unexpected system breakdowns and difficulties can be rapid and decisive.

As the private cable industry has grown so has the menu of available programming and other services. Marketing can now be targeted towards specific age, interest, cultural or ethnic groups. Furthermore, SMATV systems can offer entertainment via satellite that would otherwise be inaccessible. This diversity is even greater than that offered by most CATV systems which must cater to the majority interests of larger numbers of consumers as well as to city and local governments. There is the potential to offer a diverse range of programming. These include first run movies, sports, news, health and fitness, children's and religious programs. The more unique fare includes educational and instructional courses for the practicing professional, one-time sporting events, continuing education courses for college credit and live concerts.

Other services offered may include the management of energy control systems, public affairs programming or even a channel dedicated for use by the owners of the building complex. The advent of two-way interactive technologies, which may be structured as simply as using a telephone line to communicate back to the headend, presents other diverse possibilities. These may include teleconferencing, electronic banking, video games, in-home shopping or in-home security systems.

Private cable systems offer building owners or homeowner's associations the advantage of revenue sharing. SMATV operators are quite aware that an apartment or condominium complex represents a substantial customer base. The access to this property has real value. However, the franchise CATV operator typically does not compensate these owners or associations and has even gone to court to force access into their units. SMATV operators are clearly more than willing to share a piece of the pie with the owner. This creates a healthy business environment and a win-win business deal for both parties involved.

The Market for SMATV and Wireless Cable

The market potential for private cable is truly enormous, particularly because many consumers are beyond the reach of conventional cable TV networks. These multi-unit dwellers demand at least the same programming variety that cable TV subscribers can purchase. Even within cabled areas, many consumers are now weighing the advantages and disadvantages of subscribing to a private cable service versus CATV.

An SMATV system can be rapidly installed in response to this demand from wherever it might arise. The market includes all varieties of mobile home parks, hotels, motels, condominium projects, apartment complexes, RV parks, schools, churches, country clubs, restaurants, resorts, hospitals, nursing homes, airports, correctional institutes and retirement homes. Since targeted customers usually represent demographically distinct segments of the population, such as senior citizens, religious groups or families, private cable operators can respond with services customized to their needs, not necessarily with programming dedicated to the tastes of the masses.

For example, in Canada the number of people living in apartments and duplexes in 1976 numbered 2.4 million out of a total population of 25 million. Also, the total number of hotel rooms in the United States and Canada that could be serviced by a cable network is about 2.5 million. However, only about 300,000 are hooked up to cable. But some have realized the business opportunity and have moved rapidly. The Holiday Inn hotel chain, for example, has installed 325 dishes in 42 states. Known as the Hi-Net Network, the system supplies programming to each hotel room and also forms the hub for a revenue-generating, teleconferencing network. In addition, hospitals now have access to satellite broadcasts delivering health care information such as live operations and in-depth diagnostics. By incorporating these satellite services with an SMATV system they have the full flexibility of closed circuit TV, inter-hospital teleconferencing, security, fire detection and paging systems as well as a valuable source of educational materials. These same features make private cable systems desirable to universities, colleges, high schools, retirement homes, exclusive housing complexes and other specialized groups of people.

Wireless cable is now posed for significant worldwide growth. At the beginning of 1992, wireless had 400,000 subscribers in 80 systems. Some experts predict that up to 10 to 15 million subscribers will be on-line by the end of the decade.

NSPN

National Satellite Programming Network (NSPN) was the first company created to distribute satellite delivered television programming and to provide critical support services to private cable and wireless cable operators nationwide.

We believe in <u>professional</u> people, doing <u>professional</u> work, for <u>professional</u> people.

It's the difference between "being first" and "staying first".

National Satellite Programming Network
1909 Avenue G
Rosenberg, Texas 77471
(713) 342-9655   (800) 937-NSPN   FAX (713) 342-7016

Professionals, making your future visible.

CHAPTER 2. ECONOMICS, SALES CONTRACTS and REGULATIONS

A. ECONOMIC OPTIONS for INSTALLERS, OPERATORS and OWNERS

A number of options are available to those wishing to participate in the private cable industry. These include:

- Contracting to build a system for a property owner
- Building and then selling a system to a property owner
- Using an intermediary leasing company that would lease a system back to the owner
- Assuming full responsibility for building, owning and operating a system.

In an outright sale of an SMATV system, the purchaser could be the building owner, administrative manager(s) or a co-operative group of unit renters or condo owners. This arrangement means that all rights and liabilities would pass to the end user(s). In these cases, the builder should either separately sell or include a maintenance contract in the contract price to ensure that the system will be properly operated and maintained in the future. This serves to guarantee the reputation of the private cable builder and, if properly structured, can contribute to yearly cash flow. Such service contracts can be on either a straight time and materials or a yearly flat rate basis.

When a flat rate contracts is negotiated, the contractor typically receives 3% to 5% of the total purchase price to provide all required maintenance services. This percentage then increases by 1% each year to cover expected increases in service requirements as the system ages.

If a SMATV system is sold to a leasing company which in turn leases the system to the end user at a preset rate, a maintenance contract should also be arranged with the leasing company. It is of value to both the lessee and lessor that there are no interruptions in system operation. Again, the maintenance contract can provide additional cash flow to the operator.

When a system is simply leased to the end user, the deal should be structured so that the disadvantage of carrying the initial capital outlay is more than offset by the cash flow generated. The economics of such an arrangement can be excellent. An SMATV system typically costs about $200 to $300 per unit to construct while buy-outs often bring in $600 to $1000 per unit and more (all dollar figures as expressed in U.S. dollars throughout this book).

ECONOMICS, SALES CONTRACTS & REGULATIONS

There are many pitfalls in being a successful private cable builder/operator. It is important to obtain accurate information about residents demographics in a proposed site. Knowing the current occupancy rate is especially important. These factors as well as system quality can have a direct impact upon system value. For example, a system in Houston sold for a low $75 per unit while across town another was purchased at $400 per subscriber.

Operating an SMATV business goes well beyond calculating equipment costs. The headend, satellite antenna and distribution equipment comprises only 25 to 40% of the total capital outlay needed to begin the venture. Other on-going costs include marketing costs as well as up front and on-going monthly general and administrative overhead expenses that include the basic operating costs such as office rent, office equipment, telephone, heat, light and office staff salaries. Most SMATV systems typically require 1,500 or more subscribers to support overhead expenses. Underestimating these could mean the difference between a successful operation and outright failure. It is wise practice to have the services of a financial management professional or a knowledgeable industry attorney before a project is undertaken.

Income is generated by subscriber and service fees. These generally include a monthly fee that typically ranges from $15 to $20, plus the cost of any premium service, usually an additional $8 to $11 per month. Of course, royalty payments to the program suppliers must be deducted from this gross revenue. Basic programmers typically charge between 10 to 50 cents per month and the premium movie channels from $3.50 to $6.00 monthly. The rate depends upon the volume of the purchase as well as negotiations with the various programmers or their representatives. Service brokers such as Heifner Communications, the National Satellite Programming Network (NSPN) and World Satellite Network (WSN) can provide expert help in licensing and purchasing programming.

B. SALES CONTRACTS

There are two basic types of contracts encompassing the installation and operation of private cable systems: owner/operator and direct sales contracts. A sale to a leasing company and subsequent lease to the building owner falls into the latter category. It is crucial to clearly understand both how SMATV deals are structured as well as the content of the supporting documentation. The importance of contracts cannot be overstated.

Under the terms of an owner/operator contract, the entrepreneur assumes responsibility for installing and financing the SMATV system and subsequently for marketing the programming directly to customers. Typically, this task includes negotiating with the project owner for the rights to operate the system. Under terms of this arrangement, the private cable operator generally shares a portion of the profits with the real estate owner.

If a property owner wishes to simply purchase a private cable system, the deal would be structured around a direct sale contract that outlines all construction terms. The entrepreneur and expert then acts as a contractor to the project owner and builds the system for a predetermined price. Following its successful installation, the SMATV company may also be contracted to manage either or both the system operation and maintenance.

As in any business venture, contractual agreements are designed to eliminate confusion and to insure that everyone understands the details and limits of their responsibilities. Few private cable operators are lawyers. The SMATV expert should not be the ultimate authority on contract law. This would be an extremely risky strategy! In the same vein, the following information is presented only as a guideline. Although it is based on actual contracts and personal experiences, we strongly recommend the use of qualified legal counsel when drafting and completing the necessary contracts. To do otherwise would be foolhardy.

The Owner/Operator Contract

One of the most profitable methods of participating in the private cable industry is as a system owner/operator. The revenue streams that can be generated from this type of operation can repay system construction costs within two to three years. In addition, every paying subscriber represents consid-

ECONOMICS, SALES CONTRACTS & REGULATIONS

erable equity to the owner in the form of a continuing revenue stream. The per subscriber net worth, often exceeding a thousand dollars, also makes a buy out by other SMATV or CATV operators quite profitable for the original owner. It is clear that preparing a complete set of contracts and other documents is of central importance in protecting this valuable asset from costly disputes.

One of the principle contractual elements that an owner/operator must understand is the duration of ownership. The longer the term negotiated for operating a given project, the larger the revenue stream available. Industry norms for contract durations run from 5 to 10 years with 7 years being the average. This time component of a contract can be affected by other factors such as revenue sharing and buy out clauses. Operators should attempt to include a 3-year renewal clause in an owner/operator contract. Negotiations should concentrate on achieving a contract duration that will insure an adequate payback on investment and a profit margin that justifies the assumed position of risk.

Another important factor in the contract is the amount of revenue to be shared with the project owner. In most cases, an owner can well understand the potential profit involved and the fact that the local franchise operator does not likewise share cable TV revenues. Therefore, the deal must be sufficiently attractive to the real estate owner yet still must protect and maintain an adequate profit margin for the owner/operator. Typically, the terms of revenue sharing can be based on either the gross or net receipts from operation, and can run from 3% to at most 5% monthly. This percentage can be used as a negotiating tool to increase contract duration.

When a system becomes operational and subscriber income starts rolling in, the owner will certainly realize the healthy flow of revenue generated from his or her property. It is not uncommon, at this point, that the owner begins to develop an interest in buying out all interests in the system. By properly negotiating this point in the original contract, problems and ill will between both parties can be avoided. After all, any astute businessperson knows that customer referrals are the best form of advertising. At the time of purchase, the buy out price should be based on a formula that reflects the present value of the anticipated revenue stream over a predetermined time period. If the system is not producing the expected revenues, the contract should have a clause that allows the operator to reclaim all equipment that can be economically removed and to cease system operations. Similarly, if the system is leased to the property owner, the lease contract should contain a clause stating that if revenue is not generated as expected the lessor has the right to remove and reinstall the same equipment in another location.

All equipment installed on a given project is, in effect, utilizing the owner's real estate for the operator's personal gain. In order to protect the initial investment, the contract must specify in detail the method of compensation for use of the property. Otherwise, the real estate owner could suddenly demand excessive rental fees that would make continued operation uneconomical. In addition, a lease agreement for the property usage should be included as one component of the overall contract. Both the length of the lease and all fees expected must be clearly spelled out. If possible, this agreement should be made part of the title interest in the property to avoid any problems if the owner decides to sell the real estate.

It is also important that the initial menu of services the types of programming and the number of channels to be offered are spelled out in the contract. In addition, system design should be clearly outlined. However, option clauses must also be included to permit future changes in equipment and programming as are seen fit. This procedure gives the owner/operator the freedom to creatively tailor the system to maximize revenues.

It is important that both parties clearly understand who owns which electronic components to be installed at the site. During the preliminary survey of the project, note the type and make of all existing equipment down to fine details such as wall wiring that might be utilized during installation. The contract then should outline all existing equipment as well as any additional electronics or accessories that will be installed. The owner should also be perfectly aware of the obligation to purchase programming. The cost of decoders should be specifically stated in the contract. This is politically important and in essence is a statement of the need to operate an SMATV system legally.

When building an SMATV system into a new construction project, it is a good idea to allow the owner and his general contractor to install all underground and in-wall wiring. If equipment will be removed at some later date, this procedure can make it easier to define ownership.

ECONOMICS, SALES CONTRACTS & REGULATIONS

If the project includes the installation of new wiring on the property, the question of ownership must be dealt with thoroughly. Wiring, unlike any other part of an SMATV system, cannot be removed easily in case of a dispute. In many states and provinces it may become part of the property after a set period of time and the original owner may be left out in the cold. In any case, the owner operator should always try to have the property owner pay for the installation of the wiring. It is a direct property improvement and usually represents only a small portion of the total installation budget. If the property owner insists that the cable operator pay for the wiring, then it is important that the operator should check all local and state ordinances regarding its ownership and should try to have it registered as an utility easement.

It is also important to understand the complexity of an owner/operator arrangement. In many many cases, the associated details can rival a city franchise agreement in its complexity and terms. The need for attention to even the smallest points in contract negotiations is extremely important. Again, we highly recommend that the services of a competent attorney be utilized both during the initial drafting of documents as well as on any other occasion when the validity of an agreement is questioned. Remember, the smart businessperson uses the law to prevent the occurrence of problems, not to litigate disputes.

The Direct Sales Contract

The direct sales contract defines the terms under which the private cable firm is contracted to simply install a system. A property owner will occasionally decide to own an SMATV system outright or to pay an outside party to upgrade an existing system. When this situation arises, having skills as a contractor provides an alternative method to realize a healthy profit.

The first step in obtaining a direct sales contract is bidding the job. The bid process is extremely important and will largely determine the profitability of a project. Proper planning and attention to detail are necessary in order to prepare a successful bid. The preliminary survey provides valuable information needed to write a proposal. It is clear that the amount of effort put into preparing such a proposal will have a direct reflection on the probability of its success.

After completing the preliminary survey, a system design can be created. By doing a take-off from this design, a bill of materials is obtained and a cost analysis of all materials is generated. The final estimate is reached by adding this figure to expected labor costs. Finally, an initial bid price is generated by multiplying this sum by overhead factors and adding in a reasonable profit. Remember, the professional contractor makes a firm bid and stands by his price. This bidding process is discussed in more detail in Chapter 7.

In some cases, a cost proposal can double as a contract. If this document is sufficiently detailed, the customer merely has to sign his or her acceptance to create a binding agreement. Many contractors use a standardized form for this purpose. It is both a proposal and an agreement, and contains all the necessary items to effect a contract.

The contract document must contain details about the nature of the work to be performed, the expected time for completion, the system cost and the method of payment. When describing the work to be performed, a contractor must be sure to specify where the existing system ends and the contracted work begins. This includes giving a reasonable estimate of construction time including material deliveries as well as provisions for construction delays due to weather, for quoting the exact system cost and for specifying the method of payment. It is customary to require a down-payment, usually 10%, as well as progress payments spaced out at regular intervals during the construction process. These payments can be keyed to material deliveries and completion of certain phases of construction. On large jobs, the client may wish to retain a small percentage, in general no more than 10%, of the final payment for 30 to 60 days after system activation. This is his guarantee that any last minute needs will be properly addressed.

Legal Requirement of a Sales Contract

A contractor has certain legal rights and obligations. In some American states the first and foremost is the need to have a contractor's license. This is the legal document required both to perform any commercial work and to protect a contractor's rights. Without a license, necessary permits and city approvals needed to perform this task cannot be obtained. In addition, a licensed subcontractor would

ECONOMICS, SALES CONTRACTS & REGULATIONS

have to be hired to legally manage the construction. More details about hiring and managing subcontractors are presented in Chapter 8.

When drafting the contract document, special attention must be given to the matters of permits and planning approval. The party or parties who are responsible for obtaining any necessary forms as well as those who are required to pay the fees must be specified. In general, it will be the contractor's responsibility. However, in the case of new construction, the general contractor for the housing project may have already obtained all permits. It is also important to remember that the planning and permit process can take more time than expected and this potential delay must be factored into the contract document.

It is also important to consider the type of labor to be used on a project. On new construction jobs, there will sometimes be a demand for only union labor. If this is not taken into consideration in the bid process, unseen labor costs could erode all the profits in a job. Union labor is more expensive and harder to come by in the cable TV industry and the contractor who overlooks this fact might well soon be wishing he had never accepted a job.

Most clients will be concerned with the contractor's ability to protect them from any liability that may result from inappropriate actions. Therefore, copies of insurance policies should be submitted with the contract documents. In addition, some contracts will require the posting of completion bonds. A competent insurance agent can provide the necessary bonds and forms on an as-needed basis. Finally, the contract document must include notices to the client regarding his or her consumer rights as prescribed by law.

The Warranty

Warranty agreements must also be included in any contract. In addition to being required for certain projects, these can also be structured to provide a further source of revenue. Different types of agreements can be established, each having a potential to generate additional revenue sources.

The type of warranty offered to a customer can drastically affect the profit structure of a project. Unrealistic promises to a customer, as well as unspecified or unexpected obligations, will often result in costly service calls. In the worst but not-too-unfamiliar case, this contractual oversight could result in substantial losses and even bankruptcy.

An effective warranty provides a customer with adequate protection while minimizing a contractor's exposure to loss. All manufacturers' warranties should be utilized as fully as possible in order to cover equipment repair costs. Choosing high quality equipment in the first place and resisting the urge to go the "cheap route," saves all parties many costly headaches. A reputable manufacturer generally will go a long way towards insuring customer satisfaction.

Warranty obligations should cover the work performed on-site. This includes all construction tasks as well as installation of all equipment. It is a contractor's responsibility to ensure that equipment is installed properly and in a professional manner. It is smart practice to limit the standard initial warranty to a year or less. There are well documented cases of excessively long warranties causing the financial ruin of more than one company.

After the initial warranty period has elapsed, the system owner will still need to provide for the upkeep and maintenance necessary to protect his investment. By incorporating an extended warranty package into the original contract, not only is the investment protected but also is a continuing revenue stream created for the contractor.

An extended warranty can be simple or complex. It can be an agreement to provide needed repairs on a time and material basis or can involve monthly fees that cover all maintenance expenses. Provisions for a management contract, namely the operation of all system functions including marketing and billing, can also become a part of the agreement. The more service that a contractor provides to the system owner the greater is the potential for additional revenues.

A time and material contract provides for system maintenance on an as- needed basis. Thus, the system owner pays for repairs as they are needed at a rate determined in advance. Unfortunately, with this type of arrangement, it is very hard to predict what the income stream will be. In addition, there is nothing to stop the system owner from deciding at any time to hire a third party to do the work at a cheaper rate. However, the original contractor has an obligation to perform any preventative maintenance, while any needed parts are paid for by the system owner.

ECONOMICS, SALES CONTRACTS & REGULATIONS

A more profitable arrangement is the extended warranty agreement. Under the terms of this arrangement, the contractor guarantees the operation of the SMATV system and all its components. If a piece of equipment fails, it must be repaired or replaced. The contractor is compensated for this obligation by a fee arrangement that provides for both costs and profits. The cost of an extended warranty is based on a percentage of the original system price. A 5 to 10% yearly fee with a built-in inflationary rate increase is a typical industry payment. Contract periods should be a minimum of 3 years but can run as long as 10 years. It is important to realize that a comprehensive maintenance program will go a long way towards reducing equipment replacement costs.

The extent of a contractor's obligations must be spelled out in detail. Make sure that both parties know exactly the nature of all responsibilities and what costs the customer will incur to keep the system operational. The contractor should guarantee only the work that has been completed and the equipment that has been installed under the terms of the agreement. It is a wise choice to incur only those obligations to maintain any existing system components as specified in the extended warranty agreement.

One final warranty concern relates to programming clauses in an SMATV contract. It is common knowledge that the availability of programming is constantly changing. This rapid evolution can be used as a strength to provide for additional revenues if it is properly understood and anticipated. Attention to contract details relating to programming can help eliminate problems and can increase the level of service offered to a private cable customer.

The system owner is typically responsible for all programming arrangements and costs. It is therefore important that contract documents detail this obligation to the customer. It is also important that no guarantees regarding program availability are made. The contractor's primary responsibility is to provide a customer with a technical means of procuring programming.

The main service offered is clearly the programming. The system owner will usually find it advantageous to allow the private cable expert to handle all the paperwork and negotiations needed to secure programming. The warranty contract should be structured so that it provides the customer with an easy method of obtaining programming.

In addition to programming, provisions should be made to supply the equipment needed to receive and process the signal. Sales of decoding equipment can prove very profitable. Increased usage of Ku-band frequencies for signal delivery can also generate sales of new reception equipment. By planning in advance for the purchase of this ancillary equipment, revenue can continue to be generated long after the initial sale.

In closing, it is important to realize all the potential revenue available from warranty and service contracts. Extended warranties, programming sales and equipment purchases all add up to healthy after-market revenues. Remember, the smart entrepreneur views a warranty agreement not as a necessary evil, but as an opportunity to generate recurring revenue and to provide the customer with all the services needed to successfully operate an SMATV system.

A Sample Contract

A sample owner/operator SMATV contract is outlined in Table 2-1. It is intended as a guide only and should definitely not be used without a careful review of your situation by an attorney. Neither Baylin Publications nor the authors make no claims and cannot assume any liability connected with the legality or accuracy of this contact and with its use. Also please note that this contract does not include some important exhibits: a legal description of the property, a list of the equipment to be installed, a list of the channels to be offered and technical specifications for the system.

The importance of a properly designed contract cannot be over-stressed. This document is crucial in a successful business venture. By all means hire a lawyer who specializes in this field.

TABLE 2-1. A SAMPLE SMATV CONTRACT

AGREEMENT

THIS AGREEMENT is made this _____ day of _____, 1993, by and between

_____ of _____ (the "Operator")

and _____ (the "Owner")

RECITALS

A. Owner is the owner of certain real property located in _____, more particularly described in Exhibit "A" attached hereto and made a part hereof (the "Complex").

B. Operator desires to install certain equipment at the Complex in order to provide cable television services to the residents of the Complex.

Therefore in consideration of the mutual promises contained in the Agreement, the parties agree as follows:

1. Installation of System
Operator will install a satellite reception system at the Complex, consisting substantially of the items listed on Exhibit "B" attached hereto and part a part hereof. The system will be installed in such space as mutually agreed upon by both Operator and Owner. Owner will provide and maintain a climate controlled space and electrical service, at Owner's expense, for the headend electronics and related components. This site must meet the technical requirements specified by Operator. In addition, Operator will install whatever ground distribution systems as necessary to provide cable television services to subscribers in the Complex.

2. Services
Operator will initially offer at the Complex the channel listed in Exhibit "C" attached hereto and made a part hereof. Operator makes no warranties as to the continued availability of the programming listed in Exhibit "C" and assumes no responsibility for programming changes instituted by any program vendor. All rates and charges are subject to increases as determined by vendor rates and other financial considerations. Operator will obtain and maintain all FCC, state or local permits and utility agreements necessary to install and operate the cable television system at the Complex. Operator will, following installation, restore the grounds and buildings of the Complex to substantially the same condition as prior to the start of installation.

3. Operation of System
Operator will, at his own expense, operate and maintain the satellite reception system. Operator will respond to service requests for the system within 48 hours after notice. Agreements for individual services will be made directly between Operator and the residents. Connections, billing, installation, disconnections and line maintenance for each individual subscriber will be the sole responsibility of Operator.

4. Ownership of the System
Unless otherwise agreed to by amendment, title to the System and all equipment installed by Operator and all replacements and improvements thereto, delivered to or installed in or for the benefit of the Owner during the term of this Agreement, shall remain at all times the property of the Operator. Owner agrees to keep the System free and clear of all levies, liens and encumbrances other than created by this Agreement. There will be no charges to be paid by Owner under this Agreement. Operator will be solely responsible for the cost of the system and the maintenance and replacement thereof.

5. Length of Contract
(a) Owner hereby grants to Operator an exclusive right to install, inspect and maintain the cable television equipment, lines and facilities described in the Agreement for a period of ten (10) years from the date of this Agreement.

(b) Owner will grant operator an automatic renewal of the rights specified in section (a), for an additional five year period, upon completion of the original terms of this contract. Such renewal will be automatically granted at the end of term, unless written notice to cancel is provided to Operator by Owner six months prior to expiration of the term. Such notice may only be given if just cause is shown, as described in paragraph (e) of the section.

TABLE 2-1 continued. A SAMPLE SMATV CONTRACT

(c) The exclusive grant to Operator herein by Owner is irrevocable and shall be deemed to be an easement and covenant which shall run with the land for the term of this agreement, and the undertakings and obligations created herein shall be binding on and inure to the benefits of all future Owner(s) of the property. Owner agrees to include this contract as a title interest in the property deed, and further agrees to provide evidence of such registration in the form of a copy of the notarized filing with the appropriate authority.

(d) Operator agrees to pay Owner a fee of $10.00 per year for the use of the antenna and headend site, and all of the easements required for the installation of the distribution system. This rate will be fixed for the entire length of the contract and will apply to any renewal period as well.

(e) This contract may not be canceled for any cause, whatsoever, except for failure of Operator to build and operate the system in accordance with the specifications set forth in Exhibit "D" of this agreement. A breach of contract will occur only after Owner has notified Operator, in writing, of the failure to meet the required specification and has given Operator 20 days to correct such failure. In the event of any breach of contract, other than for just cause as defined in this agreement, Owner shall be liable to Operator for damages. The amount of such damages will be determined by the number of months left until the termination of the contract term, as specified in paragraph (a & b) of this section, times the average revenue derived from the system during the last three months of operation prior to the breach of contract.

6. Insurance
(a) Throughout the term of this agreement, Operator will maintain comprehensive general liability insurance covering Operator and Owner for bodily injury liability and property damage liability with coverage in the amount of at least $1,000,000 combined single-limit occurrence.

(b) Owner will provide, at his own expense, liability insurance in the amount of at least $1,000,000 combined single-limit per occurrence, protecting the investment of the Operator in event of the damage or loss of the system through fire, theft, flood, demolition or any other act not under the direct control the Operator. Such insurance will cover the current cost of replacement for all damaged and lost items, and provide for lost revenue, if any, during the course of repair. Owner further agrees to list Operator as additional insured on said policy.

7. Access to Property
Owner hereby grants to Operator the right to enter the complex between the hours of 9 a.m. to 9 p.m., 7 days a week, for the purpose of soliciting customers for the cable television system described herein. Owner additionally grants Operator the right of full time access, 24 hours a day, for the purpose of effecting repairs to the system.

8. Force Majeure
Operator shall not be liable in any manner whatsoever, for any delay or failure in performing any of his obligations under this agreement if such delay or failure is the result of legal restrictions, failure of Operator's vendors or suppliers, labor disputes, strike or lockout, boycott, fire, flood, revolution, insurrection, riot, war, public emergencies or any other cause beyond the control of Operator, specified above or not.

9. Entire Agreement
This agreement is the entire agreement between the parties, superseding all prior agreements, written or oral. This agreement may be amended by a written instrument signed by both parties.

10. Notices
Any notices by either party shall be in writing and delivered either personally or by certified mail to the addresses listed in this agreement.

11. Assignment
Operator may assign its rights under this agreement to any other person or entity without owner's prior consent.

IN WITNESS THEREOF, the parties hereto have executed this agreement on the date and year first above written.

by:_____
         Operator

by:_____
         Owner

E. REGULATIONS

In general, SMATV concerns are subjected to substantially fewer regulations than in the CATV industry in the United States. However, private cable is a relative newcomer and confusion still remains in some areas, especially in Canada and some European countries, nations that are generally more highly regulated than the United States.

As would be expected, in most countries there is always the need to adhere to local zoning and construction ordinances. These matters are discussed in more detail in the following chapters.

The United States

The regulatory environment in the United States is certainly less complex than that existing in Canada. There are no licenses, registration processes or fees required to receive or distribute conventional, off-air broadcasts. In addition, privately owning and operating a satellite reception system has been de-regulated since the late 1970s.

Aside from programming, commercial SMATV operations in the United States are not subject to any stringent regulations. Until mid-1992 SMATV operations were exempt from state and local franchising requirements as long as the system served only commonly owned multi-family dwellings and did not cross over or use public rights-of-way. Thus, service to mobile home parks, planned unit developments with single family homes and systems serving contiguous but non-commonly owned multi-family dwelling have been considered to be "cable systems" under federal law and required a local franchise even if the cable or communication link was not in a public street.

A United States Court of Appeals ruling issued in mid-1992 has essentially eliminated the need for franchises for operators who serve multiple multi-family dwellings owned by different parties when no public rights-of-way are used. In other words, for example, a single headend in one development may now be used to beam its signal via a line-of-sight relay to other multi-family properties. Or one SMATV headend may now be hard-wired to other properties owed by different parties in the same city block. Since one headend can now provide signals to multiple systems, costs can be reduced and smaller properties can be economically served.

As a commercial business, an SMATV operation may be subject to local laws concerning consumer protection, wiring codes, sales tax and the like. However, local zoning ordinances that target satellite dish antennas for discriminatory treatment have been pre-empted by the Federal Communications Commission (FCC).

Some states have cable access laws that allow the franchise CATV company to "overbuild" an SMATV system. While not a direct regulation of SMATV, these laws do impact system economics. Some operators respond to this situation by entering into "bulk" contracts with building owners, i.e. contracts for a flat fee paid by the owner based on the number of units in the property. As of late 1992, some courts are beginning to rule that the franchise cable operator may not overbuild an SMATV system even with an access law. This is an evolving area of the law and it would be wise to consult a knowledgeable industry attorney if this problem arises.

Any satellite antenna(s) used for receiving broadcasts need not be registered with the FCC. Only transmit facilities must be registered. In some cases it may be wise to keep track of any changes in the registration of transmit sites in the vicinity of an SMATV system. New installations could become a future source of terrestrial microwave interference.

The source of SMATV revenue is primarily the purchase and resale of pay-TV and superstation programming. Pay-TV programmers such as HBO, Cinemax, Showtime and TMC must be paid on a per subscriber per month basis. The details involved in obtaining programming either directly or via brokers are outlined in Chapter 3. Of course, private cable operators have access to a respectable number of channels offered free of charge (C-Span, CBN, etc...) with or without commercial support.

The U.S. Copyright Office also disseminates information about distribution of royalties. This can be obtained from:

Copyright Royalty Tribunal
1111 20th Street, N.W.
Washington, D.C. 20036

ECONOMICS, SALES CONTRACTS & REGULATIONS

Canada and European Nations

The regulations governing private cable systems vary from country to country throughout the world. However, the general trend has been towards a decreasing restrictions as communication technologies have evolved. The SMATV industry has experienced tremendous growth in some European countries including Sweden and Spain during the past few years.

In Canada, the relationship between SMATV operators and the law has been quite vague. The Canadian Radio-Television and Telecommunications Commission (CRTC) which regulates communications has not listed any rules aimed specifically at governing the activity of SMATV operations. The few that do exist were instated in the pioneer days of the telegraph. Nevertheless, cable companies carry a programming menu that has been essentially cleared by the CRTC and this selection has mainly been dictated by the CRTC's mandated "Canadian content" rules.

CATV operators must be licensed by the CRTC as common carriers in order to offer services for a profit. Licensing is granted on the basis of a number of factors including the financial standing of the applicant, the ability to provide uninterrupted service and the degree of infringement on local established broadcasters. A license is rarely granted to a company wishing to compete within the territory of another established concern.

However, while there has previously been no provision in Canadian law that would allow an SMATV operator to offer services for profit, the situation seems to be changing and the regulators are becoming more accepting of the SMATV industry's role.

CHAPTER 3. PROGRAMMING

> **Contributed in part by
> Barbara White Payne
> Vice-President, NSPN**

The name of the game in the private cable industry is quite simply programming. The game is "delivering entertainment via satellite to multi-family apartments, condominiums, mobile home parks, hospitals, universities, hotels, correctional facilities, military installations, retirement centers or any commercial establishment desiring to receive such programming by alternative technologies." Without easy access to the same entertainment channels offered by cable companies, the entrepreneurial private cable (SMATV) or wireless cable (MMDS) operator do not have a business.

However, legally obtaining and distributing programming to end-users for a profit has not always been an easy task for the small SMATV operator. Fortunately, the struggles of many of the pioneers in this industry have clarified and simplified this situation in the United States. Most programming that is transmitted via satellite is now legally available to private cable broadcasters who follow correct registration and purchase procedures. All conventional off-air programming has been and still is free for reception and distribution.

This situation is quite different in Canada where only a well-defined menu of services can legally be sold for profit by private cable operators. This list of "acceptable" programming has already been presented at the end of Chapter 2. Although the bulk of the following discussion therefore relates to the SMATV industry in the United States, the laws and regulations are evolving so quickly that the situation may change at any time in either country.

A. AVAILABLE SATELLITE PROGRAMMING

Many customers subscribe to SMATV systems in order to obtain premium programming which is usually transmitted via satellite. The number of options available is quite staggering. Table 3-1 outlines most of the satellite programming available to operators. This list is constantly changing. Therefore, the satellite TV guides that publish a weekly or monthly list of most currently broadcasted programs should be consulted for more current information (see Appendix D). The addresses and telephone numbers of most U.S. programmers are listed in Appendix C.

Technical Factors Relating to $2°$ Spacing

A number of new broadcast satellites launched in 1992 and following years is changing the location of numerous channels. During the month of June, 1992, most services on Galaxy 1 were transferred to the new, more powerful, 16-watt Galaxy V satellite. Reassignment of channels to Satcom C4 in late 1992 and Satcom C3 in 1992 as well as immediate

PROGRAMMING

changes are listed in Table 3-1. The movement of Telstar 303 to 125°West and its replacement at 125°West by Galaxy V is a factor in these changes.

As high powered satellites are shifted to 2° spacing, some adjustments to receiving equipment are necessary. While is has been possible to use a smaller aperture satellite reception antenna to receive Galaxy V, its higher powered neighbor would result in adjacent satellite interference. However, with satellites spaced so closely, one antenna fitted with a dual feed can simultaneously receive two satellites. These issues are examined further in Chapter 5, Headend System Design.

TABLE 3-1. SATELLITE DELIVERED PROGRAMMING and TRANSPONDER ASSIGNMENT

Program	Transponder Number	Moving	New Satellite	New Transponder Number	Date of Move
GALAXY 1					
A&E	12	Yes	Galaxy 5	23	6/1/92
Cinemax (East)	19	Yes	Galaxy 1R	?	1994
CMTV	13	Yes	Galaxy 1	2	1992
CNN	7	Yes	Galaxy 5	5	6/1/92
Headline News	8	Yes	Galaxy 5	22	6/1/92
Comedy Channel	1	No			
Discovery	22	No			
Disney	4	Yes	Galaxy 5	1	6/92
EPG, JR.	3				
ESPN	9	Yes	Galaxy 5	14	5/31/92
Family Channel	11	Yes	Galaxy 5	11	6/1/92
Galavision	20	Yes	?	?	
HBO (East)	23	Yes	Galaxy 1R	?	1994
– additional HBO (East)			Galaxy 5	15	6/1/92
Showtime (East)	5	No			
Showtime (West)	16	Yes	Galaxy 5	?	6/1/92
The Movie Channel (East)	10	?			
The Movie Channel (West)	14	Yes	Galaxy 5	?	
The Nashville Network	2	Yes	Galaxy 5	18	1992
TNT	17	Yes	Galaxy 5	17	6/1/92
USA (East)	21	Yes	Galaxy 5	19	6/1/92
WGN	3	No			
WTBS	18	Yes	Galaxy 5	6	6/1/92
WWOR	15	Yes	Satcom	?	1992
GALAXY 3					
C-Span	24	Yes	Satcom C3	?	3/93
C-Span II	14	Yes	Satcom C4	?	3/93
EWTN	10	Yes	Galaxy 1	11	1993
			Galaxy 1R	?	1994
Lifetime (East)	20	No			
MTV	17	Yes	Satcom 3 or 4	?	1993
Nickelodeon	22	?			
VH-1	15	?			
Weather Channel	13	No			

TABLE 3-1 continued... SATELLITE DELIVERED PROGRAMMING and TRANSPONDER ASSIGNMENT

Program	Transponder Number	Moving	New Satellite	New Transponder Number	Date of Move
SATCOM C1					
KCNC - Denver (NBC)	4	No			
KMGH - Denver (CHS)	6	No			
KRMA - Denver (PBS)	12	No			
KUSA - Denver (ABC)	2	No			
KWGN - Denver (independent)	14		No		
SATCOM F1					
American Movie Classics	14	Yes	Satcom C4	1	10/92
BET	20	Yes	Galaxy 5	20	6/1/92
Bravo	24	Yes	Satcom C4	?	11/92
Cinemax (West)	23	Yes	Galaxy 5	16	6/1/92
CNBC	6	No			
E! Entertainment	15	No			
HBO (West)	12	Yes	Galaxy 5	8	6/1/92
Inspiration Channel	18	Yes	Galaxy 1	17	6/92
Lifetime (West)	17	No			
Learning Channel	2	No			
QVC (Shopping Channel)	8	?			
Travel Channel	9	No			
Trinity Broadcasting	3	Yes	Galaxy 5	3	6/92
USA (West)	10	Yes	Galaxy 1	21	10/92
			Galaxy 1R	?	2/94
SATCOM F2					
WABC - New York (AB)	4	No			
WRAL - Raleigh (CBS)	2	No			
WXIA - Atlanta (NBC)	12	No			
SATCOM F4					
International Channel	12	No			
Nostalgia	21	Yes	Galaxy 5	?	1994
Playboy	24	No			
Spice	18	No			
Silent Network	24				
SPACENET 3					
KLTA - Los Angeles	15	No			
KTVT - Dallas	5	No			
WPIX - New York	9	No			
WSBK - Boston	3	No			

B. THE MECHANICS of PURCHASING PROGRAMMING

Programming created for sale to customers is protected by copyright laws. Rather well-defined and sometimes cumbersome licensing arrangements have been created to guarantee their rights. Until the early 1980s most programmers had been dealing exclusively with members of the cable TV industry who were quite familiar with the procedures necessary to legally purchase satellite broadcasts.

Private cable operators were not at all recognized before the early 1980s and their efforts to purchase programming were simply ignored. Some of these business people were subsequently sued when they began to provide and resell this programming without having formally obtained permission. Even though most satellite services are now directly offered to the private cable industry, the number of theft of service and copyright lawsuits filed by programmers against SMATV operators until recently had soared because many smaller operators were still not aware of the legal implications of purchasing and reselling programming.

Obtaining wholesale programming is not as simple as it first appears because each supplier has established a unique set of deposit and minimum purchase requirements. An SMATV operator can either attempt to negotiate with each programmer or can recruit the help of a co-op service organization.

TABLE 3-2. PROGRAM DISTRIBUTORS

Heifner Communications
4451 I-70 Drive, N.W.
Columbia, MO 65202
Telephone: 314-445-6163
FAX: 314-445-0757

National Satellite Programming Network
1909 Avenue G
Rosenberg, TX 77471-1489
Telephone: 800-937-6776
            713-342-9655
FAX: 713-342-7016

WSN
821 Marquette Avenue South, Suite 700
Minneapolis, MN 55402
Telephone: 800-367-3193
            612-339-9018
FAX: 612-371-8218

Co-operative Service Organizations

Co-operative organizations which serve the private cable operator have developed hand-in-hand with the industry (see Table 3-2). For example, the largest such group, the National Satellite Programming Network (NSPN) was founded by pioneers who participated in the struggle to convince programmers to sell to smaller operators. Today the three major organizations serving the industry are Heifner Communications, NSPN and WSN.

An authorized programming distributor is a company that provides a distribution license with the rights to sub-distribute satellite delivered programming on a national basis and ongoing marketing support to operators or commercial establishments without their having a requirement for direct ownership or management.

SMATV and MMDS operators would often be well advised to use the services of a program distributor to secure their desired channel line-up. Without such services, acquiring contracts from individual programmers may take many hours and could be quite costly. A distributor provides a single-source for all programming needs and thereby bypasses the time consuming task of placing multiple calls, negotiating, executing the often tedious and numerous contracts, meeting purchase minimums and financial commitments in the form of letters of credit and hefty up-front deposits. While operators may directly purchase services for a few pennies less, in the long run the cost of daily reporting, reconciling and check dispersement could be costly.

A distributor relieves programmers of the time consuming task of dealing with numerous small operators. Qualifying and signing up an operator with a 50 unit apartment takes as much time as signing up a 100,000 multiple system operator (MSO). Many programmers have established minimums in order to encourage operators to deal with distributors.

NSPN, established in 1982, was the first distributor to literally fight for access to programming for SMATV operators and today remains the largest offering 55 services to special markets in the United States. Their message, like that of the other service providers, has been to clearly convince the operator of the ease in dealing with a single source, and to continue their dedication to serving the industry by providing additional economical programming. They also provide a range of other services including a monthly programming guide, demographic information to support marketing efforts, an equipment financing program for qualified members and a yearly trade show. Central to their objectives is the desire to create an atmosphere of cooperation between its members and programmers. The value of the services provided to keep dealers and operators out of legal trouble also cannot be overstated.

Pay-per-view Services

Pay-per-view (PPV) programming is now the most rapidly growing segment of the satellite TV marketplace. Corporate suppliers such as Spectradyne and Hi-Net cater specifically to larger-scale hotel and motels. This technology allows a lodging operator to earn additional revenues from the sale of premium movies and other entertainment to guests or from the additional bookings provided by enticing customers with free services.

Spectradyne, originally an equipment manufacturer, is a pioneer in the PPV business having begun developing this "impulse" purchase technology in the early 1970s. It usually offers six movie channels delivered via cassette and a free programming package typically consisting of HBO, ESPN and CNN Headline News received via satellite. Nearly 400,000 rooms were served as of early 1987.

There are a number of competitive PPV firms The largest among these is the Holiday Corporation's Ku-band Hi-Net Network. While most cater solely to hotels and motels, a growing number are attempting to service other private cable market segments. For example, the NSPN leases PPV equipment to its members and then provides a pay-per-day programming package consisting of the Playboy and Nostalgia Channels. At present, technologies are being developed to allow lower-cost PPV systems to economically service systems having 100 or fewer subscribers.

C. CURRENT TRENDS and ISSUES in the U.S.

The status of programming available to a private cable or wireless cable broadcaster has changed dramatically during the past few years. When HBO finally jumped on the private cable bandwagon in late 1990 by announcing their intention to serve this industry via its distributors, only a handful of programmers who were still unwilling to participate remained. Notable among these are some of the sports channels who steadfastly refused to sell their product to either SMATV or MMDS operators. These include:

- Prime Ticket
- HSE (Home Sports Network)
- Sunshine Network
- Sports Channel (Pacific, America, Philadelphia, New York, Chicago, Florida and New England)
- Los Angeles Sportschannel
- Prime Sports Northwest
- Midwest Sportschannel
- Prime Sports Upper Midwest
- Sports News Satellite
- Prime Pittsburgh

As of this writing, only a few SMATV operators may resell such sports channels they purchase from a friendly, local cable operator. Other holdouts who will not license to special markets include:

- Encore
- Court TV
- VISN (Vision Interfaith Satellite Network)
- Sci-Fi Channel
- The Monitor Channel
- Univision
- Mind Extension University

PROGRAMMING

To date, HBO, Cinemax, Disney, ESPN, American Movie Classics, Bravo, TNT and The Nashville Network are the only services who refuse to work with distributors in serving MMDS broadcasters. They impose restrictions such as guarantees, hefty deposits and other strict criteria on MMDS firms. USA is the only programmer to charge a higher rate for MMDS than SMATV operators.

With direct broadcast, high power Ku-band (DBS) and compression technologies becoming a reality, the programming environment could change drastically. The opened question is whether the programmers will offer their services to the likes of companies such as Sky-Pix and Hubbard Communications. Even though the DBS broadcasters offer over 100 pay-per-view movies, the public may still demand the standard services such as MTS and CNN. Eventually, the DBS services should have their own news and sports channels but whether they can compete with established names such as ESPN remains to be seen. In the interim, the existing programming services and DBS could compliment one another by coexisting peacefully and working together to bring as much entertainment and information to the public as possible, by any and all technologies available. As of late 1992, Viacom Networks intended to be the first existing programmer to use video compression to offer their six services, Showtime, TMC, MTV, Nickelodeon, VH-1 and All News Channel (or more) on one transponder.

Today, there are sill some concerns regarding access to and ownership of programming services. With the continuing support, absorption and financing of programmers by MSOs, an inherent mechanism that enhances CATVs competitive barriers to competing technologies such as SMATV and MMDS is established. While financial interest and network syndication rules bar the television networks from owning the programs they distribute, no such rules exist for the large MSOs. They can own and set policy for the programmers. Certainly the combination of government regulations, restrictions and price discrimination on programming, and lack of financial resources have inhibited the development of MMDS in the United States. Regulation is generally not a benefit to any industry. However, the hampered growth of alternative broadcasting technologies resulting from the monopolistic ownership of programmers should be taken as a serious impediment to broadcasting to previously inaccessible areas via technology that may be more cost effective than traditional cable.

Even the definition of the term "private cable" is a subject of some debate within the FCC. The existing definition states that when SMATV systems interconnect multiple unit dwellings that are separately owned, controlled and managed, a franchise is required. On March 6, 1992, in Beach Communications et. al. V.FCC 91-1089, an appeal of the FCC's SMATV cable definitions was file in Washington, D.C. The court, at press time, asked the FCC for a rational basis for this definition. The court has basically stated "on the record before us we fail to see a rational basis for franchising external quasi-private SMATV, but not internal and wholly private systems." In the past, the fact that cable TV uses public rights-of-way has always been the rationale for local franchises making a distinction between quasi-private external and internal wholly private systems. This is somewhat confusing because none of these use public rights-of-way. The commission may reconsider its position before the court and declare all private property systems to be exempt from the local franchise requirement.

CHAPTER 4. THE SITE SURVEY and PLANNING

The secret to success in any venture lies in careful planning. This is especially the case for a private cable system that can be quite complex in design and construction. A detailed site survey and system design and subsequently a well conceived plan to implement the project are vital steps in successfully achieving both budgets and time schedules.

The plan must incorporate a carefully considered design, probably the single most important element in the overall project. A well-designed private cable system will ultimately save time and money compared to a poorly conceived one. Chapters 4, 5 and 6 explore details of the site survey, planning and design process.

A. THE SITE SURVEY PROCESS

The first critical step preceding the design and installation of an SMATV system is the site survey. The purpose of this procedure is to consciously and methodically gather detailed information about the resources on hand at the site. It is the most crucial input to the design process, and the skill or lack of skill by which it is implemented can lead to either the success or failure of the venture. It is essential to be as careful and as accurate as possible during the survey process.

The site survey can be broken down into five basic components which relate to the availability of broadcast signals and to the optimal method for distributing them throughout a complex. First, the particular environment impacting upon the detection of satellite signals is examined. This involves a consideration of where to install the satellite reception antennas, whether or not terrestrial interference is present and what regions of the geosynchronous arc are to be targeted. Second, a survey of off-air signals is conducted. These first two steps involve measuring the strength and quality of both satellite and off-air signals. Third, the location where the headend will be installed is studied. Fourth, the details of the proposed distribution system must be examined to determine the most fruitful course of action. Finally, all this information is assembled into a detailed design and plan of action.

The reader should note that there are a number of detailed texts available which provide more information on certain aspects that are briefly examined in this book. Some of these references are outlined following the Appendices.

B. BACKGROUND TECHNICAL INFORMATION

Basic Electrical Theory

The most basic electrical theory is needed to understand the essentials of private cable operation and design. This and the next two subsections provide the background material required to become a competent system designer. Most technically educated readers should be able to either skim through or skip these introductory sections.

Ohm's Law and Resistance

The basic law of electricity is Ohm's Law. It simply states that an electrical direct current (dc) is generated in proportion to the voltage provided. This is algebraically expressed as:

$$V = IR$$

where V is the voltage in volts, I is the current in amperes and R is the resistance to current flow in ohms. A simple analogy to the flow of water in pipes can make this equation clear. The amount of water flow (current) increases as the force driving the water (voltage) increases. If the resistance to flow increases, the current must decrease for a given driving force.

The amount of power produced by a given current flow and voltage equals the product of voltage and current. Using Ohm's Law this can be expressed in the following equivalent forms:

$$P = VI = I^2R = V^2/R$$

Impedance and Alternating Currents

When the electrical current is rapidly varying, factors in addition to resistive losses come into play. As electrical charges alternate (alternating current, ac), the higher frequency signals are more likely to leak away to ground resulting in power losses. This phenomenon is known as capacitive reactance. Inductive reactance occurs when moving electrical charges "induce" movement in nearby electrically charged particles. This is also a type of signal loss experienced in circuit elements known as coils and chokes. The combined effects of resistance, capacitive reactance and inductive reactance are lumped together and expressed in a parameter known as impedance. These losses can be appreciable in long cable runs and substantial amounts of power may be required to "drive" signals through lines.

Electrical devices have characteristic impedances. For example, coaxial cables used in SMATV systems generally have a 75 ohm characteristic impedance. The lead from the dipole of a wireless antenna to the downconverter has an impedance of 50 ohms. All amplifiers, receivers and other electronic components have characteristic input and output impedances. If the impedance of a cable does not match that of the device it feeds, there will be a substantial reflective power loss. This mismatch would be similar to sending water from a larger diameter pipe into one having a smaller diameter. The resistance to flow would be heard as knocking and seen as turbulent flow. Similarly, if the output impedance of an amplifier in not matched to the impedance of the coaxial line, reflections and losses will occur.

The Decibel Notation

The decibel notation is used throughout this book and the satellite, private cable and cable TV industries. It serves as the background to understanding and managing much of the technical information that follows. It is invaluable as a tool in the process of designing private cable systems.

The decibel scale was named after Alexander Graham Bell to allow relatively small numbers to describe the potentially enormous changes in power that occur at various stages in an electronic communication chain. It also describes the logarithmic fashion in which human senses such as hearing and vision respond to changes in sounds and light levels. Using decibels greatly simplifies calculations because components like amplifiers and antennas typically increase power levels by factors of tens or hundreds of thousands. The alternative to the decibel notation, constantly writing out such numbers, could have become very cumbersome. Very large changes in power or voltage levels are therefore translated into manageable values expressed as decibels.

Decibel (dB) changes in relative power are defined by the following logarithmic equation:

Decibel Change = 10 logarithm (signal A/signal B)

SITE SURVEY/PLANNING

For example, if signal A is 1000 watts and signal B is 10 watts, then signal A is 20 dB stronger than signal B because:

Decibel change in power = 10 log (1000/10)
= 10 log 100
= 20 dB

Working the equation backwards shows that a 3 dB change means a doubling of power, while a relative change of 30 dB means an increase of power by a factor of 1000. Some examples of decibel values are listed in Table 4-1.

TABLE 4-1. THE DECIBEL NOTATION

Number of Decibels	Relative Increase in Power
0	1
1	1.26
3	2
10	10
20	100
30	1,000
50	100,000
100	10,000,000,000

It is important to realize that the decibel scale is a relative one. Therefore, in the above example, signal A is 100 times stronger than signal B. A 20 dB change could also describe a hundred-fold relative increase in sound or electrical power.

Standard abbreviations of dBmV and dBw have been created to refer to increases relative to 1 millivolt and 1 watt, respectively. Therefore, for example, 0 dBw refers to 1 watt. Signal levels in private cable calculations are usually measured in microvolts (millionths of a volt). The standard level used for comparison is 1000 microvolts which equals 1 millivolt or 0 dBmV measured across 75 ohms of impedance.

It is important to understand how and why the method to calculate decibels relative to one millivolt differs from that used to measure decibels relative to one watt. The equations used are:

$$dBm = 10 \log (\text{signal in watts}/1 \text{ watt})$$

and

$$dBmV = 20 \log (\text{signal in millivolts}/1 \text{ millivolt})$$

The voltage levels in dBmV are defined as twice the power changes as expressed in dBm. This difference stems from Ohm's Law which shows how the power varies as the square of the voltage level. If a given voltage doubles, power increases by a factor of 4 which equals 6 dB. In order that the decibel changes in power and voltage are consistent, the definition for dBmV includes an additional factor of 2. The following example should further clarify the factors underlying the difference in definitions.

Compare the difference in decibels between 15 millivolts and 60 millivolts impressed across a 75 ohm resistance. The powers are:

$$P = V/R = 15/75 = 3 \text{ watts}$$

and

$$P = 60/75 = 48 \text{ watts}$$

Therefore, power and voltage differences in dBm and dBmV, which are expected to be equal, are given by:

Change in power = 10 log (48/3) = 12.0 dB

Change in voltage = 20 log (60/15) = 12.0 dB

The decibel scale incorporates an extremely useful feature based upon one of the mathematical properties of logarithms. When amplifiers, attenuation pads or other electronic components are cascaded, the changes in power can be calculated by simply adding the gains or subtracting the attenuations as expressed in decibels.

An example of this point is quite illustrative and should be clearly understood. If an antenna with a gain of 35 dB (a factor of 3,162) feeds into an amplifier with a gain of 10 dB (a factor of 10), the resultant gain is 3,162 times 10 which equals both 31,620 and 45 dB, also the sum of the individual gains expressed in decibels. If a signal of -46 dBmV (0.0050 millivolts or 5 microvolts) is fed into this system, the output signal is -46 dBmV plus 45 dB which equals -1 dBmV (0.89 millivolts). It is important to understand that input signal is expressed as a voltage in dBmV, the amplification as a relative increase in decibels and the output signal again as a voltage in dBmV.

All MATV and SMATV design calculations, including antenna and amplifier gains, lines and splitter losses, insertion losses and isolations are expressed in decibels. The decibel notation allows sim-

SITE SURVEY/PLANNING

ple additions and subtractions to be used so that otherwise complicated design calculations become easy. For convenience, a conversion from dBmV to microvolts is provided in Table 4-2.

TABLE 4-2. dBmV to MICROVOLT CONVERSION TABLE

dBmV	uV	dBmV	uV	dBmV	uV
-40	10.00	0	1000	41	112200
-39	11.22	1	1122	42	125900
-38	12.59	2	1259	43	141300
-37	14.13	3	1413	44	158500
-36	15.85	4	1585	45	177800
-35	17.78	5	1778	46	199500
-34	19.95	6	1995	47	223900
-33	22.39	7	2239	48	251200
-32	25.12	8	2512	49	281800
-31	28.18	9	2818	50	316200
-30	31.62	10	3162	51	354800
-29	35.48	11	3548	52	398100
-28	39.81	12	3981	53	446700
-27	44.67	13	4467	54	501200
-26	50.12	14	5012	55	562300
-25	56.23	15	5623	56	631000
-24	63.10	16	6310	57	707900
-23	70.79	17	7079	58	794300
-22	79.43	18	7943	59	891300
-21	89.13	19	8913	60	1000000
-20	100.0	20	10000	61	1122000
-19	112.2	21	11220	62	1259000
-18	125.9	22	12590	63	1413000
-17	141.3	23	14130	64	1585000
-16	158.5	24	15850	65	1778000
-15	177.8	25	17780	66	1995000
-14	199.5	26	19950	67	2239000
-13	223.9	27	22390	68	2512000
-12	251.2	28	25120	69	2818000
-11	281.8	29	28180	70	3162000
-10	316.2	30	31620	71	3548000
-9	354.8	31	35480	72	3981000
-8	398.1	32	39810	73	4467000
-7	446.7	33	44670	74	5012000
-6	501.2	34	50120	75	5623000
-5	562.3	35	56230	76	6310000
-4	631.0	36	63100	77	7079000
-3	707.9	37	70790	78	7943000
-2	794.3	38	79430	79	8913000
-1	891.3	39	89130	80	10000000
0	1000	40	10000		

Note: 0 dBm = 1000uV Reference Level

Three rather simple but illustrative examples of using decibels for system design calculations are presented below (see Figures 4-1, 4-2 and 4-3).

Example I

In this example, an off-air antenna receives a 200 microvolt (0.2 millivolt) signal. This voltage is passed into a preamplifier and subsequently into a television set via a coaxial cable line (see Figure 4-1). The input to the set is calculated by first converting the signal to dBmV as follows:

$$\text{Input signal (dBmV)} = 20 \log (0.2/1)$$
$$= -14 \, \text{dBmV}$$

Then the signal level delivered to the television is found as follows:

Antenna output signal level =	−14 dBmV
plus preamp gain =	19 dB
less cable loss =	−3 dB

Therefore, signal level to TV = +2 dBmV

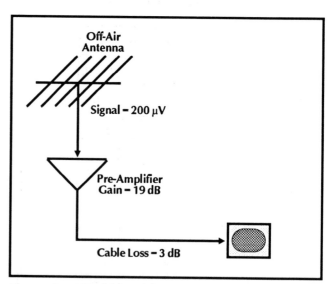

Figure 4-1. Working with Decibels - Example I. *This simple example consists of an off-air antenna, preamp, cable and television.*

Example II

In this second case, two televisions are fed from one off-air antenna which delivers a 1000 microvolt or a 0 dBmV signal (see Figure 4-2). The preamp gain is 17 dB, losses in the line to the splitter are 6 dB, the loss in the splitter is 3 dB and the subsequent cable loss is 8 dB. Then the signal level available to the television set is:

Antenna output signal level =	0 dBmV
plus preamp gain =	17 dB
less cable loss to splitter =	–6 dB
less splitter loss =	–3 dB
less cable loss to television =	8 dB

Therefore, signal level to TV = 0 dBmV

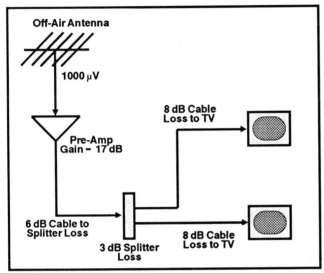

Figure 4-2. Working with Decibels - Example II. *This second example also incorporates a 2-way splitter having a 3 dB loss.*

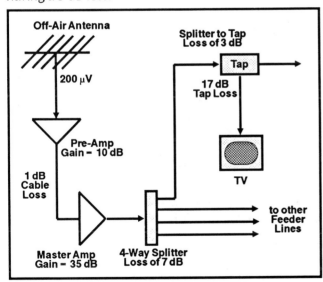

Figure 4-3. Another Example of Using Decibels. *Here a 200 microvolt signal, which equals –14 dBmV, is increased in level to –4 dBmV by the 10 dB preamp gain. Following a 1 dB cable loss, a 35 dB line amplifier gain, a 7 dB splitter loss and 3 dB cable loss, the signal reaching the tap is found by addition and subtraction to be 20 dBmV. When the 17 dB tap isolation loss is subtracted, 3 dBmV finally reaches the television set.*

Signal-to-Noise Ratio and Picture Quality

The concept of signal-to-noise (S/N) ratio is central to understanding television broadcast systems. S/N is the ratio between the signal and noise power expressed in decibels. The performance of an electronic device is determined by the input S/N ratio and its internal design.

Television Picture Quality and S/N

Experimenters have determined that television picture quality is directly related to the input signal-to-noise ratio. In 1959, the FCC authorized a task group to study the problems associated with television broadcasting and reception. This group was known as the Television Allocations Study Organization, or TASO. The results of a picture quality study were included in their final report. In this test, a standard TV set was viewed by numerous people who were asked to grade the quality of the picture as S/N ratio was varied. The results are shown in Table 4-3.

TABLE 4-3. SUBJECTIVE RATING of PICTURE QUALITY vs. S/N RATIO (TASO Study)

S/N Ratio (dB)	Subjective Picture Quality
above 45	Excellent quality. As good as could be desired
40	Good quality and providing enjoyable viewing. Some interference perceptible
33	Passable quality. On-screen interference not objectionable
28	Marginal. A picture of poor quality with somewhat objectionable interference
22	Poor. A just-watchable picture having definite objectionable interference
below 22	Unusable. Pictures too bad for viewing.

SITE SURVEY/PLANNING

A second similar study was conducted by the National Cable Television Association. Their results show more clearly the statistical and subjective nature of arriving at judgments of television picture quality. For example, while 50% of viewers judged picture quality to be excellent when the S/N ratio exceeded 43 dB, 40% more or 90% of the subjects judged pictures to be excellent only when S/N ratio was greater than 49 dB. These results are listed in Table 4-4.

TABLE 4-4. SUBJECTIVE RATING of PICTURE QUALITY versus S/N RATIO (NCTA Study)

RATING	90%	50%
Excellent	49 dB	43 dB
Fine	42 dB	35 dB
Passable	37 dB	30 dB
Marginal	30 dB	25 dB
Inferior	23 dB	19 dB

Noise

All signals are subjected to the influence of random electrical signals known as noise. Noise is caused by the endless motion of the molecules that compose matter. These small, vibrating charged particles generate electromagnetic fields that can mask the organized signal sent by man-made devices. Noise is present in all matter at temperatures above absolute zero degrees Kelvin, the temperature at which all molecular motion ceases. There is no temperature colder than absolute zero, $0°K$, which equals both $-459.69°F$ and $-273.16°C$.

Noise is generated by internal heat in power supplies, amplifiers, receivers and other electronic equipment. Noise which is detected by all satellite and off-air antennas also stems from a combination of man-made and natural sources. Man-made noise can be generated by numerous devices including electric motors, neon lights, coronal discharges from power lines and vehicle ignition systems. Natural noise sources include galactic noise, solar noise, heat and even distant lightning discharges (see Figure 4-4).

Noise power increases with both temperature and signal bandwidth. This makes sense because, as bandwidth increases, more noise is detected along with the signal. Noise power is given by:

$$\text{Noise Power} = kTB$$

where k is Boltzman's constant equal to 1.38×10^{-23}. T is the temperature in degrees Kelvin which equals degrees C plus 273, and B is the bandwidth in Hertz.

The approximate noise power contained in a television signal is now easily calculated. The bulk of information in this signal is contained in a 4.2 MHz band. At room temperature, approximately $290°K$, the noise power is found to be 1.6×10^{-14} watts. If this power is fed into a 75 ohm impedance, the voltage generated is calculated by Ohm's Law to be 1.1 microvolts or -59 dBmV. Note that the 75 ohm value is chosen here because most antennas, amplifiers, receivers and electronic components used in private cable systems are designed to have 75 ohm input and output impedances to match the impedances of standard coaxial cables.

Amplifier Noise

Amplifiers increase both noise and signal levels presented at their inputs. Therefore, for example, if -59 dBmV were fed into a 20 dB amplifier, the output noise power would be increased to -39 dBmV. However, amplifiers are not perfect devices and also add some noise as a signal passes through their internal circuits. In the above example, if the noise power were measured at the output and found to be not -39 dBmV but -33 dBmV, then clearly 6 dBmV of noise was added by the amplifier.

Amplifier noise figures range from extremely low values of 0.4 dB, less than $30°K$, for LNAs (hence the name low noise amplifier) to a range of 2 to 15 dB for more conventional amplifiers. Off-air antenna preamps typically have noise figures that fall at the lower end of this latter range. Preamps are usually mounted right at the off-air antenna for best low-noise reception. Amplifiers with higher noise figures are usually designed for relatively high signal levels so that the precise value of the noise figure is not as critical.

The concept of equivalent noise figure is important in private cable design calculations. If amplifier gain is known and noise power is measured, the equivalent noise figure is the measured noise power less the amplifier gain. Using the same numbers as in

SITE SURVEY/PLANNING

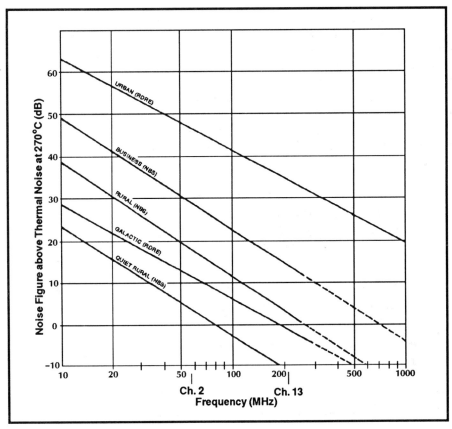

Figure 4-4. Typical Antenna Noise Levels. *The expected noise figures at the temperature of an average summer day are plotted versus channel frequency for a number of situations. Noise levels are higher for the lower frequency channels.*

the previous example, if the output noise power is −33 dBmV and the gain is 20 dBmV, the equivalent noise figure is −53 dB.

Amplifiers, along with headend processors, are the two main sources of noise in an SMATV system. While the noise generated by amplifiers is generally flat across the frequency spectrum, noise from the headend is most often related to the frequency being processed. A useful rule of thumb to keep in mind is that the noise figure of a system increases by 3 dB every time the number of downstream amplifiers is doubled. Therefore, for example, if 2, 4 and 8 amplifiers are used in a distribution system, noise levels will increase by 3, 6 and 9 dB, respectively, at the output of these components.

Calculating Television S/N Ratio

The necessary ingredients are now in hand for determining the S/N ratio at the input to a SMATV system. The calculations are simple. The S/N ratio is given by:

S/N Ratio = Signal power
less
Equivalent amplifier noise figure

For example, if the signal power is −12 dBmV and the equivalent noise figure is −53 dB, the S/N ratio is calculated as −12 dBmV minus −53 dBmV which equals 41 dB. This signal-to-noise ratio is sufficient to provide a good quality television picture.

As the signal progresses through a headend and distribution system it undergoes various amplifications and attenuations and, in the process, extra noise is added. However, if the S/N ratio at the pre-amp output is sufficiently high, maintaining this level throughout the cable network is fortunately a simple task.

The S/N ratio at the output of a satellite receiver must also be adequate to properly drive all televisions connected to the distribution system. While an outline of these calculations is presented in Chapter 5 and Appendix B, for more information the reader should refer to either the C- or Ku-band satellite manuals referenced at the end of this book.

SITE SURVEY/PLANNING

C. THE SATELLITE PROGRAMMING SURVEY

The first step in conducting a detailed site survey is an assessment of the available satellite channels. A list of the programming desired for distribution must then be created. This step leads to a determination of the satellites to be viewed. Table 3-1 in Chapter 3 has outlined much of the programming currently available to SMATV operators in North America. This menu is changing from month to month as the satellite broadcasting industry grows and evolves, so this list should be considered an approximation. Please see Appendix D for a listing of Satellite TV Guides from which further information can be obtained.

D. THE TVRO SITE SURVEY

Once the satellite menu has been chosen, a detailed site survey must be performed to determine if there are any technical constraints or other impediments to detecting the required satellites and transponders.

A well-conducted site survey is the critical step before installing the satellite reception portion of a private cable system. Three important tasks should be completed. First, a location with a clear view of the communication satellites to be viewed must be found. Second, a test is conducted to determine whether or not terrestrial interference (TI) is present at this location. In those cases where TI is a concern, its level and source must be determined. Third, the installation of the satellite antenna and cable routes to the headend must be planned. Important facts such as what type of equipment will be used, where to place the dish to protect it from wind loading and interference, how much concrete is necessary, whether or not to use an az-el or polar mount, how to protect the system from lightning strikes and how to route cable runs must all be decided well in advance of submitting a bid and installing the system.

Ensuring a Clear View of the Arc

The satellite antenna must have a clear view of each satellite to be detected because any obstruction can absorb or reflect microwaves and subsequently lower the S/N ratio and hinder reception. Water in deciduous trees is a particularly strong absorber of microwaves and the signal losses it causes can make pictures unwatchable.

The two instruments required to locate any communication satellite are an inclinometer and a compass (see Figures 4-5 and 4-6) Each satellite can be targeted by knowing its azimuth and elevation angle (Appendix B presents the formulas to calculate these angles). The azimuth is measured in degrees of rotation from true north in a plane parallel to the horizon and the elevation is measured in degrees above the horizon.

Figure 4-5. Site Survey Tool. *The compass is one of the basic tools required by a satellite TV installer. It is used to locate and aim at the geosynchronous arc.*

34

SITE SURVEY/PLANNING

Figure 4-6. Inclinometers. *An assortment of inclinometers is pictured here. The rounded instruments display the reading on an external scale. The smaller device is sighted through to the target. The instrument in the rear is a level/inclinometer with a digital read-out.*

When choosing the antenna site a number of other factors must also be considered. The dish should usually be hidden from view to improve the cosmetic appearance of the site and to protect equipment from vandalism. It must be installed in a location where wind loading is not be a major factor and where protection from lightning is easier. Both these factors suggest that it should be mounted as closely to ground level as possible. The site should also be in relatively close proximity to the headend equipment to minimize cable runs.

E. AVOIDING TERRESTRIAL INTERFERENCE

Terrestrial interference (TI) occurs when a satellite reception system detects unwanted microwave signals. It is the major cause of reception problems in SMATV systems. The effects on picture quality can range from a mild case of "sparklies" or "snow" to complete wipe-out. TI is present in most urban environments and has been increasing every year as new microwave links are established. An estimated 20% off all new TVRO installations now experience some form of TI.

Nevertheless, TI can easily be understood and avoided in order to prevent or minimize reception degradation. Its elimination should not be a long and tiresome process. It is vital to be as complete as possible in identifying and solving any potential problem. There is nothing worse after an installation is completed to discover that an antenna must be moved or replaced to deal with a TI problem. Such a mistake can be costly and can lead a client to seriously doubt an operator's competence.

If TI is detected, the nature, amplitude, frequency as well as the elevation and azimuth of the TI source must be determined. These parameters will determine what countermeasures can be taken. In rare cases, it may even be wise to avoid bidding the job.

Fortunately, as Ku-band broadcasts assume a larger portion of the private cable menu, concerns with TI will decrease. There is only a rather remote chance for interference to occur in this higher frequency range.

Sources of TI

Satellite broadcasts which are transmitted in the 3.7 to 4.2 GHz C-band region of the frequency spectrum are not alone. Most local telephone companies also utilize this band for their line-of-sight, earth-based microwave relays. In addition, many businesses use microwave point-to-point services for data transfer. Microwave towers beam and receive the full range of communications including computer data, network TV and audio feeds and telephone transmissions (see Figure 4-7).

Interference originating from any signals sharing the band of frequencies which are used at any point in a satellite TV receiving system can potentially cause problems (see Table 4-5). This ranges from the microwave C-band (3.7 to 4.2 GHz) and above to as low as the audio and video baseband range (0 to 6 MHz). Thus, for example, if an unwanted 70 MHz signal leaks into a receiver which processes satellite channels on the same frequency, pictures can deteriorate. Or, if a nearby communicator is using a 980 MHz relay, the 950 to 1450 MHz range generated

SITE SURVEY/PLANNING

Figure 4-7. Line-of-Sight Relay Tower.
Thousand of towers such as this one relay C-band microwaves for many terrestrial communicators such as common carriers. (Courtesy of Microwave Filter Company)

by most LNBs and block downconverters might be affected.

The interference band can be broken down into two segments: the ingress interference band below 1 GHz; and the antenna interference band from 1 GHz to approximately 8 GHz. The difference between these two bands is that, unlike microwave signals in the 1 to 8 GHz range, the lower frequencies cannot enter via the antenna/feedhorn/LNB route.

The Ingress Interference Band

TI of frequencies lower than about 300 MHz will probably enter through poorly grounded or improperly connected equipment. For example, a local TV station might be received along with the satellite TV signal if a ground connection is "floating" and has not been properly connected. At frequencies above 300 MHz, wavelengths are short enough that signals can leak in through poorly shielded equipment cases or into openings of sufficient size (greater than about 2/10th of an inch at 300 MHz).

TABLE 4-5. POTENTIAL SOURCES of ANTENNA INTERFERENCE

Frequency (GHz)	Nature of Potential Offender
0.960-1.350	Land-based air navigation systems
1.350-1.400	Armed Forces
1.400-1.427	Radio astronomy
1.427-1.435	Land-mobile: police, fire, forestry, railway
1.429-1.435	Armed Forces
1.435-1.535	Telemetry
1.535-1.543	SAT–maritime mobile
1.605-1.800	Radio location
1.660-1.670	Radio astronomy
1.660-1.700	Meteorological radiosond
1.700-1.710	Space research
1.710-1.850	Armed forces
1.990-2.110	TV Pick-up [2]
2.110-2.180	Public common carrier
2.130-2.150	Fixed point-to-point (non-public)
2.150-2.180	Fixed omnidirectional
2.180-2.200	Fixed, point-to-point (non-public)
2.200-2.290	Armed forces
2.290-2.300	Space research
2.450-2.500	Radio location
2.500-2.535	Fixed, SAT
2.500-2.690	Fixed point-to-point (non-public) Instructional TV
2.655-2.690	Fixed, SAT
2.690-2.700	Radio astronomy
2.700-2.900	Armed forces
2.900-3.100	Maritime radio navigation
2.900-3.700	Radio location [3]
3.300-3.500	Amateur radio
3.700-4.200	Common carrier (telephone) Earth stations [1]
4.200-4.400	Altimeters [3]
4.400-4.990	Armed forces
4.990-5.000	Meteorological /radio astronomy
5.250-5.650	Radio location (coastal radar)
5.460-5.470	Radio navigation – general
5.470-5.650	Maritime radio navigation
5.600-5.650	Meteorological ground based radar
5.650-5.925	Amateur
5.800	Industrial and scientific equipment
5.925-6.425	Common carrier and fixed SAT [2]
6.425-6.525	Common carrier
6.525-6.575	Operational land and mobile
6.575-6.87	Non-public point-to-point carrier
6.625-6.875	Fixed SAT
6.875-7.125	TV pick-up
7.125-8.400	Armed forces
8.800	Airborne Doppler radar

Interference frequencies are listed in order of occurrence
The most likely sources of TI come from communicators operating in or near the 3.7 to 4.2 GHz C-band range of frequencies.

1. Telephone carrier spectrum co-located with TVROs.
2. Widely distributed common microwave carriers.
3. Seldom occurring frequencies close to TVRO Band.

The Antenna Interference Band

Whereas ingress interference usually arises when an installation is not "perfect," TI can enter through the front door, the satellite antenna, even in the most professionally installed system. The antenna interference band can be subdivided into two distinct regions: in-band TI centered on 3.7 to 4.2 GHz; and out-of-band TI spanning the 1 to 8 GHz range but excluding C-band frequencies.

In-Band TI

The source of in-band TI is nearly always the common carrier, repeater stations that share the C-band with satellite TV transmissions. These communication relays carry voice, video and data traffic for a fee via the familiar microwave antennas situated on large towers scattered across the country. AT&T, Sprint, MCI, and Allnet are among the better known telcos but many regional and local common carriers also are in service.

These terrestrial C-band networks have proliferated as telephone, data and other communication needs continue to grow. It is estimated that new license applications and requests for network modifications have been submitted in the early 1980s at a rate of 30 to 50 every month. In total, the growth rate has been at least 10 to 20 percent yearly. Today, such networks crisscross the United States like a giant spider web centered on the major metropolitan areas but spanning rural regions. In-band TI accounts for about 95% of all interference problems. Fortunately, this situation will slowly be improving as fiber optic networks continue to expand. Fiber optics trunks have far greater communication capacities than the existing microwave lines and operate with light which cannot cause interference.

Each antenna on a repeater station can carry up to 6 different frequencies of either vertical or horizontal polarization. Each tower can have multiple antennas. In some areas, only one frequency is used; in others, all 25 possible common carrier channels may cause serious problems for TVROs receiving satellite TV broadcasts.

The FCC has wisely allocated these "Ma-Bell" transmissions different center frequencies than those used by satellite TV broadcasts (see Table 4-6). The allocated frequencies lie 10 MHz above and below the center frequency used by each C-band satellite channel. For example, channel five is centered on 3800 MHz, so that a common carrier relay using frequencies of 3790 and 3810 MHz can be potential sources of interference. A total of 25 frequencies are assigned to common carriers; 23 between each channel and one each above and below the satellite range.

TABLE 4-6. COMMON CARRIER CENTER FREQUENCIES

Transponder Number	Satellite Center Frequency (MHz)	Common Carrier Center Frequency (MHz)
		3710
1	3720	
		3730
2	3740	
		3750
3	3760	
		3770
4	3780	
		3790
5	3800	
		3810
6	3820	
		3830
7	3840	
		3850
8	3860	
		3870
9	3880	
		3890
10	3900	
		3910
11	3920	
		3930
12	3940	
		3950
13	3960	
		3970
14	3980	
		3990
15	4000	
		4100
16	4020	
		4130
17	4040	
		4150
18	4060	
		4170
19	4080	
		4190
20	4100	
		4210
21	4120	
		4230
22	4140	
		4250
23	4160	
		4270
24	4180	
		4290

SITE SURVEY/PLANNING

Common carriers usually relay signals having a 3 to 5 MHz bandwidth (see Figure 4-8). Both the amplitude and bandwidth of the interfering signals increase as telephone traffic increases. These narrow-band transmissions which are typically used for telephone links can often be easily filtered or "notched out" from the signal detected by a satellite receiver. Narrow-band transmissions are a relatively small portion of the typical transponder bandwidth which ranges from 20 to 32 MHz. However, wide-band formats often used for videoconferencing links and data service networks have become more commonplace. Such 5 to 30 MHz bandwidth signals severely interfere with satellite broadcasts and cannot be eliminated with notch filters without destroying a great deal of the picture information. More expensive and sophisticated phase cancellation methods are required to overcome this form of TI.

Out-of-band TI

Out-of-band TI have carrier frequencies that generally lie in the 1 to 3.7 and 4.2 to 8 GHz bands. There are four major communication bands accounting for most of the out-of-band TI. These are listed in Table 4-7.

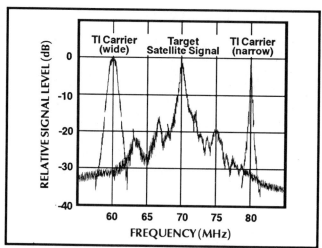

Figure 4-8. Wide versus Narrow Band TI. *TI detected from line-of-sight transmitters can have center frequencies ± 10 MHz from the satellite carrier. In this case, the IF of the satellite receiver is 70 Mhz. The wide band interference illustrated here has a bandwidth of approximately 5 MHz.*

TABLE 4-7. MAJOR SOURCES OF OUT-OF-BAND TI

1.990 - 2.110 GHz	TV studios to relay towers & between metropolitan areas
2.110 - 2.180 GHz	Common carrier bands for relaying data
6.425 - 6.525 GHz	
5.925 - 6.425 GHz	"Fixed SAT" for carrying remotely originating programs to uplinks for satellite broadcasts

Effects of TI on Satellite TV

Ingress Interference

Ingress interference, namely interference not entering the system via the antenna, has the recognizable effect of causing common problems on all satellite channels. This is observable as a characteristic picture defect or as an audio buzz.

Out-of-band TI

Out-of-band TI also has recognizable effects on satellite television reception. The most dramatic disturbances are usually seen at either end of the satellite band and often on channels of one polarization. Occasionally, the presence of interference is observed as a gradual change from "bad" to "good" channels across the satellite band. An out-of-band interfering signal affects those satellite channels closest in frequency. Thus, for example, if a strong 3.5 GHz signal were detected by an antenna and transmitted into a receiver it would most strongly impact upon the lower frequency channels. If this interfering signal happened to be either horizontally or vertically polarized, it would most strongly affect channels having the same signal polarization. This type of picture deterioration is generally the same on all affected channels.

It is rare that weak to moderate out-of-band antenna interference below approximately 2577 MHz will affect a satellite receiving system. The WR-229 waveguide flange on all C-band LNA inputs strongly attenuates signals having frequencies below 2577 MHz. However, powerful out-of-band signals can and do sometimes overpower the system and interfere with reception.

SITE SURVEY/PLANNING

In-band TI

In-band TI also has characteristic signs. Repeater stations transmit up to 6 carriers of a given polarity spaced 80 MHz apart. Thus, if interference from one relay were intercepted, 6 satellite channels spaced 2 channels apart would be affected. If more than one transmitting antenna were used and if the satellite antenna also detected those signals received by the repeater station having different polarity or frequency, all 24 channels could be affected. Usually, patterns of good and bad channels are seen when in-band problems arise (see Figures 4-9a, 4-9b and 4-10). In general, even if all 24 channels were affected, the degree of interference would vary between channels.

The effect that in-band TI has on satellite television depends upon its power relative to the satellite signal (see Table 4-8). If it is 18 dB below the desired signal then it is not noticeable on a TV screen. As TI powers rise to about −10 dB, sparklies or snow begins to appear on the screen. These increase until about −3 dB when the dots reach a "flurry" level. At 0 dB, the "blizzard" level, the picture is nearly lost. Above this power, TI begins to detune the receiver away from the satellite broadcast; the automatic frequency control (AFC) circuit begins to track the TI instead of the satellite signal. At +3 dB the screen has no detectable picture and a coarse appearance. At +5 dB the screen is blank and uneven. Above +10 dB the screen is completely whited out and has a fine, even texture.

If the AFC circuit is disabled, usually with an external or internal receiver switch, the satellite frequency can be manually tracked. Then a watchable picture may be seen even with TI levels as high as 10 dB. Note that if a receiver has frequency synthesized tuning it must have a very "stiff" AFC circuit which deviates less than 3 MHz from the selected frequency. If not, it may track even lower levels of TI than a receiver normally would.

Both wide-band and narrow-band TI have similar effects on satellite picture quality. But wide-band interference carries more power and therefore has more pronounced negative effects. Note that TI may be noticed only at certain times of day depending upon the telephone or data transmission traffic load. It is wise to conduct a site survey both during peak business hours and later in the day when telephone traffic is heaviest.

TABLE 4-8. RELATIVE TI LEVELS AND SYMPTOMS

TI Relative to Satellite TV Signal	Symptom
Less than −18 dB	No Problems
−18 to −10 dB	None to Light Sparklies
−10 to −3 dB	Heavy Sparklies – Picture Barely Watchable
−3 to 0 dB	Lines or Random Pattern – No Picture.
Greater than 10 dB	Complete Wipe-out

KEY TO FIGURES 4-9a and b

- Light Interference
- Medium Interference
- Heavy Interference
- Wipe-out
- ↓ Vertical Interference
- ↓ Horizontal Interference

SITE SURVEY/PLANNING

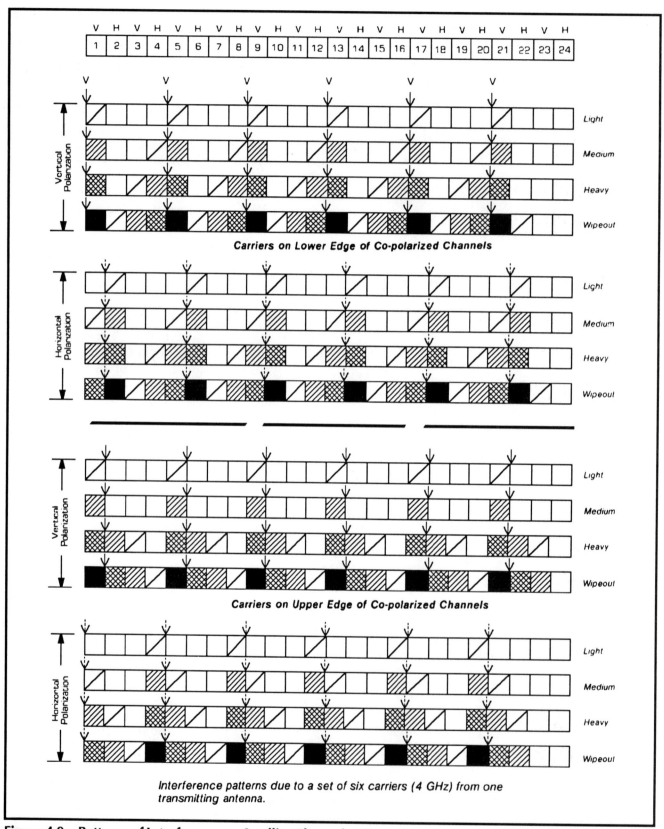

Figure 4-9a. Patterns of Interference on Satellite Channels. *Six carriers from one earth-based transmitting antenna show a characteristic regularity in patterns of channel disturbance. The key to reading both of these figures is shown on the left. (Courtesy of Microwave Filter Company)*

SITE SURVEY/PLANNING

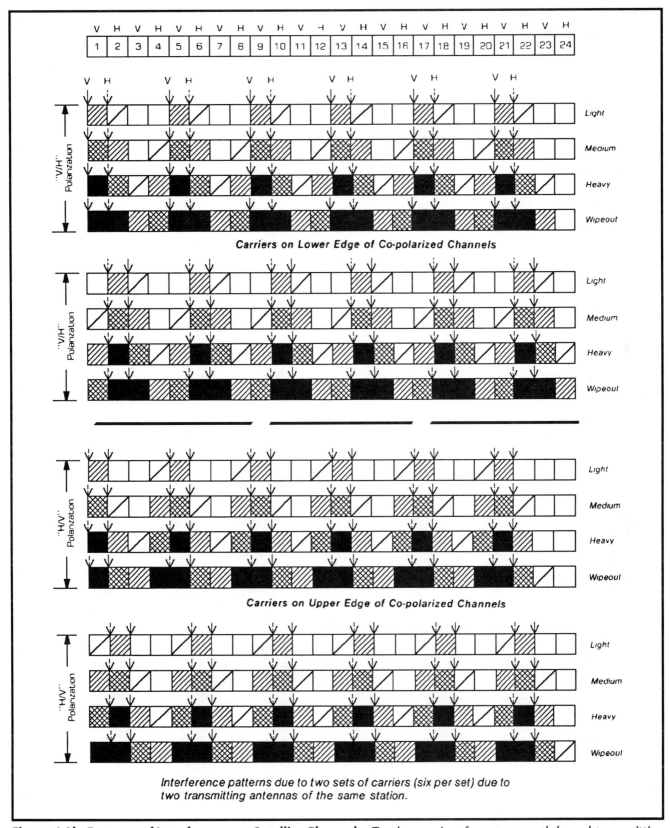

Figure 4-9b. Patterns of Interference on Satellite Channels. *Twelve carriers from two earth-based transmitting antennas show a characteristic regularity in patterns of channel disturbance. The key to reading this figure is shown two pages earlier. (Courtesy of Microwave Filter Company)*

SITE SURVEY/PLANNING

Figure 4-10. The Effects of TI on Picture Quality. *Picture quality deteriorates as the power of an interfering carrier increases relative to that of the incoming satellite signal. The interfering carrier in all these cases is at 10 MHz below the satellite center frequency. (Courtesy of Microwave Filter Company)*

SITE SURVEY/PLANNING

Using a Portable Satellite System to Detect TI

A portable satellite system can be an effective first cut method to test for TI at the proposed installation site. A small aperture antenna is inexpensive and may be easily transported and quickly set up at the site (see Figure 4-11). Many satellite technicians already own either a portable system that consists of a 6 or 7 foot diameter dish mounted on a trailer.

The survey should be conducted in the approximate area and, if possible, at the same location where the permanent antenna will be installed. All satellites to be detected should be targeted since TI can be very directional. Interference may not show up on Galaxy V but can be strongly disruptive on Satcom IV at the other end of the arc or on nearby Telstar 303. Sometimes rotating an antenna just a few degrees can cause an antenna side lobe to point at a source of interference where none was detected before. It is also wise to conduct tests as closely as possible to the same height where the permanent antenna will be placed. Moving down only a few feet may suddenly recruit a fence as a good artifical screen.

All transponders on each satellite should be checked since some forms of interference often affect only a limited number of channels. In addition, these channels should be observed a number of times for at least two days during business hours. TI is usually more prevalent during the work day and measurements made at night or on a weekend can be misleading. If TI is present it may be quickly cured by moving the dish a few feet laterally or by placing it closer to the ground or in a small depression in the ground. If a TI- free location cannot be found, bandpass or notch filters can be inserted to test their effect (these are discussed in more detail in Chapter 5). Remember, it is important not to confuse a weak signal caused by the antenna being blocked by an obstruction with interference. Also, in those cases where picture quality has deteriorated in an existing installation, do not confuse obstructions caused by the growth of new leaves on trees with TI. This can be easily diagnosed because all channels will be equally affected.

If TI is not detected with a portable antenna then all is clear. However, if TI ranges from moderate to severe levels, those results obtained with the portable system should be supplemented by more sophisticated methods described in Chapter 10 where use of the spectrum analyzer is explored. A small dish may simply not have sufficient gain, low enough noise temperature and directivity compared to a commercial grade antenna. Even though complete wipe-out may occur on some transponders when viewed with a portable small-aperture system, the permanent antenna might require only moderate filtering in order to obtain good quality pictures. Nevertheless, when TI does occur, knowing the gain of the small dish and being able to estimate the TI level can generally give a good indication of the required size of the permanent antenna.

An alternative to testing with a portable small-aperture system is to use a hand-held LNB/survey-feedhorn assembly connected to an LNB, receiver and TV set or monitor. In this case, it is best to have a spectrum analyzer or, at least, a signal strength meter that is more sensitive than the one often built into a satellite receiver. The test unit is then scanned in all directions across the sky, not only at the satellites, because TI coming from off-axis directions can often be detected by antenna side lobes. The LNB/feedhorn should be oriented in both the horizontal and vertical planes so any interfering carriers of both polarities can be located.

While watching both the signal strength meter and the TV screen, if the random white noise or snow on the TV remains unchanged and if there is little or no reading on the signal strength meter, then no problem exists. If the screen changes and goes blank, black or white, or if black bars or lines appear and if the meter gives a positive indication, then some form of TI is present. If the feed is covered up all these symptoms should disappear. If not the test

Figure 4-11. Portable Antenna for Site Checks. *A small portable antenna should be used if there is any uncertainly about obstructions blocking the field of vision of a dish or if TI has been detected by other methods. A 1.5 meter dish is pictured here. (Courtesy of DH Antenna)*

SITE SURVEY/PLANNING

equipment is malfunctioning. If a TV set or monitor is not used, the signal strength meter can be employed to detect moderate to heavy interference as indicated by a continuous fluctuation in its read-out. However, this method is not very sensitive to light to moderate levels of interference.

If TI is detected, the feed/LNB should be scanned back and forth to find the direction where the meter reading is highest and the feed should be oriented in all planes. This will locate the direction of the TI source and identify its polarity.

If TI has been detected, the same test can be conducted at another potential site, perhaps one better shielded by natural barriers. A logical first choice would be a location where there is an obstruction between the new site and the source of TI. If the interference is coming from behind the antenna, it will not have as great an effect because the dish will block out most of the microwave signal. If no location with a clear view of the arc of satellites and free of interference can be found, a reasonably good quality portable antenna could be used to examine the effect of inserting microwave and IF filters.

The use of a spectrum analyzer to obtain records of the TI levels and sources is outlined in more detail below. Professional engineers must have quantified information before proceeding with a satellite reception system design.

It is important to understand that all these methods of testing for interference have their limitations. The information gathered applies only to TI sources that were operating during the site survey. The 1 P.M. to 5 P.M. time period especially on Friday afternoons is usually when microwave traffic is heaviest. But none of these methods will protect against common carrier relays that may be activated at some point in the future. The only protection in this case is to register and license the installation in order to be informed about any planned microwave project. Incidentally, it might be good protection to include a clause in the SMATV sales contract which includes payment for insertion of filters if they are necessary at some future date.

F. THE OFF-AIR SITE SURVEY

Probably the most crucial step in a site survey is to determine the strength and quality of local off-air signals. Private cable systems should receive some of the available terrestrial broadcasts because most subscribers want to view local news, sports and weather. In some cases, the solution is ready-made because local signals may already be available at the site in an existing MATV system. But, in general, a new off-air antenna array must be installed. Many SMATV designers have discovered to their dismay that the most difficult system design task often lies in providing high quality, local broadcasts. This challenge and the problems associated with combining off-air and satellite signals can sometimes be so frustrating that numerous private cable companies simply refuse to bid on installations that are either additions to MATV systems or are projects requiring a number of off-air channels.

Nevertheless, once the fundamentals are understood, only a modest degree of design ability and a minimum of test equipment are necessary to successfully tackle any private cable project. Presenting this knowledge in a clear and understandable fashion is one of the principle aims of this text.

Off-air signal levels must first be either measured or estimated before any design issues are considered. The received signal levels vary with a number of factors including the transmitted effective radiated power generated at the broadcast tower, the distance between the transmitter and the site, the frequency of the particular channel, details of both the interleaving and surrounding terrain, seasonal and daily variations in weather conditions, the gain characteristics of the off-air antenna, antenna height and losses in the download. The download is the cable linking the antenna/preamplifier to the off-air processors. When required, a much shorter cable is used to connect antenna output to preamp input. Although there are rather complex computer simulations and other methods available for estimating signal levels, the best policy is to either measure the power levels at the site or, as a second best choice, to use estimating tools.

SITE SURVEY/PLANNING

Measuring Off-Air Signal Levels

In those cases where there is either no existing MATV system or where the existing one has deteriorated to the point where it is inoperative, a portable antenna should be used to measured signal levels at the site. The information gathering in a site survey is a crucial step to be taken before an SMATV system can be properly designed. The site chosen for installing off-air antennas should be as close to the head-end equipment as possible.

The minimum equipment necessary for a site survey includes a quality field strength meter (FSM), at least 20 feet of masting, a color portable TV and a moderate gain broadband antenna or an antenna constructed from half-wave dipoles cut to the correct length to maximally receive each individual channel. Some contractors also include a 40 foot, crank-up tower and rotator to facilitate measuring off-air signal levels near the top of a proposed tower (see Figure 4-12). The test antenna(s) should be just large enough to receive local signals so that a cumbersome device need not be hauled around for site surveys. It should be lightweight, collapsible, inexpensive and easy to handle as it probably will not endure a long time under site survey conditions. This antenna can be supported by a short mast and temporarily mounted in a pipe clamp close to where the permanent device will be located.

The gain figures for the VHF-low (2 to 6), VHF-high (7 to 13) and other bands to be employed as well as that for the downlead should be either known or calculated. Such specifications can be either found on the manufacturer's data sheet or can be supplied directly from the source. The typical gains for a six to eight element broadband antenna are 1 to 2 dB on the VHF-low band, 4 to 6 dB on VHF-high and 5 to 9 dB on the UHF band. (More background details about off-air antenna design and construction are examined in detail in Chapter 5).

Signal strength measurements can be made with either a portable spectrum analyzer or a field strength meter. The levels in dBmV should be recorded in a logbook for future reference. Given the test antenna gain and the downlead attenuation, the gain of the permanent antenna relative to a dipole required to deliver the desired signal can then be easily computed. First subtract the antenna gain and add the cable losses from the measured signal level. Then subtract this result from the required permanent antenna output signal level to obtain its required gain. This output power is determined by available signal levels as well as by what type of preamplification and processing will be employed.

While scanning the skies with the antenna and field strength meter, note any possible interfering signals that may be present. Particular attention should be paid to FM radio stations that are generating signals as strong or stronger than those of adjacent TV channels. Wild swings in the reading obtained from the FSM are probably due to mobile radio sources such as taxi dispatchers or pagers. These will have to be filtered or trapped out later in order to avoid on-screen interference.

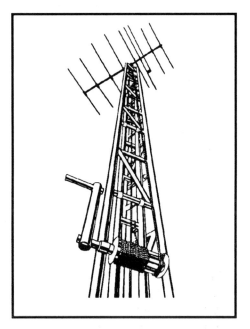

Figure 4-12. Portable Crank-Up Tower. *This tower comes with a fully insulated 4' x 6' x 6' enclosed shelter and trailer. The top view shows the fully extended tower. (Courtesy of Aluma Tower Company)*

SITE SURVEY/PLANNING

Carefully note the condition of the picture on the portable color television. Snowy pictures are not a problem as the weak signal causing this outcome will be managed in the design stage. It is important to look for types of interference such as ghosting or audible noise from man-made sources such as a nearby power line. These will probably not show up on the FSM. If ghosts do exist, the antenna should be moved to another location. If there are electrical noises, note if they are steady or intermittent. Identifying and curing such problems is covered in more detail in Chapter 10.

The off-air signal levels should be measured over a period of a couple of days. Doing so will identify any transient reception problems that may occur at odd hours or under particular weather conditions.

When using the existing antenna to perform this survey, particular attention should be paid to its type and make. If this antenna is later removed, it should be replaced with one that has the same gain characteristics. In an urban area that has strong off-air signal levels, an approximate replacement will usually work perfectly well. However, in those locales where broadcast signals are weak, it may be critical to use an antenna that duplicates the performance of the one used for the survey. Note that any existing preamplifiers used on the antenna to boost input signal will also affect the results of these conclusions.

Estimating Off-Air Signals

In those cases where the strength of off-air signals cannot be easily measured, levels can be estimated based on grade A and B signal level contour maps for any particular location. The indicated fields strengths on these maps are defined in terms of microvolts per square meter for an antenna located 30 feet above ground. Those areas between the two contours, known as the grade A and grade B contours, have a 50% chance of exceeding the estimated signal indicated on the map. The values of these contours for the standard NTSC channels are listed in Table 4-9. Typically, grade B contours range from 30 to 70 miles from a television transmission antenna. These coverage maps can be obtained by request from local television stations or from some published texts. For example, the *Television and Cable Factbook* includes all the grade A and B contours for TV stations in the United States.

Once the field strength at a particular site has been estimated, standard equations can be used to calculate the expected signal level in dBmV. This voltage is related to field strength, channel frequency, antenna gain and the height of the antenna above ground. For example, an 11 dB antenna mounted at 120 feet above ground receives a signal of approximately 6.4 dBmV on channel 12. The same assumptions were used to calculate the expected signal levels for VHF channels 4 and 12 and UHF channel 72. Results are presented in Table 4-9.

TABLE 4-9. GRADE A and B CONTOUR LEVELS

Grade of Service	TELEVISION CHANNEL		
	Low-Band	High-Band	UHF
	Field Strength (microvolts/meter^2)		
A	68	71	74
B	47	56	64
	Signal Level (dBmV) *		
A	12.2	7.4	1.4
B	11.2	6.4	0.4

* Please see text for discussion of underlying assumptions.

TABLE 4-10. TYPICAL SIGNAL STRENGTHS at TV ANTENNAS

DISTANCE from TRANSMITTER (Miles)	AVERAGE SIGNAL LEVEL (dBmV)
0 to 20	More than +20
20 to 40	0 to +20
More than 40	Less than 0

SITE SURVEY/PLANNING

These estimated signal levels can be used in conjunction with contour maps to guess at what would be expected at a given site. The fact that every doubling of antenna height typically gives a 6 dB increase in signal strength allows some changes to be made in the underlying assumptions. However, the exact level depends on many unpredictable factors such as the type of terrain and the nature of obstructions such as hills or buildings lying between the transmitting and receiving antennas. As a result, estimated signal levels are just that, estimates. These maps are not completely accurate and should not be relied upon in areas of questionable reception. By far the most reliable option is an on-site measurement. This eliminates guesswork.

A further aid in estimating off-air signal strengths is presented in Table 4-10. These typical signal levels are only intended as a guide. If the transmitter powers are different from those assumed, multipliers can be used to obtain approximations for the situation at hand.

Standard NTSC Channel Frequency Assignments

A listing of standard NTSC and PAL channel frequency assignments, is presented in Table 6-5 as an aid in this survey of off-air channels. Note that in countries where the North American NTSC standard is not used, the frequency assignments, both channel bandwidths and center frequency locations, are substantially different.

G. OFF-AIR INTERFERENCE and RECEPTION PROBLEMS

Interference can be as annoying a problem in receiving off-air broadcasts as with satellite transmissions. It must be understood and managed from a position of strength.

Two categories of interference can be identified. Co-channel interference arises from signals occupying the same frequency band as the desired broadcast (see Figure 4-13). Out-of-band interference such as adjacent channel interference is simply referred to as interference. All unwanted signals except for those originating from co-channel sources can potentially be eliminated by filtering. Although co-channel interfering signals can be phase canceled, this is not usually necessary (phase cancellation is explored in Chapter 5). It is rare that television stations within 170 to 220 miles of each other are licensed on the same channel. Therefore, co-channel interference is generally experienced only when a receiving site is located far from a metropolitan area and is in between two relatively distant transmitting antennas. In this case, the relative levels of the effective radiated power of each transmitter and the receiving antenna's gain pattern determine the magnitude of the problem.

Similarly, adjacent channel interference is relatively rare because licenses for adjacent channels are allocated to stations at least 60 miles apart. Again, a receiving antenna located in a rural area somewhere between two transmitters can potentially have problems with both unduly low signal strengths and adjacent channel interference.

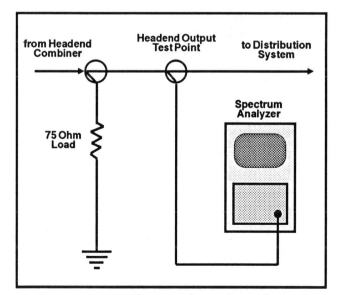

Figure 4-13. Measuring Co-Channel Interference.
This illustrates a simple test for co-channel interference just following the headend combiner. If it is not measured at the headend, it will only enter the system via ingress into the distribution lines.

SITE SURVEY/PLANNING

In a private cable system that carries signals ranging from channel 3 to channel 12, strong local FM stations or mobile radios can produce interference. While FM broadcasters only occupy the out-of-band area between the channels 6 and 7, mobile radios are licensed on almost all the frequencies within the television mid-band. Even though mobile radios transmit vertically polarized signals, these eventually end up with the same horizontal polarization as generated by TV transmitters because of the way an RF signal interacts with and is polarized by the surface of the earth. Mobile radio, mid-band interference is particularly annoying because transmissions come and go at random times through the day.

The site survey is again the essential first step in headend design. Signal levels on all channels that will be distributed should be measured, even if there is no known local channel that is broadcasted on that frequency. Measurements should be taken at a number of times during business hours to determine the level of mobile radio activity. If an adjacent channel is present, its level should be measured and its hours of operation noted.

In those rare cases when a visit cannot be made to the site, the published grade A and B contour maps can be used to estimate the potential for interference from both local and distant broadcasters. Remember, this is an estimate that does not necessarily identify interfering signals from mobile radios, FM broadcasters or television stations having unusual signal propagation characteristics. Some companies have also developed theoretical computer survey programs based on the location, height and power levels of all transmitters in the area, the local geography and the receive antenna location, height and gain. Again, these are theoretical approximations to real world measurements and should be taken with the appropriate grain of salt.

Another site survey aid does exist. In most areas, there are electronics stores that sell antennas and hardware for the community. The sales staff in these outlets often have a great deal of knowledge about local reception problems. Therefore, an ongoing relationship may be established with such people in order to avoid many unseen pitfalls. Most major antenna manufacturers such as Winegard and Channel Master also provide detailed product and local area application sheets to their customers. A constantly updated file of this information can be very valuable.

The Bottom Line

If a signal level of zero to 10 dBmV is measured, this should be adequate to clearly receive most television channels. However, a preamplifier can produce usable pictures from levels as low as −5 dBmV if no adjacent channels are present. In addition, any interfering signals should be at least 55 to 60 dB below the level of the desired channels. Again, it makes sense to always use the best available resources on hand. In an area having extremely low power, video broadcast signals or very high levels of interference, the expertise of an local antenna contractor could prove to be invaluable for the pioneer SMATV projects.

The off-air processing equipment available to properly receive television broadcasts and to avoid trouble with interfering signals is examined in detail in Chapter 5.

H. HEADEND SURVEY

An SMATV headend should be located in a secure, air-conditioned enclosure near the logical center of the planned distribution system and as close as possible to both the satellite and off-air antennas. However, the actual situation may differ from the optimal design. Whether or not an existing system is being retrofitted, numerous questions must be answered before a headend design is implemented. These include:

- What is the best location for the headend? It should be in a secure, climate controlled environment. Usually a headend requires a small walk-in room This decision will be affected by the number of satellite and off-air channels desired, by the choice to use or not use an on-site computer controller and by the type of electronic components selected.
- Will other services such as a security system, VCR programming or locally originated channels be included?
- Has a previous CATV operator already installed a quality distribution system that places a restriction on the headend location?
- Must the headend equipment be installed in a cabinet or an opened rack?
- Will one or more off-air UHF channels be converted to VHF signals? Are some local VHF channels so strong that they have to be converted to other VHF channels to prevent ingress interference?
- Is the system going to have adjacent channels on its output?

The answers to these and numerous other questions should become clear as this text unfolds.

I. DISTRIBUTION SYSTEM SURVEY

After determining which signals will be received, it is time to complete the signal distribution plan. Accomplishing this task requires accurate information regarding the physical characteristics of the building project and a knowledge of the condition of any existing distribution plant. The distribution survey is designed to provide this information.

Distribution systems can assume many forms. They can be very simple or extremely large in span and complex in design. However, they all function on the same principal, namely, the delivery of a specified signal level to all outlets or drops on the system. Quite clearly, the design and type of materials used in an SMATV system determines its suitability for multi-channel distribution.

Home-Run vs. Loop-Run Systems

There are two principle types of distribution systems, home-run and loop-run (see Figure 4-14). In a home-run system a cable runs from each subscriber's outlet into a distribution box. This type of system is preferred because it allows an operator to control each individual's access to the system. Therefore, a subscriber can be turned on and off without affecting anyone else. In contrast, the loop-run or loop-tap system utilizes a single cable that runs from outlet to outlet and employs a tap in each unit to feed subscribers. Although it minimizes the use of coaxial cable and can be relatively inexpensive, if a cable should break all subscribers would lose service. In the past, this type of installation was the predominate method. Today it is avoided if at all possible. In general, it does not allow for control of individual subscribers and, as well, is prone to extensive service problems. So the home-run system has been the method of choice used by most cable companies.

Some addressable distribution systems are designed along the lines of computer "local area networks" (LANs) which use loop-run designs. In an LAN, the addressing information is transmitted along with the signal and serves to authorize and control services to subscribers. In these special cases, a newly installed loop-run network can be a very effective distribution solution.

SITE SURVEY/PLANNING

Loop-tap systems are found in many older buildings, hotels and most institutional distribution systems. Although not the technology of choice, such systems are still often installed because of their low initial cost. Most experts agree that all new systems should incorporate home-run distribution except in cases where bulk purchase contracts have been signed or in hotels, motels and other institutional systems where addressable technology is to be employed. Installing a loop-tap system will reduces the sales value of an SMATV system.

The first task in the distribution survey is to determine what type of system, if any, has been installed, home-run or loop-run. This can be accomplished by removing the face plate in a number of units and then by examining its internal construction. A loop-run tap can easily be identified because it has two cables connected and normally has a few resistors and capacitors on the rear of the wall plate. A home-run tap has just one cable screwed onto a single fitting that extends through the plate (see Figure 4-15). Most home-run cables are routed to a junction box or panel located on the building or in a utility closet.

Some installers may argue that if an existing distribution system has no outlets that deliver snowy pictures or ghosting due to direct ingress or impedance mismatches, it should be left intact, regardless of its design. However, this choice may limit the future installation of a more advanced addressable or pay-per-view system.

Trunk and Branch Wiring

Trunk and branch wiring is typically used in a larger-scale CATV network and, occasionally, in larger private cable installations. A main trunk line first routes signals to major independent areas such as large buildings or individual floors of large buildings. Then the cable branches into a home-run cable arrangement. This method both saves on coax and incorporates the advantages of home-run type architecture. Often home-run systems have elements of the trunk and branch wiring configuration.

Checking System Condition

After identifying the system wiring design, the coaxial cable must be examined to determine its general condition. A number of questions must be answered. Is it faded and brittle from being exposed to the elements? Does it have many breaks and or splices? How many foil shields and wire braids make up its jacket? Is it copper or aluminum? These are all questions that will help to determine its suitability for use in the proposed system.

The distribution network is the backbone of an SMATV system and must operate perfectly in order to deliver strong signal levels to all subscriber drops. If there are any serious inherent problems, it must be rebuilt to insure system reliability.

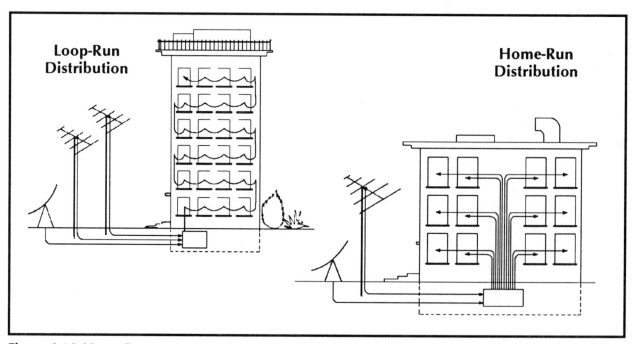

Figure 4-14. Home-Run versus Loop-Run Systems. *The home-run system is generally a more highly desirable SMATV signal distribution configuration. (Courtesy of Private Cable Magazine)*

SITE SURVEY/PLANNING

The type and size of wire used is an extremely important factor in system design and performance. All signals are distributed on 75 ohm coaxial cable. If the project under review has any other kind of wire, such as 300 ohm flat lead or 50 ohm communication cable, it must be replaced with 75 ohm coax. Coax comes in many sizes and styles. In general, the larger the cable, the less signal is lost due to cable attenuation. If there are cable runs between buildings of more than 100 feet, RG-11 or 500 cable should have been used. Higher loss RG-59 and RG-56 cables are normally used for subscriber drops.

If the existing system will be incorporated into the new design, its level of performance must be determined. The type and size of all cables, fittings, boxes and tap values should be noted. If the property has been previously served by a cable company, the point or points of entry to the property should also be catalogued. Then a system map with all the existing components and measured signal levels must be drawn. These levels should be measured at all proposed channel frequencies. Any major variances from these predetermined voltages can indicate a problem with the cable or system equipment. Problem areas must be located and the amount and types of repairs needed to make the system operational should be determined. Sometimes it may be cheaper to build a new system rather than to repair an old one. In addition, a new system requires less maintenance and can have greater customer appeal.

Surveying for New Construction

When installing new wiring, the next step is to determine where and how the cable will be placed. The new outlet should be located as close as possible to the existing one. Sometimes this will not be possible because of installation constraints and a line may have to be looped around the room to reach the television set. In this case, the space between the wall and the carpet tack strip can provide a useful route.

One of the key aspects of installing drops is the method chosen for connecting the coax from a subscriber box to the wall outlet. It is necessary to choose between exterior or interior wiring to facilitate wire placement. Exterior wiring can be less costly and more rapidly installed. But it is also exposed and therefore subject to weathering and vandalism. In contrast, interior wiring is completely protected and is less unsightly. Most property owners will prefer interior wiring is possible.

The presence of stacked closets can make installing interior wiring a simple task. If a building has the same floor plan on all levels, chances are excellent that a stack of closets exists from the attic to the bottom level. In this case, the drop can be run down through the closets in some form of molding or conduit. However, when the floor plans are different on each level, wires have to be installed on the building exterior in areas where they are not noticeable, such as under trim molding or along drain pipes. Conduit or molding should always be used and the final job painted to match the building. These items are discussed in more detail in Chapter 8.

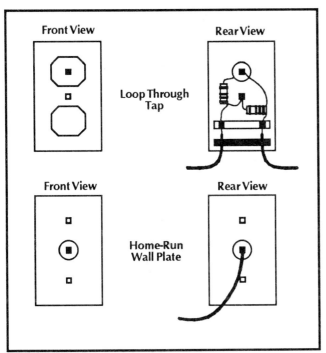

Figure 4-15. Home-Run and Loop-Through Wall Plate Configurations. *The differences between the wall plates used for home-run and loop-run systems is quite obvious upon inspection.*

Final Evaluation

The final examination requires re-inspection of all details. The condition of the existing system and cable should be noted. Is the cable it cracked and crinkled from exposure? Is it the right size for the distance it has to run? Is it directly buried in the ground, or is it in conduit? Any problem areas must be catalogued and plans made for corrective actions. In addition, an inventory of all electronics must be completed to determine which components may have to be replaced.

When building a new system, the existing one may be able to serve as a guide for cable placement. Note any existing conduits and cable raceways that can be used. The route that existing cables take in crossing under sidewalks and driveways should be mapped. A footage wheel can be used to measure all cable runs and pavement crossings. In general, following the route of the existing system improves the chances of minimizing the impact of any new construction. Upon completion of all phases of the site survey, all the data to prepare a detailed plan for building the private cable system will have been gathered. It is clear that a complete and accurate set of data can ensure the success of the venture, while haphazardly collected information can easily lead to unexpected problems, delays and unforeseen additional costs. The importance of performing a thorough and accurate survey cannot be over-stressed. The success of a private cable venture will be directly related to the effort expended in this planning stage.

CHAPTER 5. HEADEND SYSTEM DESIGN

Data collected from the system survey provides the necessary background to complete the preliminary headend system design. This design then provides the details and information needed to bid a private cable system. It is important that this plan is thoroughly prepared with great attention being paid to even the smallest details. The fact that a well designed system will operate smoothly and profitably but that a poorly planned one is an invitation to trouble cannot be over-stressed.

The first system designs attempted may seem difficult and somewhat mysterious and could take a seemingly tremendous amount of time until the process becomes more familiar. However, in time, even the most complex plans will proceed smoothly and rapidly. In general, it is good practice to create two or three different designs for each job being bid. This should include devising a number of hypothetical situations so that each would require a different set of equipment and construction techniques.

In time, the process of designing, bidding and installing SMATV systems becomes nearly secondhand nature. An experienced private cable designer can rapidly and accurately estimate construction costs as well as the time needed to implement each phase of any project.

An alternative to designing a system in-house is to employ one of the many companies offering design and planning services. Most of the major equipment manufacturers including General Instruments, Scientific Atlanta, Winegard and Channel Master offer free or low cost support based on site survey information. Such professional help can be very useful when tackling a large or complicated system having a unique set of design problems.

A. THE BASICS of HEADEND DESIGN

Every private cable system is composed of two principle components, the headend and the distribution system. The headend is where all the programming, received from either satellite, off-air or MMDS broadcasts, is processed and modulated onto one or more carrier signals. The programming is then transmitted via a distribution system from the headend to all end-users. The basic rule to follow is straightforward. In order to deliver excellent quality signals via the distribution system to all customers, the same quality must be produced in the headend.

All headend designs can be divided into two major categories, the off-air and the satellite processing groups. Each of these sections employs its own special antennas and equipment to receive and electronically prepare the desired signals for distribution. The off-air group is composed of antennas, channel separators, off-air processors, strip amps, pass band filters, UHF-to-VHF converters and mixing networks. The satellite group is assembled from a dish, feedhorn/LNA or LNB, satellite receivers, channel modulators, premium channel descramblers and mixing networks. Each component serves

HEADEND SYSTEM DESIGN

a specific function to accomplish the overall task. All these components are explored in detail below.

The operational tasks of every headend can be divided into four major categories. First, signals are detected and, if necessary, amplified. For example, an LNB is always used to amplify the very weak satellite transmissions in a satellite reception system while off-air amplification is not always necessary. Second, individual channels are separated. Third, these signals are processed and "cleaned up." This may be a simple task in some cases and a rather complex and painstaking endeavor in other situations. Finally, all channels are mixed and balanced for input into the distribution network.

Signals fed into a distribution system can be drawn from sources as diverse as FM broadcasts, building security systems or VCR movie playbacks as well as from satellite, off-air and wireless broadcasts. All rules which apply to the signal levels of either video or audio sources used in isolation must be applied to feeding signals into an SMATV headend. Satellite broadcasts must be sufficiently above receiver threshold in order to ensure excellent picture quality. Off-air signal strength must also be continuously maintained. Even brief downtimes are generally not acceptable in SMATV facilities.

Headends can range from being very simple to having extremely complex designs. Any combinations of local off-air television channels, signals from distant TV stations and satellite originated broadcasts are possible. Headends can process anywhere from a few to over 60 channels of programming. However, most private cable systems are typically designed to offer 12, 24 or 36 channels. Twelve channel systems are popular in smaller installations because they use only the available VHF channels which can be tuned in by any television set. This eliminates the need for a set-top converter in each subscriber's residence. Larger systems generally are required to provide more levels of programming in order to be competitive with local cable TV franchise operations.

When designing both headend and distribution systems it is important to be able to interpret and judge specifications published by manufacturers of the various SMATV components. The material here is designed to provide such a background.

Balancing Video Signals

Whenever signals are modulated onto two channels occupying adjacent frequency bands, there is always the possibility for interference to occur. Government agencies such as the FCC in the United States and the DOC in Canada have wisely allocated sets of non-adjacent, off-air channels in each region or metropolitan area of their respective countries. For example, in the United States, the city of Denver has channels 2,4,6,7 and 9 in use while Philadelphia area TV stations broadcast over channels 3,6,8,10 and 12. An explanation of why channels 6 and 7 are not adjacent is discussed below.

Most SMATV systems are designed to relay at least all 12 VHF television channels. Headends therefore must usually process adjacent channels for distribution and must be designed so that each one is received without interference from all others. Although the UHF band is also very occasionally used because a greater number of channels is available, cabling losses are substantially higher than in the lower frequency VHF band. Therefore, private cable designers first use the available 12 VHF and midband channels and then the progressively higher frequency super and hyperbands to distribute programming.

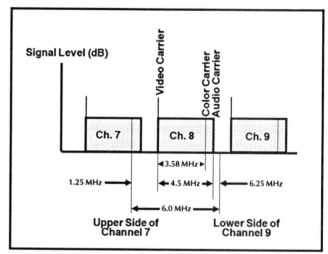

Figure 5-1. NTSC Audio and Video Channel Format. *Each NTSC television channel is organized to have a video carrier centered at 1.25 MHz above the lower side of the channel, a color subcarrier centered at 3.58 MHz above the video and an audio signal of 0.25 MHz centered below the upper edge of the channel. When mixing or combining adjacent channels it is very important to select modulators and have adjustable audio and video gain controls that filter out the lower sideband which can interfere with reception of the lower adjacent channel.*

HEADEND SYSTEM DESIGN

Each NTSC, PAL or SECAM television broadcast has its video, color and audio information organized as shown in Figure 5-1. If the audio subcarrier power is too high, it can bleed into the upper adjacent channel video signal and place a cross-hatch or pattern over the television picture. If the video bandwidth has not been properly filtered so that unwanted signal falling below the lower channel edge has not been eliminated, it can bleed into the lower adjacent channel and cause scratching sounds and crosstalk on its audio output.

The audio subcarrier can interfere with the upper adjacent channel picture carrier by beating with it to produce, in an NTSC broadcast, a 1.5 MHz difference signal. Off-air television broadcasts typically have their sound levels set at only 6 to 10 dB below the picture carrier frequency (see Figures 5-2 and 5-3). This reduction in level is not sufficient for a cable distribution system using adjacent channels because the beat or cross-modulation would be strong enough to be observable. If the level of the audio subcarrier is reduced to less than 15 dB relative to the video carrier, the beat is not detectable on-screen. However, if the level falls below –20 dB, the upper video carrier begins to interfere with the audio subcarrier and causes a noise known as sync buzz. The audio subcarrier is usually adjusted by either using a trap known as an aural carrier reducer or by internal circuitry in the headend processing equipment.

In light of these considerations, whenever adjacent channels are being distributed, three important rules must be followed. First, the signal must be properly restricted by a bandpass or SAW filter to protect bleeding onto the lower adjacent channel. Second, the audio level must be maintained at 15 dB below the video carrier. Third, power levels of the video carrier of each adjacent channel must be nearly equal to prevent the stronger signal from overloading and bleeding onto the weaker channel. The signal levels of all channels should be maintained within ±1 dB of each other. For best results this balancing should be accomplished somewhere between the off-air antenna and headend amplifier or, for satellite signals, between the modulator and headend launch amplifier. Although some newer televisions can even manage nearly 10 dB differences between adjacent channels, most sets require inputs that are within 5 dB of each other so that no noticeable degradation is visible on the weaker channel.

Channels having adjacent numbers are not necessarily adjacent in frequency. Table 5-1 outlines NTSC channel frequency allocations. Because there is a large space between the VHF-low (2 through 6) and the VHF-hi channels (7 through 13), channel 6 does not have an upper adjacent and channel 7 does not have a lower adjacent signal. Fortunately for SMATV system designers, there is also a 4 MHz guard band used to transmit marine radio communications between VHF channels 4 and 5.

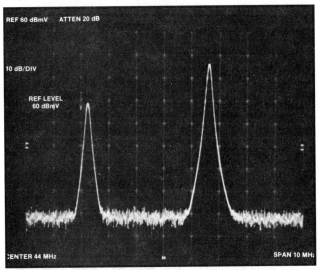

Figure 5-2. Audio and Video Signals. *This analyzer display shows an inverted audio and video television signal. The video carrier to the right is 4.5 MHz above the audio signal.*

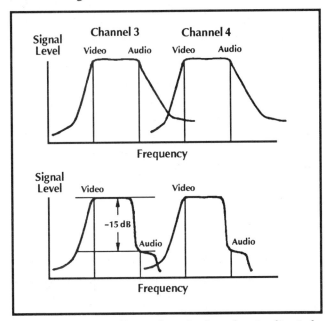

Figure 5-3. The Effect of Reducing the Audio Subcarrier. *By reducing the audio subcarrier on channel 3 the potential for beat interference on the upper adjacent channel 4 is minimized. In general, this reduction in level should be about 15 to 17 dB.*

HEADEND SYSTEM DESIGN

Table 5-1 provides some very important information for the headend designer. For example, assume that an SMATV distribution system feeds two satellite channels to each room in a small hotel along with local off-air broadcasts. If channels 2, 4, 7 and 9 are occupied with the local broadcasts, satellite signals could be modulated onto channels 11 and 13 with no concern for interference. But if a television station is already relaying a broadcast on channel 11, a good choice would be to modulate the satellite signal onto channel 6 which has no upper adjacent channel as well as onto the free channel 13. This sensible choice would save time and money in correctly designing a headend for a small distribution system. This would allow use of less expensive modulators and signal balancing techniques than would be the case if channels 5 and 6 had been chosen.

Although signal levels must be equalized before being fed into a distribution system, precise balancing is not nearly as important when using non-adjacent channels. Therefore, an SMATV system having less than 12 channels should be designed, if possible, with non-adjacent channels as targets for modulation. When adjacent channels are used, however, high quality processors, modulators and other headend components are required to clean up the audio and video signals and to provide filtering and amplification as necessary.

TABLE 5-1. VHF CHANNEL ALLOCATIONS

Channel Number	Frequency Range (MHz)	Upper Adjacent	Lower Adjacent
2	54-60	3	none
3	60-66	4	2
4	66-72	none	3
5	76-82	6	none
6	82-88	none	5
FM Band	88-108		
7	174-180	8	none
8	180-186	9	7
9	186-192	10	8
10	192-198	11	9
11	198-204	12	10
12	204-210	13	11
13	210-216	none	12

B. OFF-AIR ANTENNAS

Television signals are collected from both local and distant stations with off-air antennas. These are critical components of an MATV system because no amount of expensive headend equipment will compensate for an inadequate front-end that produces a substandard signal.

The main objective in all MATV system designs is to use durable, reasonably-priced antennas to obtain pictures that are free of interference and snow. For smaller installations where low cost is one of the central objectives, broadband antennas that simultaneously detect either a group or all of the desired off-air channels can usually be employed if signal levels are strong and consistent. For larger scale jobs, an antenna tuned to each channel is recommended in order to obtain greater flexibility in balancing signal levels and in controlling interference.

Antenna Designs

Physics dictates that an off-air signal can be efficiently detected by a conductor whose length equals one half of its wavelength. Wavelengths range from 13 inches for the highest UHF frequencies to over 17.0 feet for the lowest VHF channel frequencies (see Appendix E and Table 6-5). The most elementary antenna is a "half-wave" dipole. This design consist of two quarter wavelength elements that are insulated from each other. Each element is connected by a transmission or feed line to the input terminals on a receiver. The radiation pattern for a simple dipole is shown in the Figure 5-4. When a resonant signal, namely a signal whose wavelength precisely matches that of the antenna, is received, antenna gain varies from zero to its maximum broadside to the dipole. A simple dipole has an impedance of about 72 ohms.

HEADEND SYSTEM DESIGN

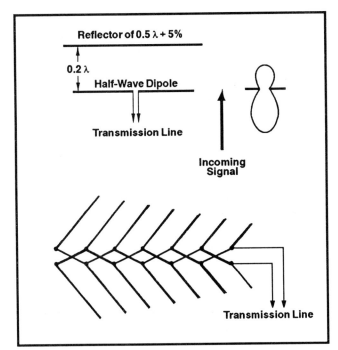

Figure 5-4. Basic Yagi and Log Periodic Antennas. *The top configuration is the basic Yagi antenna that consists of a half-wave dipole connected to a transmission line and a reflector element located 0.2 wavelength behind the active element. In contrast, the log periodic arrangement has all its active elements wired to the downlead.*

In general, more elaborate antennas and arrays of antennas are various combinations of dipole elements. The design objectives are to optimize each antenna's front-to-back ratio, i.e. increase reception from the source and to decrease reception from undesired directions, and to maximize gain while using a manageable number of elements. Numerous creative designs have been developed to simultaneously accomplish both of these goals.

A folded dipole antenna consists of a dipole and a half-wave conductor joined at their ends. This antenna has an impedance of 300 ohms. If a reflector element is located one fifth wavelength behind a half-wave dipole, it serves to reinforce signals arriving from the front but discriminates against reception of signals impinging from the rear. The reflectors are typically about 5% longer than the dipole. Adding the reflector element decreases the impedance of a simple dipole to about 36 ohm and that of a folded dipole to about 150 ohms.

Yagi Antennas

Incorporating additional elements in front of the dipole serves to intercept and re-radiate the incoming signal so that it arrives in phase with the signal detected by the dipole. The result is an increase in its overall gain. When these elements are not electrically connected to the dipole, they are known as parasitic elements. A Yagi antenna is composed of multiple parasitic directors, one reflector and an active half-wave dipole (see Figures 5-5).

Yagi antennas are relatively high gain devices designed to tune to a rather narrow range of frequencies. This type of narrowband device generally features 5 to 9 directors for reception of low-band channels and 10 to 12 directors for high-band operation. Both types use one reflector element.

Log Periodic Antennas

A log periodic antenna is designed to have a moderate gain and to detect a broadband selection of channels simultaneously coming from one general direction. All elements are active and are connected together with the longest at the back (see Figure 5-4). Each adjacent element is then decreased in length by a fixed ratio from the previous one. The distance between elements also becomes shorter by a constant factor, typically 35% of a quarter wavelength, in progressing from element to element towards the back of the antenna. Log periodics have the advantage of having a 75 ohm impedance that matches that of standard coax.

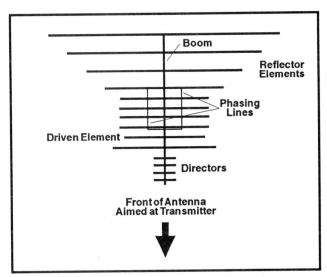

Figure 5-5. Basic Yagi Details. *The director elements in front initially detect the signal. Then the reflectors bounce it back into the driven elements. The downlead is connected to driven elements through the phasing lines. The boom supports the whole structure.*

HEADEND SYSTEM DESIGN

Aperture Antennas

Aperture antennas refer to the class of devices normally used for reception of UHF or higher frequency broadcasts. These are constructed from various types of dipoles and reflective screen and are substantially smaller than VHF antennas because the detected signals have higher frequencies (see Figures 5-6 and 5-7).

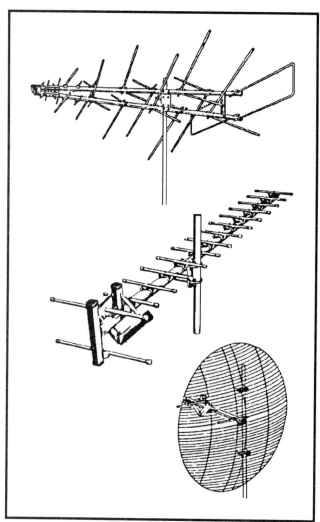

Figure 5-6. Antenna Types. *The top antenna is a log periodic model J-283X, the middle is a standard Yagi model J-275D and the lower one is a parabolic variety model D-1138-BB designed for higher frequency use. (Courtesy of Delhi Antenna Products Division)*

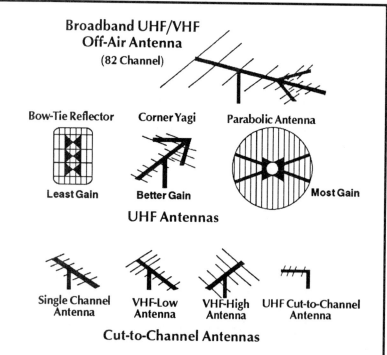

Figure 5-7. Off-Air Antenna Types. *A wide variety of antennas are diagrammed here.*

58

HEADEND SYSTEM DESIGN

Antenna Performance Parameters

Gain

Off-air antennas are characterized by a number of performance parameters including gain, bandwidth and directivity. There are two ways of expressing the gain, either referenced to an isotropic antenna or to a dipole. An isotropic antenna, a theoretical device, radiates and detects power equally in all directions. In contrast, a dipole has a more directional reception pattern and consequently has a gain when detecting signals along its "boresight" that is 2.2 dB higher than that of an isotropic antenna (see Figure 5-8 and 5-9). When gain is referenced to an isotropic antenna, it is followed by the abbreviation dBi; when referenced to a dipole the abbreviation is dBd (see Table 5-2).

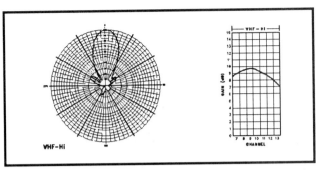

Figure 5-8. Antenna Characteristics. *The plot on the left shows the receptivity pattern and the graph on the right the channel gain characteristics. Both of these specifications are for the model JD-275D shown in the previous figure. (Courtesy of Delhi Antenna Products Division)*

Bandwidth

Antennas are designed to perform their best over a specific range of frequencies, the operational bandwidth. Yagis which are "cut-to-channel" or tuned to one TV channel can suffer from some problems specifically relating to antenna bandwidth. The lower frequency channels are potentially the most sensitive to these problems because in this range the 6 MHz channel width is a larger percentage of the carrier frequency. Antennas with bandwidth problems could exhibit picture distortions such as ghosting, smearing or poor color definition. Signal strengths may give an indication of bandwidth difficulties, especially in cases where the sound carrier is discovered to be equal to or stronger than the picture carrier.

Log periodic, off-air antennas do not generally have the same sensitivity to bandwidth problems because they are broadband in design. However, if a problem does occur on one channel, it can often be traced to the narrowband element which detects that channel frequency. The element length can easily be found by referring to the wavelength/frequency chart in Appendix E.

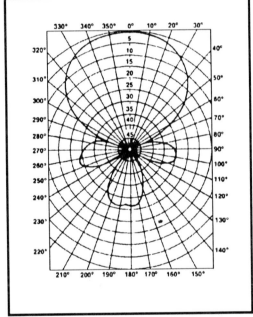

Figure 5-9. Typical Yagi Antenna Pattern. *This pattern is typical of most Yagi-type antennas. This particular variety has two side lobes and one back lobe. If reflected signals happened to be arriving from the side at 90 degrees off boresight, they would be detected at 30 percent relative to those arriving directly from the target.*

TABLE 5-2. TYPICAL OFF-AIR ANTENNA GAINS

Channel Range	Type of Antenna	Gain Relative to Dipole (dB)
Low-band	5 element Yagi	6-8
	10 element Yagi	9-10
	Broadband	2-7
High-band	5 element Yagi	8
	10 element Yagi	9-11
	Broadband	5-8
UHF	Broadband	7-18

HEADEND SYSTEM DESIGN

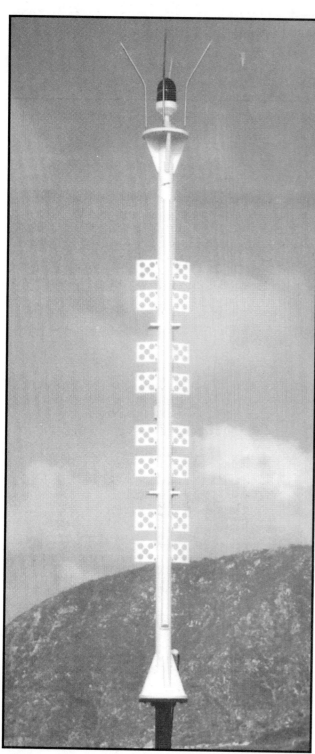

Figure 5-10. Television Transmission Antenna. *This antenna used to broadcast from a television station would be omni-directional except for the pattern directors which are used to modify the azimuth radiation patterns. (Courtesy of Private Cable Magazine)*

Directivity and Front-to-Back Ratio

Each antenna has a characteristic radiation pattern (see Figure 5-10). The simplest of these is the dipole pattern. Comparisons can be made by inspecting the polar plot of their radiation patterns. This plot is generated by rotating an antenna a full 360° while measuring its output signal level relative to a fixed transmitter. Clearly, the signal should be at a maximum when pointed at the transmitter and should decrease to one or more minimums.

Directivity refers to the ability to concentrate all the incoming power into a single lobe, known as the front lobe. This lobe pattern, which is typically different in the vertical and horizontal planes, bestows upon antennas the ability to reject both signals impinging from non-source directions as well as reflected signals which could cause ghosting. A good quality antenna is able to largely ignore most interfering signals coming from sources such as citizen band radios, pagers and electrical equipment.

The width of the main lobe is defined by a parameter called antenna beamwidth. This is the angle between points where the ability to detect incoming signal has dropped to half of its maximum. Half-power points occur where the voltage is 0.707 times the maximum. This beamwidth measurement permits a comparison of the directivity of various antennas.

The front-to-back ratio, expressed in decibels, is found by dividing the relative gain of the front lobe by that of the back lobe. The ability of an antenna to reject unwanted signals impinging from its rear largely determines how well it operates in locations where co-channel interference is a problem. This parameter is defined by:

Front-to-back Ratio =
20 log (Front Lobe/Back Lobe Voltage)

For example, the back lobe for the Winegard JD 275D, shown in Figure 5-11, is 0.045 of the front lobe maximum voltage of 1.0. Therefore:

Front-to-back ratio = 20 log (1.0/0.045)
= 27 dB

In general, larger antennas have more directional vision, namely, a higher front-to-back ratio and a more narrow beamwidth. This parallels the fashion in which beamwidth decreases as the size of a satellite antenna increases.

HEADEND SYSTEM DESIGN

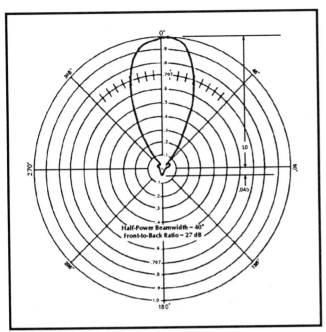

Figure 5-11. Front-to-Back Ratio. *The front-to-back ratio can be calculated from the polar plot. The ratio of the front to rear lobes is 1.0/0.045 which equal to 27 dB. (Courtesy of Winegard Corporation)*

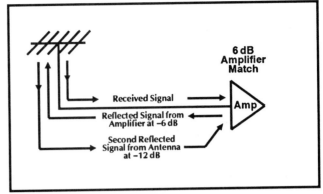

Figure 5-12. Ghosting and Return Loss. *Ghosting in this case is caused when signals are reflected back and forth between a poorly matched antenna and preamplifier. The signal reflected back from the amplifier is 6 dB below the input signal from the antenna. The re-reflection from the antenna is 12 dB below the level of the original signal and so on. This situation is explored in more detail in the text.*

Impedance Matching and Ghosts

Off-air antennas must be impedance matched to the interconnected equipment. MATV systems typically use 75 ohm cables and equipment and sometimes electronic components with 300 ohm input impedances. If a transmitting antenna were perfect, all the energy fed into this device would be radiated. In practice, the antenna reflects some of this energy back to the transmitter. Similar mismatches occur in receiving antennas. This "mismatch" is quantified by a parameter named the voltage standing wave ratio, VSWR for short.

Figure 5-12 illustrates a situation where both the antenna and preamplifier have a 6 dB mismatch. This means that half of the signal coming from the antenna to the preamp is reflected back. This reflected signal is 6 dB below the original level. In turn, part of this signal is itself reflected by the antenna. Its level is now 12 dB below the original voltage. This returning signal is delayed by the extra time spent in traveling three times down the coaxial line. This delayed signal will appear as a ghost on the TV screen. The longer the time delay, the wider the spacing of the ghost from the original picture. The greater the mismatch, the brighter the ghost appears on-screen. In general, all equipment used in MATV should have mismatches of no greater than 14 dB to avoid problems with ghosts as well as signal loss.

Antenna Configurations

Various narrowband Yagi and broadband log periodic antenna configurations have been created to satisfy off-air reception needs. For example, a quad array of four log periodic antennas has reduced wide-angle pickup and better front-to-back ratios than a single antenna of this type (see below for additional details). This performance improvement reduces the chances of detecting interference from co-channel sources.

Broadband antennas detect all channels whose center frequencies fall within their specified frequency range. They are used in areas where all channels are broadcasted from one site and where these signals are nearly equal in power. This is the case in most urban areas. These varieties of antennas work well in locations where there is no co-channel interference caused by signals broadcasted from nearby cities. Broadband antennas also have the advantages of being less costly to use than multiple, single channel Yagis, of occupying less space and of requiring less mounting hardware (see Figures 5-13, 5-14 and 5-15).

Single channel or cut-to-channel Yagis are tuned specifically to one particular band of frequencies. They have high gain on the specified channel but detect other channels at a much lower level. High gain, single channel antennas are used to receive distant stations or local stations which are transmitting from a different location than the majority of local chan-

HEADEND SYSTEM DESIGN

Figure 5-13. Close-Up of Broadband Antenna's UHF Elements. *These elements detect the UHF signals. (Courtesy of Channel Master, Division of Avnet, Inc.)*

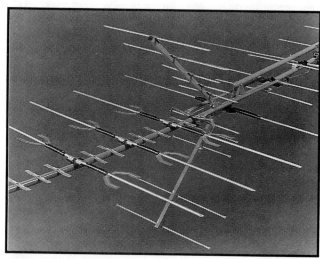

Figure 5-14. Close-Up of a Broadband UHF/VHF Antenna. *(Courtesy of Channel Master, Division of Avnet, Inc.)*

nels. This type can also function side-by-side with a broadband antenna in order to receive distant signals, but only when these distant channels do not correspond to those frequencies relayed by local broadcasters.

UHF antennas can also be of broadband or single channel design. Sometimes they are shaped to only detect a specific portion of the UHF spectrum, such as the range between channels 14 to 20. UHF antennas are usually separate components although combination UHF/VHF (or U/V) antennas are available. But such hybrids usually prove inadequate except in areas having high signal levels.

Another common configuration uses two broadband antennas each of which is tuned to half of the VHF band. One receives low frequency VHF channels; the other the high frequency stations. The purpose of this design is to give a very uniform gain across the whole frequency range frequencies and to facilitate easier separation of signals for processing in the SMATV headend.

Purchasing an Off-Air Antenna

The objectives in designing and installing an antenna array are to provide a high gain and good frequency selectivity on each desired channel. Unwanted ghosts and interfering signals must be properly rejected. These antennas must also be strong, durable and able to withstand occasionally severe wind loads.

Commercial-grade antennas should always be used in private cable systems. Varieties manufactured for consumer use do not deliver consistent performance and generally have shorter lifetimes. The difference lies in the fact that commercial-grade antennas are constructed with seamless, anodized aluminum tubing of a thicker gauge, have rust-resistant hardware and feature heavier grade mounting brackets.

The best practice is to purchase an excellent quality, commercial off-air antenna. When an antenna that serves a large number of customers fails, the price to pay in lost revenue and replacement is quite high. In the long run, a long-lasting, reliable system is the best and least expensive advertising available.

For the record, the most common approach to designing an antenna array for off-air reception in areas between cities is to install a separate Yagi for each channel desired. These antennas and the associated preamps then are all mounted on a high tower near the site. In urban areas, experienced technicians typically install one or two broadband antenna to cover the frequency range of interest. Nevertheless, a wide choice of antenna designs are available to the private cable designer. Note that most antennas are designed to have a 75 ohm output impedance. Some varieties having 300 ohm outputs must be installed with a 300-to-75 ohm matching transformer.

HEADEND SYSTEM DESIGN

Obtaining Required Signal Levels

In general, a signal level of +10 dBmV is considered to be the minimum input required by off-air signal processors, although experienced designers can make do with 0 dBmV or less. Some home viewers even tolerate levels as low as –20 dBmV, but usually at the expense of poor video quality. Note that more expensive full-performance heterodyne processors may function adequately at signal levels which are often considered too low for commercial use.

It cannot be over-stressed that before an off-air system is designed and components are purchased, a site survey should have determined the level of available signal at the antenna site. This defines the necessary gain as well as type of antenna or antenna arrays required.

In locations where the off-air signal level is too low designers have two options, to increase antenna gain or to increase the tower height. None of the other parameters such as transmitter power or distance, weather conditions and terrain characteristics are under control of the SMATV system design engineer.

Increasing Antenna Height

Increasing antenna height to increase gain can be fairly straightforward and effective. However, it can become expensive. In general, a doubling of tower height usually results in roughly a 6 dB and 12 dB increase in signal power on VHF and UHF channels, respectively. This usually holds true provided the starting height is more than three wavelengths above ground level. When mounting on top of buildings, the main reason for raising the antenna is to reduce the effects of the roof on its performance. This height is measured relative to the surrounding terrain. Doubling the height of an antenna mounted on top of a high rise building would prove very difficult. This rule of thumb can be seriously violated if unusual terrain or other obstructions and buildings alter the immediate environment. For example, in cases where the installation site is in line of sight with the transmitting tower, there would be little change in signal strength with change in antenna height. Or raising the antenna too high in a strong reception area may place it in a position to detect unwanted signals reflected from nearby buildings or terrain so that ghosting is caused.

Stacking Antennas

Multiple antennas can be stacked to improve reception directivity or to increase overall gain. Stacking horizontally improves the directivity in the horizontal plane while stacking vertically increases the directivity in the vertical plane. Reductions in beamwidth of up to 45% are typical.

Two or more high quality Yagi or log periodic antennas can also be stacked to increase input signal levels. A rule of thumb is that a double or a 3 dB increase in gain can be obtained by stacking two antennas. For optimal gain, vertical spacing should be half the signal wavelength and horizontal spacing should be approximately 0.9 times the wavelength. The optimal spacing for gain can be traded off for improved directivity. This is most effectively accomplished by adjusting the horizontal not the vertical spacing of multiple antennas. Note that in general a good deal of useful information on stacking both Yagi and log periodic antennas can be obtained from their manufacturers.

Examples of stacking antennas are shown in Figure 5-15. Single channel antennas are spaced approximately one wavelength apart to avoid any interactions between the received signals. This spacing should be maintained for the lowest frequency, longest wavelength channel when broadband antennas are used. In order for stacked antennas to work optimally, signals from each should arrive at their common mixer output in phase. This allows the two output powers to be additive. Combining out-of-phase signals results in a reduction in combined signal levels.

In order to ensure that signals are added in phase, the antennas are mounted on a common mast and a directional coupler is used to combine the signals. The downleads from each antenna have the same length within a tolerance of approximately half an inch. This method can be used to combine both broadband and cut-to-channel inputs.

A quad stack of antennas can provide a 6 dB increase in gain (see Figure 5-16). A low-loss phase combiner can be used to combine the outputs of a 4-bay stack. Generally no more than four antennas are stacked for SMATV use because the increase in weight and expense does not justify the increase in gain. In reality the overall gain obtained from stacking antennas is slightly less than would be expected because of the 2.5 and 5.3 dB attenuation in the splitters for 2-bay and 4-bay stacks, respectively.

HEADEND SYSTEM DESIGN

Figure 5-15. Stacking Antennas. *Antennas can be vertically or horizontally stacked. It is important to keep the downleads of equal lengths. These are also called phasing lines because if they do have unequal lengths, phases of the off-air signals coming from each antenna will differ. It is important to make the horizontal supports of non-metallic materials such as redwood or cypress.*

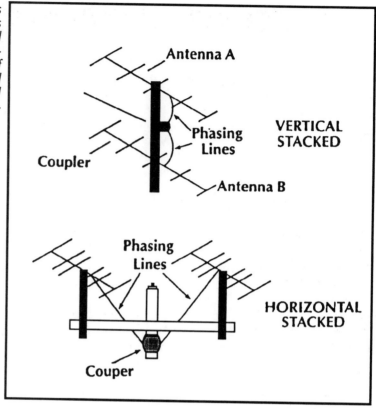

Figure 5-16. Quad Stacked Antennas. *This headend receives signals from a satellite antenna and a quad stacked array of four off-air antennas (Photo Courtesy of Simone Carlysle-Smith)*

HEADEND SYSTEM DESIGN

C. OFF-AIR ELECTRONIC COMPONENTS

All the electronic components required to construct an MATV system are examined in this section (see Figure 5-17). These include preamplifiers, traps, channel separators, strip amps, demod/remod systems, processors, mixers, splitters, filters and traps.

Preamplifiers

MATV amplifiers are classified into various categories such as preamps, single-channel strip amps, broadband distribution amps or line extender amplifiers. These components are all designed to amplify the incoming signal within a specified band of frequencies while introducing as little noise as possible. Often internal filtering is provided. Many similar characteristics apply to all types of amplifiers.

Preamps are generally used in conjunction with antennas to amplify relatively weak off-air signals (see Figure 5-18). They are designed to have low noise figures and usually are enclosed in weather resistant, antenna-mounted cabinets. Their proximity to the antenna minimizes cabling losses. Preamps generally have a separate ac power supply located in the headend building. This provides 18 to 24 Vdc along the coax that carries the signal from the preamp to the headend.

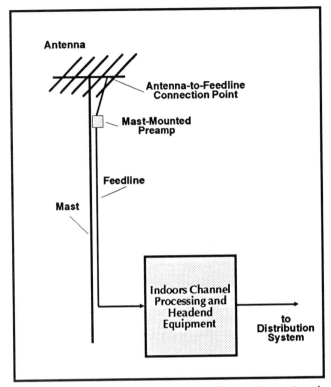

Figure 5-17. Off-Air Signal Processing. *Once signals are received by an off-air antenna they are typically routed into a preamp. Subsequently they follow a cable called a downline or feedline to channel separation and processing components indoors at the headend.*

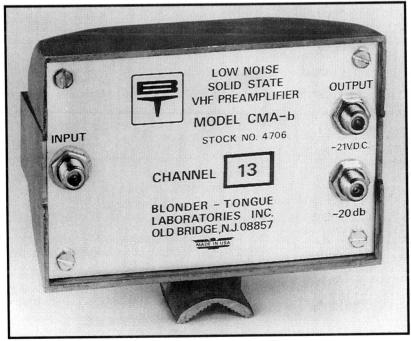

Figure 5-18. Typical Preamplifier. *This unit, model CMAb, is packaged in a weather resistant case and has a mast mounting flange. (Courtesy of Blonder-Tongue Laboratories)*

HEADEND SYSTEM DESIGN

Preamps, like off-air antennas, are designed to either be broadband or channel specific. Broadband preamps are available in many different ranges including VHF-low, VHF-high, broadband VHF, partial or full UHF and all channels. But preamplifiers are not necessarily the same even though they may span the same frequency band or bands. Some varieties may incorporate one amplifier covering the entire band. Others may have two or three amps, each tuned to a smaller portion of the frequency band. As would be expected, the manufacturer's specifications provide all the information necessary to correctly choose a preamp for any given situation.

Gain and Noise Figure

A preamp must have adequate gain so that the signal level is increased enough to overcome the lead-in cable losses. The output signal-to-noise (S/N) ratio must also be adequate so that the noise figure of the remaining electronics can be virtually ignored. Typical gains range from 10 or 12 dB to a maximum of 30 dB.

The noise figure is the most important preamp specification. Its importance parallels that of selecting a low noise LNB for a satellite reception system. The preamp noise figure determines how much noise is added to the MATV system while the very weak input signal is being amplified to an acceptable S/N ratio. Typical noise figures for preamp range from 6 to 8 dB. A minimum value to be expected in a top quality unit is, at best, 3 dB. However, if preamps were built with lower noise figures, this improvement would be unacceptable since it is usually achieved at the expense of amplifier attenuation matching.

Bandwidth

The bandwidth specification, in addition to providing information about which channels a preamp amplifies, suggests how this device is constructed. For example, if the bandwidth is listed as 54 to 890 MHz, there is probably one amplifier for the entire band. Such preamps are usually the least expensive and are well suited for most MATV tasks. However, these units are prone to overloading from signals transmitted by out-of-band communicators such as ham radio enthusiasts. In addition odd harmonics generated inside the amplifier from moderately strong VHF signals can potentially ruin reception of some UHF channels.

Also, for example, if the bandwidth of an amplifier were listed as 54 to 216 MHz and 470 to 890 MHz, it could mean that a bandpass filter could have been used to eliminate the intermediary portion between 216 and 470 MHz. Alternatively, if two amplifiers each covering one band were used, harmonics bleeding from the VHF into the UHF band would not be a concern. But input bandpass filters that would cause some signal losses and that reduce the sensitivity by increasing the noise figure would be required. The difference in this case is about 0.5 to 1.0 dB, an extra bit of noise that could make a difference when operating in weak signal areas. A similar type of reasoning would also apply to other multi-range preamps such as one having a bandwidth in three ranges, 54 to 88 MHz, 174 to 216 MHz and 470 to 890 MHz.

Input and Output Specs and Distortion

All amplifiers suffer from a phenomenon known as compression. Beyond a certain signal strength, the input signal saturates an amplifier and a further increase in input power does not cause an increase in output power, namely the amplifier has a non-linear response. The lower voltage signals then "catch up" with the saturated peak voltage signals and distort the output. The limiting output factor for a single channel amplifier is the point at which the television synchronization pulses, the highest voltage points on a television signal, begin to saturate the amplifier (see Figure 5-19). At this point when the horizontal and vertical sync pulses are lost, pictures will begin to "tear" and "roll." The 0.5 dB compression point specifies the maximum input that an amplifier can manage.

Single channel preamps can also be affected by a form of distortion known as intermodulation. This occurs when the picture and sound carriers in the television signal interfere with each other and mix with each other. This undesired interaction is know as beating.

Broadband preamps can produce another form of distortion known as cross modulation. This occurs when the carrier from one channel bleeds onto and mixes with that of an adjacent channel. The sign of cross modulation is bars that move back and forth on the screen similar to the movement of windshield wipers. The greater the bandwidth of a preamp, the greater the potential for cross modulation. The maximum output capability of a broadband preamp is rated at a certain level of cross mod when the device is carrying a specified number of channels. In those cases where signal levels have not been properly balanced before transmission, the input from one

HEADEND SYSTEM DESIGN

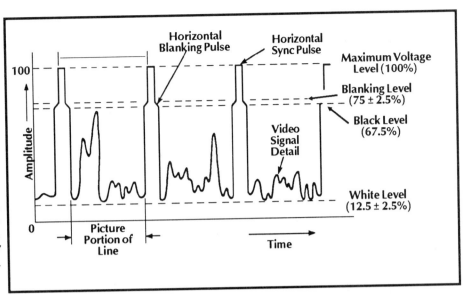

Figure 5-19. The Television Signal. *This illustrates the structure of three lines of a television video signal. Above the 67.5% level, the black level, the beam is shut off. The "blacker than black" level falls between 67.5% and 100% of the maximum voltage level. The picture information is relayed between the black and white levels. Therefore, during the horizontal blanking interval, which contains the blanking pulse and sync pulse, no illumination is produced. The sync pulse, the highest voltage point and point of potential amplifier saturation, commands the TV receiver when to begin each scanning line.*

channel may overdrive the amp while another channel may be well below the amplifier's maximum input signal handling capacity. An under-power input level tends to deteriorate the S/N ratio and fails to take advantage of the maximum designed output capability of an amplifier. A single channel preamp also may occasionally suffer from cross modulation as a result of an unduly strong adjacent channel (see Chapter 6 for a more detailed discussion on amplifier distortion).

Bandpass Filters and Traps

There is seldom a situation where an MATV system does not use a number of bandpass filters and traps. A bandpass filter allows only those carriers within a specified frequency band to pass but attenuates higher or lower frequency signals (see Figure 5-20). A trap is a filter that removes a narrow portion of the frequency spectrum but leaves higher and lower carrier frequencies untouched.

The most common application for filters occurs when separate antennas are used for each television channel. For example, if channel 7 and 8 signals are detected by two narrowband antennas and then mixed, a problem arises because a portion of the signal from each channel can leak into the opposite channel antenna (see Figure 5-21). The unwanted signal, essentially interference, entering the wrong antenna can be distorted by that antenna's off-channel performance. The resulting phase mismatched signal, when mixed with the original broadcast, can cause on-screen ghosting, smearing or poor color representation. Typically, a separate bandpass filter

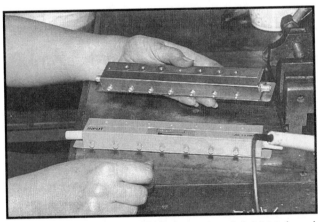

Figure 5-20. Bandpass Filter. *The model 3303 bandpass filter comes in range of single channel passbands that include channels 2 - 6, A2 - I, 7 - 13 and J - W. It has a return loss of 14 dB and an insertion loss that ranges from 3 to 5 dB across this frequency range.. (Courtesy of Microwave FIlter Company)*

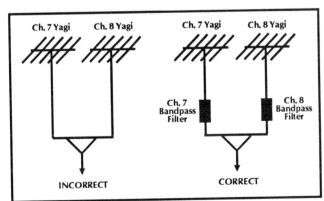

Figure 5-21. Combining Cut-to-Channel Antennas. *The incorrect and correct methods to combine two single channel Yagi antennas are shown here. The bandpass filters are necessary to prevent interactions between the received signals so that color smearing and other distortions are not produced.*

67

HEADEND SYSTEM DESIGN

or a combination preamplifier/bandpass filter can isolate the input from each channel to just that cut-to-channel antenna.

Band rejection filters are also available. A common variety is a channel elimination filters that is used to remove a channel from a headend system so that another may be inserted in its place (see Figure 5-22). FM band rejection filters are employed to prevent strong FM signals from overloading the processing equipment and distribution amplifiers. Crossover filters or band separation filters are used to separate UHF and VHF signals from broadband antennas (see Figure 5-23).

A trap can be narrowly tuned to eliminate just a portion of the frequency band. They are used in some cases instead of bandpass filters to prevent strong adjacent channels from interfering with weak ones. The frequency range over which substantial attenuation occurs can have rather sharp boundaries. In contrast, filters have rather wide "skirts." High "Q" traps can filter out the adjacent channel without adversely affecting the desired signal. Single channel interfering sources such as spurious signals and conversion beats can also be effectively removed with traps. In addition, they are commonly used to reduce the sound carrier level within a TV channel without affecting the video level. Figure 5-24 illustrates the frequency response for different types of traps and filters. In general, traps are inexpensive devices providing up to 50 dB isolation. They are inserted via F-connectors, are weatherproofed and can be used over a wide temperature range. As example of a high Q resonant cavity type of channel deletion filter is shown in Figure 5-26.

Performance Characteristics

The performance of filters and traps is characterized by their insertion loss, match and selectivity. Insertion loss defines how much a signal is attenuated as it passes through the device. Generally high quality brands have insertion losses that are under 1.5 dB for VHF signals and under 3 dB for UHF signals. Attenuation is generally higher at higher frequencies because capacitive signal leakage increases.

Match describes how well the impedance of a filter or trap is matched to that of a 75 ohm coax. Match, which is measured by the relative reduction in power of reflected signals, typically should be less than −8 to −9 dB. Note that both insertion loss and match are used to describe other SMATV components including taps, splitters and mixers.

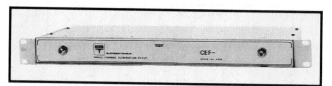

Figure 5-22. Channel Elimination Filter. *This filter removes a single video channel. (Courtesy of Blonder-Tongue)*

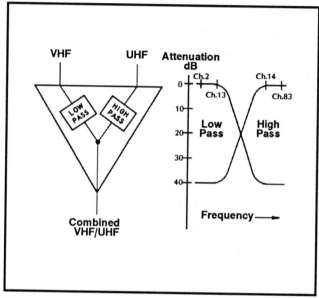

Figure 5-23. VHF/UHF Combiner. *The basic elements of a UHF/VHF combiner or a combiner used in reverse, a splitter, consists of a low pass and a high pass filter. The attenuation characteristics that make this possible are shown on the right.*

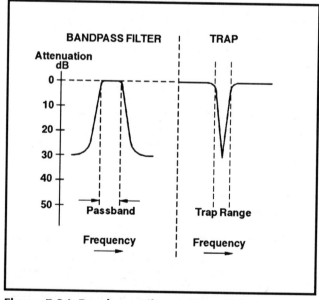

Figure 5-24. Bandpass Filter & Trap Characteristics. *The attenuation versus frequency of both a bandpass filter and trap clearly show how these circuit elements function.*

HEADEND SYSTEM DESIGN

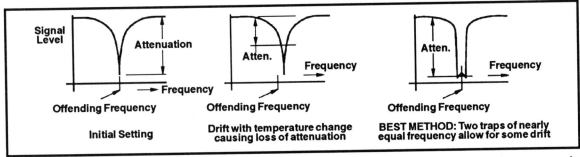

Figure 5-25. The Effect of Temperature on Trap Frequency. *The operational center frequency of a trap can drift with changes in temperature. If two traps of nearly equal frequency are used, drifting can occur in either direction with increases or decreases in temperature, and the offending frequency will still be trapped. This drifting is another good reason to have temperature stabilized headends.*

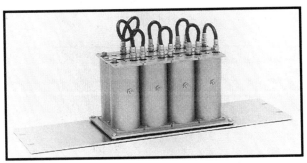

Figure 5-26. Channel Deletion Filter. *This filter is designed to remove hyperband channels 37 to 61 by use of a high Q resonant tuned cavity. It can be rack mounted. (Courtesy of Microwave Filter)*

The selectivity of a filter or trap defines how well the device removes only the desired group of frequencies. Traps should have the ability to reject the targeted signal by over 40 dB relative to surrounding frequencies. Attenuation for a bandpass filter is often measured relative to semi-adjacent channels. Therefore, for example, a bandpass filter for channel 9 should reject channels 7 and 11 by at least 12 to 15 dB. If distant channels are being observed in the presence of strong local broadcasts, the semi-adjacent channel rejection should be at least 40 dB.

The center frequency of a trap can drift with changes in temperature (see Figure 5-25). This can be a problem in a headend that is subjected to temperature extremes.

Channel Separators

After off-air antennas detect the desired signals, these are subdivided into individual channels by a channel separation network. This allows each channel to be individually processed and subsequently recombined at whatever levels are necessary for clean signal distribution. In those cases where a narrowband, single channel Yagi is used, the same is simply accomplished by bandpass filtering.

Channel separation can be managed in two different fashions. If there are only a few channels of relatively equal strength to separate, a basic splitter and bandpass filters are all that is needed (see Figures 5-27 and 5-28). For example, inserting the signal from an off-air antenna into a four-way splitter results in four identical outputs for filtering, amplification and processing. Often such splitters include built-in bandpass filters so that channels are separated in one component. This approach works well, for example, when separating a few UHF channels for conversion to VHF signals.

Such band separators are used to divide the off-air input into UHF and VHF bands. Then the VHF signal is usually separated into the VHF-high and VHF-low channels and any FM inputs are separated from the VHF-low band. Each of these groups is then filtered and inserted into processing devices such as single-channel strip amplifiers.

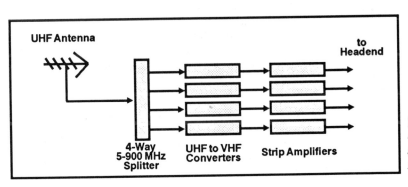

Figure 5-27. Simple Channel Separation. *A 4-way splitter with UHF-to-VHF converters and strip amplifiers can be used to easily separate UHF channels.*

HEADEND SYSTEM DESIGN

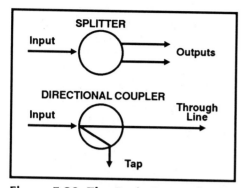

Figure 5-28. The Basic Tap and Splitter. *A tap and splitter accomplish the same task in different ways. A splitter divides a signal equally. A directional coupler taps off a variable signal while sending most of the power downstream in the through line.*

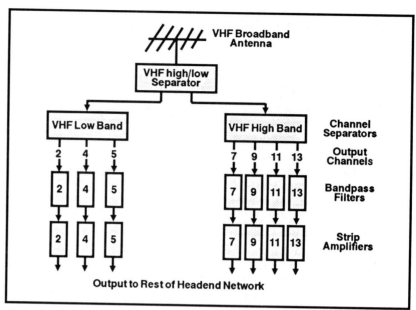

Figure 5-29. 7-Channel VHF Separation Network. *After a high/low separator, the individual low and high-band channel separators do their job. They are capable of tuning to a specific channel and can provide individual variable level adjustments.*

When processing a large number of unequal channels received from different directions, a separation/mixing network that allows more control than a simple splitter is required. Figure 5-29 illustrates a typical separation/mixing network. The separator, which is capable of tuning to a specific channel, provides a variable attenuation to allow adjustment of individual signal levels prior to processing. This particular unit has seven balanced outputs each tuned to a different frequency. These outputs can then be combined with signals from other off-air antennas for further processing in the headend.

In a small MATV system, these same tunable separators can be used in reverse to combine the balanced signal for broadband processing (see Figure 5-30. This technique can be an inexpensive method to balance the signal while delivering superior performance even in a small system.

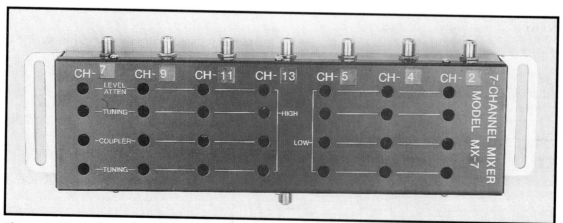

Figure 5-30. 7 Channel Splitter/Mixer with Attenuators. *In a small MATV system, the same type of components used to separate and balance these seven channels can be used to mix them for input to the headend launch amplifier.(Courtesy of Pico Macom)*

Off-Air Processors

After the channels are separated, the signals must be "cleaned up" and amplified in a processing system. This headend section has three principle functions. First, these components increase the relative level of the off-air video carrier. Second, they must reject all co-channel and other spurious signals to a level that is 55 to 60 dB below the desired channel. Third, they must ensure that the level of the in-channel audio subcarrier is reduced to about 15 dB below the video carrier power and that all channel levels are balanced. Although this is the optimal level on a cable system, most off-air broadcasters transmit the audio subcarrier at only 7 to 13 dB below the video carrier. These electronic components also have other functions depending upon the situation. The processor section must often provide automatic gain control or re-modulate an off-air channel onto a different frequency slot.

The three typical systems employed for processing off-air signals are single channel, "strip" amplifiers, on-channel processors and demodulation/re-modulation (demod/remod) systems. These processors differ in the complexity of the methods used but have numerous elements in common.

Strip Amplifiers and On-Channel Processors

Strip amplifiers are the most commonly used processing components in SMATV installations (see Figure 5-31). They amplify and equalize the signal level of individual television channels. These cut-to-channel amplifiers generally feature input and output level adjustments, good selectivity, a thru-output test jack, automatic gain control (AGC) circuits and, in most cases, are capable of producing a relatively high 66 dBmV (1 volt) output so that several hundred feet of coaxial cable can be driven. Strip amps are generally rack mounted and interconnected one below the other in order to combine all outputs. They also have the advantage of being inexpensive and usually operate relatively trouble-free when correctly installed.

When using strip amplifiers in a co-channel environment, i.e. where multiple channels are present, either internal or external bandpass filters are usually required. Most channel separators unavoidably allow some portion of the signal from an adjacent channel to be distributed downline with the primary signal. When this "dirty" signal reaches a strip amp, the amplification circuits also boost its power and cause intermodulation distortion or "beats" in the system. A bandpass filter, mounted in-line before the strip amp, eliminates this interference by filtering adjacent channels to a low enough level so that such problems do not occur.

The name on-channel processor is often applied to strip amplifiers used as off-air processors (see Figure 5-32). Most, but not all, on-channel processors have internal, input bandpass filters to prevent

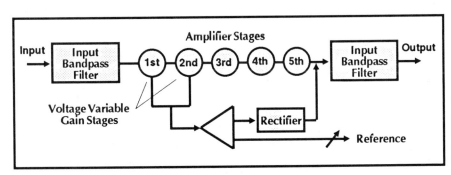

Figure 5-31. Schematic of a Strip Amplifier. *The heart of a strip amplifier consists of input and output bandpass filters, an AGC control and a series of amplifiers.*

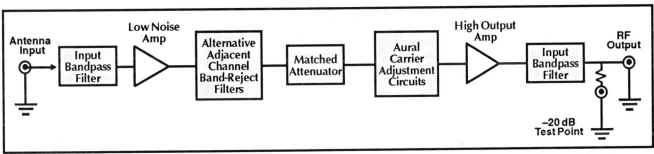

Figure 5-32. Schematic of an On-Channel Processor. *This on-channel processor is similar to a good quality strip amplifier but always has input and output bandpass filters and other features such as an aural carrier level control.*

HEADEND SYSTEM DESIGN

strong out-of-band signals from overloading the strip amp. Some brands also feature an output bandpass filter to further clean up spurious signals which may either have passed through or have been created by the system.

The signal level received by off-air antennas that are located more than 30 miles from a television transmitter are often subjected to signal fading. As this distance increases so does the potential fading, usually up to a maximum of 20 to 25 dB. The AGC circuits are designed to keep the output level constant in such environments. Strip amplifiers are rated according to maximum output capability, gain and AGC range (see Figure 5-33)

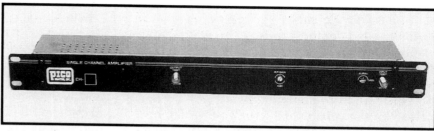

Figure 5-33. Cut-to-Channel Amplifier. *This single channel amplifier has an output of 62 dB below AGC. Its maximum output level at the 0.5% sync compression point is 72 dBmV. The AGC ranges from 4 to 40 dB while the front panel range is 15 dB. The aural trap can be adjusted from 0 to −10 dB. It has a −20 dB test point on the input and a −30 dB test point on the output. It is available in low, mid, FM, high, superband and UHF ranges. (Courtesy of Pico Macom, Inc.)*

Therefore, for example, a strip amp might have a maximum potential output of 50 dBmV, a maximum gain of 35 dB and an AGC range of 20 dB. The gain of this strip amplifier could then vary from 15 to 35 dB. The input signal would have to be 15 dBmV for this amplifier to achieve its full output. If the input increased by 10 dB, the AGC could then compensate by lowering gain. But if the signal decreased, the output would fall because the gain would have already been at its maximum value. The solution to this problem would be to design the system so that the average input signal is 25 dBmV and the output is 50 dBmV. A rise or fall of the input by up to 10 dB in either direction could then be managed by the 20 dB AGC range. In reality, signals often fade down to 15 dB below the average signal level, but rarely rise more than 3 to 5 dB above this norm. Therefore, a safe operating point for a strip amplifier is at an input signal that would force the unit to operate at 5 or 6 dB below its maximum gain. Note that this AGC circuit should preferably lock onto the peak level of the horizontal sync pulse, not the average brightness level of the television signal.

Broadband amplifiers rarely feature AGCs because this circuitry has no way of deciding which channel to "lock onto." It is unusual for every channel to periodically go off the air. However, if absolutely necessary, this design difficulty can be solved by injecting a pilot carrier at the headend to be used as an AGC reference level.

Using strip amplifiers for signal processing can cause a number of problems. For example, if the channel being processed is not broadcasted during certain time periods when the adjacent channels are on-the-air or when co-channel interference is detected, the AGC circuitry can cause the power levels to increase to the point of overload and cause interference with these adjacent channels or spurious outputs, respectively. In addition, in weak signal areas where adjacent signal problems can occur, the channel separators and bandpass filter may not be able to cleanly isolate the signals to allow clear reception. Noise calculations show that using a bandpass filter at the front-end of a strip amplifier has little effect on S/N ratio except in weak signal situations. In these cases, it may be better to use a processor or a demod/remod system.

Demodulation/Re-Modulation Systems

A demodulator acts much like a receiver. It takes an off-air signal and removes its carrier to provide the baseband (unmodulated) audio/video output. A channel modulator is then used to create a new carrier so that the baseband output is remodulated onto the same or a new channel. This system generates a clean signal and also allows for greater control over image quality than a strip amp system.

A high quality demodulator employs a surface acoustic wave (SAW) filter in its circuitry to provide sharp filtering and a clean signal. The resulting baseband output contains only the signals for the channel being processed and, unlike a strip amp, will not pass any co-channel information. In addition, many good demodulators are frequency agile, i.e. able to switch to any desired frequency, while most strip amps have to be factory ordered with desired channels specified. This frequency agility makes the demod/remod process very desirable for UHF-to-VHF conversion and also helps to minimize the need to stock spare parts. A discussion of frequency agility and its benefits follows below.

Demod/remod systems have recently become less costly. They can provide the operator of a large system with a very professional method to process signals at a reasonable cost. Many operators have realized the benefit of these components because of the inherent degree of signal quality control and have chosen to use demod/remod equipment for all aspects of channel processing.

However, there are some difficulties associated with using demod/remod components. When using this system for on-channel processing, it is important to be sure that there are no outside signals leaking into the system, i.e. no ingress interference. The synchronization pulses generated by the internal modulator must lock-up with the sync from the off-air system. However, if this modulator signal detects an unexpected signal leaking into the system, this may not occur and the co-channel interference could ruin picture quality. When using an existing system located in a region having high levels of off-air interference, some signal ingress should be expected. However, a new, well-constructed system usually should eliminate ingress and the associated problems.

Heterodyne Signal Processors

Full-performance heterodyne signal processors are generally far more complicated in design than on-channel processors and are therefore correspondingly more expensive (see Figures 5-34 and 5-35). These components do all that either on-channel processors or demod/remod systems accomplish. They amplify the off-air signal, clean it up to remove extraneous signals and output it onto any desired frequency.

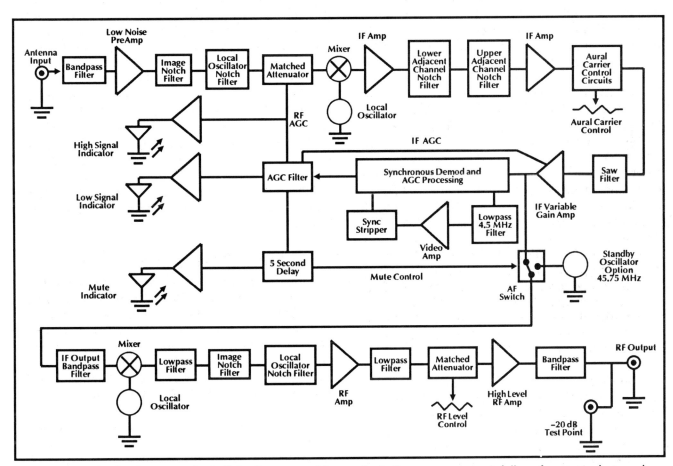

Figure 5-34. Block Diagram of a Full-Performance Heterodyne Processor. *A full-performance heterodyne processor downconverts input signals to an IF of 45 MHz and uses good quality SAW filters to maximize interference rejection. (Courtesy of Nexus Engineering)*

HEADEND SYSTEM DESIGN

The major advantage that a heterodyne processor offers is the ability to clean up each channel by removing interfering signals more effectively than the other systems. These components convert the input signal to a 45 MHz IF frequency range where low-cost, high-performance SAW (surface wave acoustic) filters are available to cleanly separate and remove unwanted signals. Good quality SAW filters have very sharp frequency boundaries and therefore have very high channel selectivities. These units also use several additional stages of filtering to achieve excellent processing. A comparison between typical on-channel and heterodyne processors is presented in the accompanying figure.

In most heterodyne processors, the input and output modules can be changed to output any desired channel. In addition, the synchronization signal can be phase-locked to the off-air sync pulses in order to prevent ingress interference problems.

Heterodyne processors allow ultimate control over signal quality (see Figure 5-36). Pioneer more expensive model were used almost exclusively by franchise cable operators in very large headend systems. Today reasonably priced heterodyne processors are the first choice in all but the smallest SMATV operations.

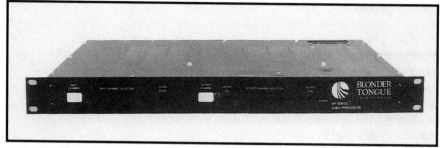

Figure 5-35. Agile Heterodyne Signal Processor. *In this processor, the input signal is converted to an intermediate frequency which is then converted to the desired frequency for insertion into the headend mixer. Both the input and output channel frequencies can be selected at will. Heterodyne signal processors are usually larger, more complex and three to four times more expensive than on-channel processors. (Courtesy of Blonder Tongue)*

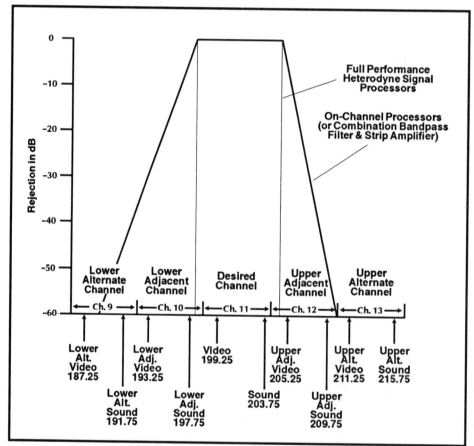

Figure 5-36. Comparison of On-Channel and Heterodyne Processors. *This attenuation versus frequency graph clearly shows why a heterodyne processor can properly handle a much wider range of off-air input signals than a standard on-channel processor.*

HEADEND SYSTEM DESIGN

UHF-to-VHF Conversion

Most SMATV systems include some local UHF stations in their off-air channel menu. Incorporating these broadcasters can provide a low cost means to expand channel capacity while, at the same time, offer additional quality programming to customers. However, the distribution of UHF channels can prove to be a technical headache because signals are attenuated more at these higher frequencies than are VHF channels. Rather than installing expensive UHF line equipment, it is much easier to convert each UHF channel to a matching, unused VHF channel for insertion onto the distribution network.

UHF-to-VHF converters exist in many different configurations and price ranges (see Figures 5-37 and 5-38). In those cases where the available input signal level exceeds 0 dBmV, a rather simple, crystal controlled UHF-to-VHF converter with built-in bandpass filters is often selected. These devices are relatively low cost items available from many equipment suppliers. Such on-channel converters/processors are channel specific and must be ordered from the manufacturer with the desired conversion pre-installed. In many regions, distributors will stock equipment preset for those channel conversions most frequently used.

A demod/remod system or a full-performance heterodyne processor can also be used for UHF-to-VHF conversions. These devices provide high quality channel conversions that are very stable. In addition, demod/remods allow some frequency conversions that are not technically possible with standard crystal converters. Although demod/remod systems are more costly than crystal converters, they do provide superior quality. Modern demod/remod systems have evolved to the point that now they are only slightly more expensive than a combination converter/strip amp system.

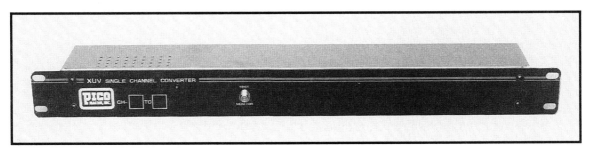

Figure 5-37. Fixed UHF-to-VHF Converter. *This crystal and PLL controlled unit has a fixed UHV channel input and VHF channel output. (Courtesy of Pico Macom)*

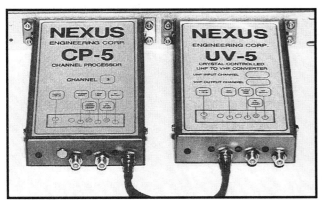

Figure 5-38. Channel Processor and Channel Converter. *The top photo shows the low-cost Nexus UV-5 crystal controlled UHF to VHF converter. After conversion, its output is fed into a channel processor which filters and amplifies the signal to be combined into the mixing system for distribution. (Courtesy of Nexus Engineering)*

HEADEND SYSTEM DESIGN

Prohibited Conversions

Conversion between some channel frequencies is not allowed because of the resultant beat frequencies that cannot be filtered out of the final signal (see Table 5-3). To understand this concept it is useful to consider how a typical UHF-to-VHF converter functions. The desired UHF channel is filtered and then mixed with a pure frequency signal generated by a local oscillator (LO). The LO and channel signals "beat" together producing sum, difference and harmonic frequencies. The LO frequency is chosen so that the UHF channel frequency minus the LO frequency equals the desired VHF frequency. Any unwanted, spurious signals produced are blocked from exiting by use of a bandpass filter.

An example can make this process clear. If UHF channel 41, having a picture frequency of 633.25 MHz, is being converted to VHF channel 13, at 211.25 MHz, the local oscillator frequency is chosen to be the difference between these two numbers or 422 MHz. But this conversion can cause problems. In addition to the channel 13 frequency being produced by the difference between the LO and channel 41 frequencies, the second harmonic of channel 13, which equals 2 times 211.25 MHz or 422.5 MHz is also created. This beats with the channel 41 UHF frequency of 633.25 MHz to produce 210.75 MHz. This frequency then combines with the VHF channel 13 of 211.25 MHz and produces an annoying 500 kHz beat. This channel 41 to 13 conversion is therefore considered forbidden because of the possibility for interference.

Incidental and Self-Conversions

There is potential for trouble even when allowed conversions are employed. "Incidental conversion" can occur when simultaneously downconverting two UHF signals. For example, if channels 28 and 30 were both being converted to channels 8 and 10, respectively, unwanted beats could result. This is because the bandpass filtering at the converter inputs is not sharp enough to completely reject the nearby channel. Therefore, some of the channel 30 signal which enters channel 28 is subsequently downconverted to channel 10. But the channel 30 converter is also producing channel 10. These two signals interact and produce interference. The same result occurs if channel 28 ingresses into 30. In this case, a better choice would be to convert channels 28 and 30 to channels 10 and 8, respectively.

"Self-conversion" occurs when there is a 12 channel spacing in the UHF spectrum. An example is when channel 29 at 561.25 MHz is being converted to channel 4 at 67.25 MHz in the presence of a strong channel 17. The LO frequency will be set at the channel difference of 494 MHz. However, if channel 17 is strong enough to enter the channel 29 input filter it also acts as a local oscillator. Its sound carrier at 493.75 MHz when subtracted from channel 29 video carrier produces a difference signal at 67.5 MHz. The resulting 250 kHz beat between this frequency and the channel 4 signal at 67.25 MHz can produce problems on channel 4. By selecting a different conversion target, this potential problem can be avoided.

In practice, investigating possible downconversion interference difficulties is rather straightforward. First choose a possible conversion, then subtract the VHF from the UHF picture carrier frequency to find the LO frequency. If this frequency falls within the band of another channel, then there is a potential problem. Also, examine the difference in frequency between the UHF channel and the LO frequency. Problems may also occur if this falls within the band of a desired channel. Note that difficulties do not necessarily always occur in the forbidden transition zones. But the possibility is always there.

TABLE 5-3. FORBIDDEN CHANNEL CONVERSIONS

UHF Channel	VHF Target Channel
14-16	5
18-22, 36	6
22-25, 51-55, 81-83	7
25-28, 55-59	8
14, 28-31, 59-63	9
16, 17, 31-34, 63-67	10
19, 34-37, 67-71	11
21, 22, 37-40, 71-75	12
24, 40-43, 75-79	13

VHF Channel	UHF Target Channel
3	8, 9, 10
7	2
8, 9, 10	3
4	12
12	4

Also Forbidden in General

VHF-low to VHF-low
VHF-high to VHF-high
Subchannel to subchannel

HEADEND SYSTEM DESIGN

Mixing and Combining Signals

After off-air and satellite signals have been processed, they must be mixed into a multi-channel format for distribution into the cable plant. To accomplish this, several different methods and configurations of active or passive combiners or directional couplers can be used.

Basic Requirements for Signal Mixing

A headend combiner takes the individual channel outputs from satellite modulators, off-air processors and audio FM modulators and combines them onto a single output cable for input to the headend launch amplifier. Either combiners, which are splitters used in reverse, or directional couplers, also known as taps, can be used to mix signals (see Figures 5-39 and 5-40). Although either of these passive devices has acceptable performance, splitters are easier to use and produce more reliable installations because fewer cables and connectors are generally required.

A passive mixer functions much like a large 8 or 16-way splitter. It is constructed from a series of modified tap-type circuits. This type of internal design increases the isolation between ports and improves the simultaneous mixing of many channels. In larger SMATV systems, outputs from a number of mixers may be mixed via a series of cascaded 2 or 4-way combiners to facilitate adding a large number of channels together onto a distribution network.

There are four parameters which characterize a combiner. These are (1) the degree of isolation between input ports, (2) the balancing of headend output channels, (3) frequency response and (4) insertion losses (see Figure 5-41).

When a splitter or any other device has a high isolation rating, a signal fed into any port will be so severely attenuated that it will not be detected in ordinary practice at any other port. But even when combiner isolation is high, if this device is not properly matched to its output cable, input port-to-port, cross-talk can become unacceptable. Note that the importance of isolation in combiners is directly related to the quality of the signal produced by the processing equipment. In some cases, a very "clean" processor output on each channel can create an excellent headend output even if the combiner isolation is near zero. Nevertheless, port-to-port isolations should generally be at least 16 to 20 dB or better.

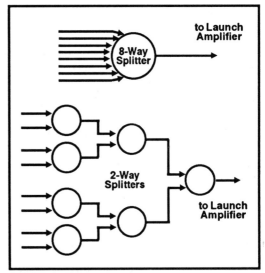

Figure 5-39. 2-Way versus 8-Way Splitters. *Either six 2-way one 8-way splitter can accomplish the same task. When this type of mixing is employed, it is better to use an 8-way splitter having good isolation between ports to minimize the number of connectors and cables that must be used.*

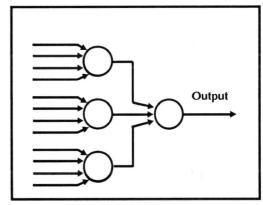

Figure 5-40. 12-Channel Combiner. *The theoretical per channel loss in this twelve way combiner is 11 dB. In practice, each channel loses 13 dB before reaching the output port.*

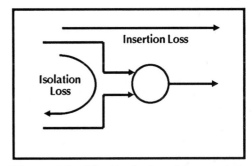

Figure 5-41. The Basic Combiner. *A 2-way combiner must have good inter-port isolation and a reasonably low insertion loss.*

HEADEND SYSTEM DESIGN

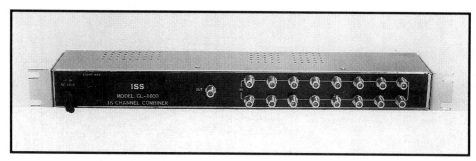

Figure 5-42. 16-Channel Active Headend Combiner. *This unit combines and amplifies the low outputs from lower cost modulators and signal processors. (Courtesy of International Satellite Systems)*

All channels must be properly balanced at the output of the mixing network. This topic has been explored earlier in this chapter. The frequency response of the components used to combine the various television signals can directly affect this balancing. If the combiners and other devices used cannot manage the full range of channel frequencies, output levels will not be easily balanced.

Insertion losses must always be factored into headend or distribution system calculations. The objective of the combination network is to provide a set of balanced channels to the headend launch amplifier.

Channel Separators/Mixers

In more simple MATV systems, the same unit utilized to separate channels can also be used for mixing. This method is used when the only requirement is to balance all channels for distribution. Instead of amplifying or processing each channel, one step of balancing all levels coming out of the separator is required. These channels are then fed in reverse into a second separator and subsequently into a broadband amplifier for distribution to the remainder of the system. This method can also work well when mixing a few satellite channels into an existing system that does not process channels individually.

A commonly encountered headend device is the VHF/UHF mixer. This unit contains both a low pass and high pass filter. VHF signal would be prevented from entering the UHF antenna by the high pass filter and therefore would not cause distortion. This same protection is provided against UHF signals ingressing into the VHF line. This simple unit can also be used to separate VHF and UHF inputs to a television set.

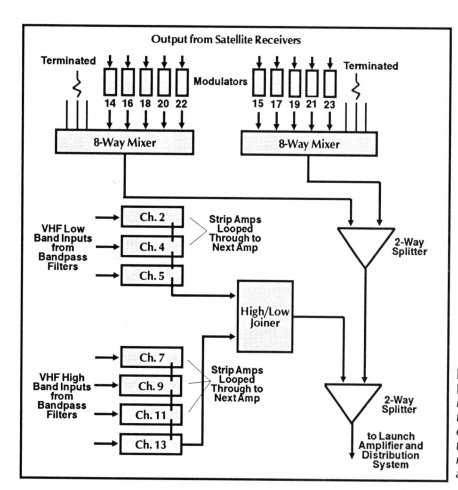

Figure 5-43. Mixing Techniques - Example I. *Splitters and mixers can be used for either separating or mixing television signals. In this 17-channel example note that all unused ports are terminated. Note that adjacent channels are not looped through from strip amp to strip amp.*

78

HEADEND SYSTEM DESIGN

Active Mixers

An active mixer is basically a large 8 or 16-way splitter having a built-in broadband amplifier (see Figure 5-42). This design compensates for the signal loss that occurs in all passive splitters and mixers. In theory, this design is a valid idea. However, practical experience suggests that these devices have shortcomings.

Most active combiners on the market are "low-end" products. They are used primarily to overcome the low-output characteristic of certain strip amps and modulators. Many of these active combiners are noisy and cannot be used to their full rated capacity. However, they are occasionally useful in relatively small systems that have a minimum number of channels.

Simple Examples of Mixing Techniques

There are many different ways to combine off-air signals. The method chosen can drastically affect the quality of the resulting signal. Figure 5-43 and 5-44, which detail two typical 17 channel mixing schemes, are referred to below.

In the first two examples, no adjacent channel is mixed in the same component. The odd channels are combined in one unit and the even channels in another. Then output signals from these mixers are combined together to produce the final output. The only deviation from this procedure is the mixing of channels 4 and 5 that are not adjacent in frequency.

In the first design (see Figure 5-43) the off-air channels are looped from one strip amp to the next. Many strip amps provide a loop-through port to help facilitate channel mixing. These ports should not be used for adjacent channel mixing because of the possibility of interference.

Finally, the outputs from the looped channel 5 and 13 strip amps are mixed together in a high/low combiner that incorporates a frequency separator that removes the high-band from the low-band channels. In general, it is better to use this combination method whenever possible because the associated signal loss is much lower than when using a splitter. However, the configuration shown in Figure 5-44 simply amplifies the VHF channels, combines them in a splitter and then routes them to a high/low combiner.

Figure 5-45 outlines an example of combining VHF, FM and UHF signals. Frequency separators are available in a number of configurations. UHF/VHF separators are also extremely useful for combining outputs from separate antennas onto one downlead.

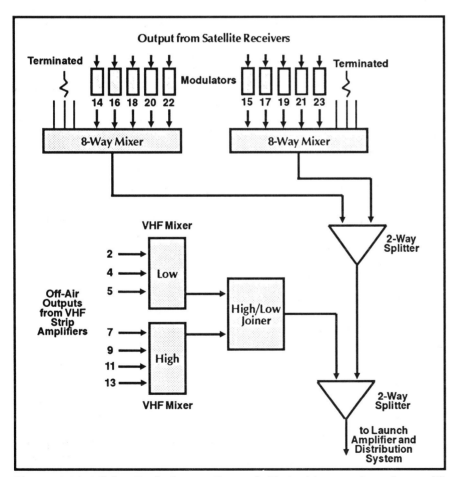

Figure 5-44. Mixing Techniques - Example II. *In this case, the strip amplifiers are looped through. Notice that no adjacent channels are looped through as this would potentially cause interference.*

HEADEND SYSTEM DESIGN

Figure 5-45. Frequency Separation/Mixing. *This simple system combines VHF-low, VHF-high, FM and UHF signals onto one line.*

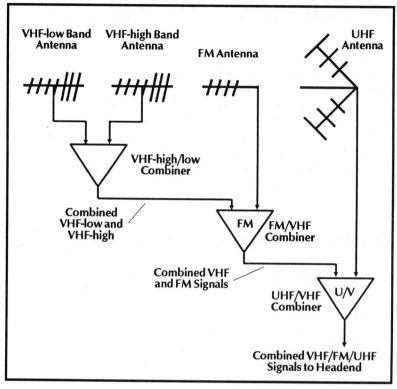

Launch Amplifiers

Once all the headend processing has been completed, the combined signal is routed into a distribution or launch amplifier and subsequently into the distribution system (see Figure 5-46). Past this point, it is essential to maintain the S/N ratio at a reasonable level. The strategy is to minimize the number of distribution amplifiers while maintaining the input to each amplifier above a minimum level. This process is outlined in more detailed in the chapter on distribution systems that follows.

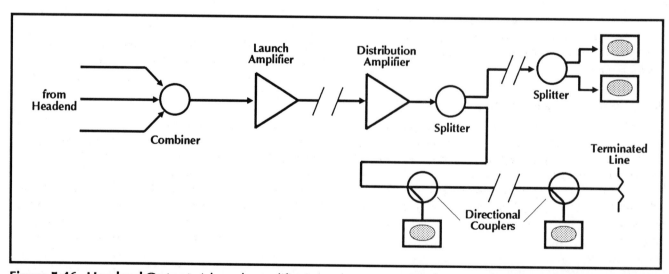

Figure 5-46. Headend Output. *A launch amplifier is used to output signals onto a distribution network. A typical output level is about 55 dBmV.*

D. SOLVING OFF-AIR SIGNAL LEVEL PROBLEMS

Off-air signals are subjected to an interesting range of transmission difficulties that require creative solutions to ensure good quality television reception. In some systems, a weak signal may vary considerably in level. Another set of problems also arises when an SMATV system is too close to an off-air transmitting antenna and therefore receives a signal having too much power.

Fading and Weak Off-Air Signals

Fading is usually a more serious problem when detecting lower power signals. These variations in signal power are more pronounced as the signal path increases, precisely the situation in which off-air transmissions are weak. The headend processing electronics must also be more sophisticated to handle fading when signal levels are marginal.

Fading is dependent upon numerous factors including signal frequency, the time of day or year, terrain roughness or slope, antenna heights, reflections along the propagation path and weather. Unusual atmospheric conditions can increase fading by causing signals to bend or duct along their paths to the receiving antenna. It is clearly very difficult to predict when fading will occur.

Signal fade can be reduced to some extent by increasing antenna height to be "above" terrain and ground conditions. This strategy usually also has the benefit of increased antenna gain. Signals tend to fade less on lower frequency channels because RF transmissions tend to be less directional and more "spread out" and averaged over terrain as frequency is decreased.

AGC Design Considerations

Unless fading is properly corrected by automatic gain circuits (AGC) in the processing components, distributed signals may not be correctly balanced and interference between adjacent channels becomes more likely. In general, when an installation is beyond the boundary of the grade B contour, processors having good quality AGC circuits are necessary.

It is important to select processing components that have well-designed AGCs. In some cases, lower quality AGC circuits can actually cause more problems than they cure. The three most important design failures are inadequate AGC range or window, insufficient immunity to certain types of interference and lack of muting triggered by pre-set signals.

SMATV system designers should understand that signals fed into processors and amplifiers having AGC circuits must fall at the proper location within their AGC range. This means that average signal levels during normal atmospheric conditions should be set at about 10 to 20% above the AGC midpoint. This allows a processor to compensate for increased signal power as well as for fade which is more common and usually is more variable than more powerful signals. Setting the correct incoming level requires knowledge of the signal power at the antenna, downlead, cable and filter losses as well as the preamplifier gain. An example of this method has already been presented in the discussion of preamps.

If the AGC range is insufficient or if the input signal is not set at the correct point then, as the power increases or decreases the range may be exceeded and the output signal would not be properly controlled. This could cause the noise component to compress the signal and "catch up" with the peak components. This amplified noise could interfere with adjacent channels. The synchronization pulses on the television signal, which are at the highest voltages, may be compressed and lost, resulting in a picture that tears or rolls. The AGC level detection circuits should determine the required gain correction by locking onto the horizontal sync pulses which are the only component in a video signal which is set at a constant reference voltage.

Problems can also arise if the AGC circuits are susceptible to noise or interfering signals. Some types of "impulse" noise from sources like distant lightning strikes can produce short voltage transients. If the AGC mistakenly tracks these signals, the processor will attempt to maintain the signal level based on noise rather than the television signal. A high quality processor must have sophisticated circuits which can discriminate between noise, interference and the desired television signal.

HEADEND SYSTEM DESIGN

AGC circuits must also have a method to mute a processor's output if the input signal levels drop below a predetermined value, for example, −25 dBmV. If not, an AGC will attempt to "find" a signal and probably lock onto noise or interference. This is a very real concern. Many television transmitters are shut down at night to conserve electricity or to reduce wear to the active transmitter components.

Overloading

Attempting to detect distant, very weak signals increases the chance of inadvertently amplifying adjacent channels. This class of problems known as overloading is more likely to occur when headend designers install preamplifiers before processors. These preamps are generally broadband devices that amplify both the desired signal and adjacent channels which, at times, could be much stronger. If these weak signals are first fed into a high quality processor that uses selective bandpass filters, adjacent channels and interfering signals can be eliminated before they become a problem.

A simple precautionary design practice is to install a high quality antenna from the start rather than to try to use sophisticated electronic processors to overcome problems after the fact. A high gain, narrow beam antenna is mounted as high as possible can go a long way towards increasing gain, reducing the effects of fading, ghosting and overloading, and rejecting interfering signals.

Ghosting and Stacking Antennas

Ghosts that appear on-screen as multiple faint images to either side of the intended television picture occur when a signal follows two or more paths of different lengths to reach an off-air receiving antenna (see Figure 5-47). This delayed reception results when a portion of the incoming signal is reflected by buildings or other obstructions.

Ghosts also can occur when there are excessive signal reflections in a distribution system or when signals directly ingress into the system at a point removed from the headend. Such ingress ghosting results because signals travel at a lower speed in the coaxial distribution network than along the direct off-air route. Other than the obvious solution of eliminating signal ingress, when channel space is available this problem can be solved by converting the input channels to a second set of unused channels. Thus, for example, if off-air channels 3, 5 and 9

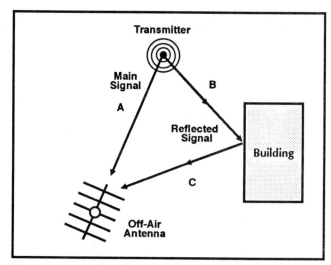

Figure 5-47. Ghosting. *Ghosting results when the same signal arrives at the television set via an off-air antenna at two different times. The signal reflected off the building has to travel a longer distance, B + C, than the direct path, A, from the transmitter.*

were translated to channels 4, 6, and 10 ingress ghosting would not occur.

The most straightforward method to potentially avoid ghosting lies in choosing an antenna which has a narrow, highly directional field of view, i.e. low side lobes and a narrow beamwidth, and aiming it well. The beamwidth can be reduced by increasing the number of antenna elements or by installing a correctly designed array of antennas.

If ghosting does occur, the first step is to attempt to eliminate the problem by slightly rotating the antenna while viewing the television picture. With some luck, this may serve to aim the antenna so that the interfering signal is detected at a null in the antenna reception pattern. However, if the angular separation between the incoming signals happens to be less than the antenna beamwidth, the phase cancellation cure described below is the only solution. Ghosting cannot be eliminated by filtering because each identical portion of this signal occupies the same frequency band.

In the more intractable cases, ghosting can often be cured by installing a stacked antenna array and using phase cancellation techniques. The horizontal and vertical staggering techniques described below are quite effective and can provide at least 20 dB of interference rejection.

HEADEND SYSTEM DESIGN

Phase cancellation is based on a precise knowledge of the signal paths. Figure 5-48 illustrates the concept of phase cancellation and Figure 5-49 a typical solution for eliminating ghosts. Both antennas are pointed at the desired signal. But the spacing between the two antennas is arranged so that the interfering signal arrives at the combiner 1/2 wavelength later than the other. Since the interfering signals detected by the two antennas are 180° out-of-phase, they cancel and eliminate the ghosting. The angle of arrival of the interfering signal determines the horizontal antenna spacing. This phase cancellation method will generally work when the principle and interfering signal arrive from directions between 5° and 160° apart and when the desired signal is at least 10 dB stronger than the weaker one.

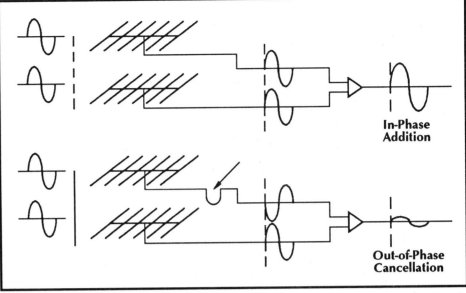

Figure 5-48. Phase and Signal Combining. *When signals are in-phase they add together. When they are out-of-phase they cancel. The phase difference in the bottom diagram is caused by the extra time it takes the signal to travel through the delay line.*

A simple equation can be used to determine the difference in downlead length required to ensure that incoming interference is properly phase canceled. One test antenna is first used to determine the incoming directions of both signals relative to north, and the angle between the two signal is obtained by subtraction. The following formula is then used to obtain antenna spacing:

$$H = \lambda / 2 \sin \theta$$

where λ is the signal wavelength, H is the horizontal antenna spacing and θ is found from the angle between the interfering and direct signals. Both λ and H are expressed in the same units. Figure 5-50 presents a graphical description of this angle and a numerical example.

In order for the interfering signals to properly cancel, the two out-of-phase signals must have nearly equal strengths. If the difference as measured at the output of each antenna is greater than 0.5 dB, both antennas should be moved while the relative separation is maintained. A location where the signal strength are nearly equal can usually be found by this method.

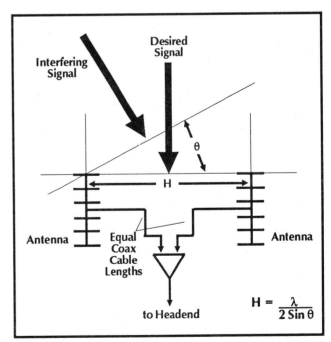

Figure 5-49. Ghost Cancellation. *Ghosts can be effectively canceled by using two identical antennas separated by a distance chosen so that the desired and interfering signals arrive a half wavelength apart and phase cancel. The distance, H, is given by the formula which is explained in the text and in the following figure.*

HEADEND SYSTEM DESIGN

Once the necessary 2-bay array has been installed and signal levels are nearly equal, any remaining ghosting can be eliminated by making small rotations to the final array position while watching a portable color TV.

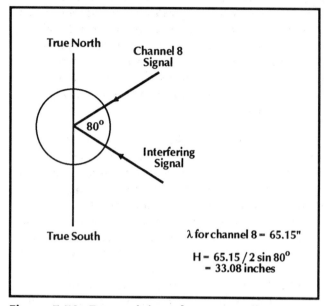

Figure 5-50. Determining Phase Cancellation Distance. *An example of determining the correct horizontal distance required between two antennas to cause complete phase cancellation is diagrammed here. The angle between the desired and interfering signals is 80°. Note that this formula is useful only when the interfering signal does not arrive from within approximately 20° of the rear of the antennas.*

In situations where the separation between main and interfering signals is between 45° and 90°, the antenna spacing may have to be so narrow that elements begin touching. One solution is to increase the inter-antenna spacing to an odd multiple of the calculated horizontal spacing. Therefore, for example, if the spacing is found to be 3 feet, a spacing of 9 or 15 feet may be adequate to phase cancel the interference. Difficulties can also arise when the angular difference between the incoming signals are less than 5° or 10°. Then the required spacing, particularly for the low frequency channels, may be impractically large.

When the interfering signal arrives from within 170° to 190° degrees of the transmitter boresight, a different stacking method must be employed. An example is presented in Figure 5-51. The vertical spacing is set at one wavelength. Then an additional horizontal stagger distance of 1/4 wavelength is included in the 2-bay configuration. Signals coming directly from the television transmitter arrive at the top antenna 1/4 wavelength or 90° ahead of the bottom one. This is compensated for by using a top antenna downlead that is 1/4 wavelength longer. Therefore, both antennas detect the principle signal in phase. The situation is reversed for interfering signals arriving from behind. In this case, the top antenna detects the reflected signal 1/4 wavelength behind that of the bottom antenna. Then the downlead contributes another 1/4 wavelength delay. The sum of these delays is 180° so that when the same interfering signal reaches the combiner, phase cancellation occurs.

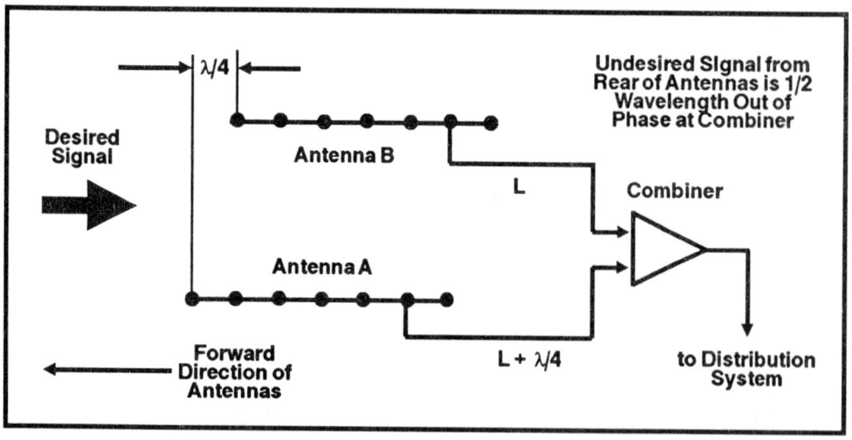

In order to calculate the difference in cable lengths, the wavelength of the signal traveling through coax is required. This is shorter than the "free-space" wavelength because signals travel more slowly within cables. The correction, the "velocity of propagation factor," can be obtained from cable manufacturers. Solid and foam dielectric cables typically have factors of 0.66 and 0.82, respectively.

Figure 5-51. Back Staggering for Rear Incoming Interference. *When interfering signals are arriving from behind off-air antennas, ghosts can be eliminated by staggering the antennas one quarter wavelength apart. For example, for channel 4 having a 67.25 MHz video carrier frequency, the wavelength in air is 175.58 inches. The wavelength in this particular brand of RG-6/U must be corrected by 0.83 so that its channel 4 wavelength is 145.73 inches. The quarter wavelength displacement is therefore 36.43 inches. Please refer to the text for further details.*

Processing Strong Signals

Ghosting and overloading can also occur when incoming signals are too powerful. It is not uncommon that an antenna in proximity to a strong local transmitter could detect a signal in excess of 20 dBmV. This can occur even when using a low gain, inexpensive off-air antenna. Such high level signals can overload a processor and cause sync compression leading to picture tearing and rolling as well as intermodulation distortion on adjacent channels. Fortunately, the technical fix to excessively strong signals is quite simply a matter of inserting an appropriate attenuation pad in the downlead from the antenna to the channel processor.

Powerful off-air signals can also ingress directly into television sets or a cable network. These signals, that do not enter the system through the off-air antenna, arrive at the television set either before or after signals coming from the headend and cause ghosting. Large cable TV systems are very susceptible to such ingress ghosting because the miles upon miles of distribution cables are potentially excellent antennas. Signals can ingress through faulty connectors or cable-shield breaks. In addition, a city-wide cable network usually has some subscribers in close vicinity to a local transmitter so that ghosting through ingress into their televisions can also occur.

In contrast, most private cable systems are highly localized and therefore such problems are much less likely. However, when a strong local transmitter is close at hand, it is important to use higher quality matching transformers and to tighten up all cabling connections. Occasionally the problem might be severe enough to force a conversion to another VHF channel. In this case, the local channel that is in the vicinity of the powerful transmitter generally must remain unused.

Two simple tests can be conducted to determine where the ghosting interference is entering a system. If a subscriber's set is disconnected from the cable and ghosting persists, then the signal is directly entering the set. In most cases, direct pick-up at the TV set can be solved by adding a set-top converter to the TV. If the incoming signal is connected into to the converter that is operating on an unused channel, usually 3 or 4, and the ghosting still persists, then the ingress is occurring in the distribution system. If the ghost is to the left of the main picture, it is probably entering the system at some point after the headend. If the ghost appears to the right of the picture, it is entering the system at the antenna.

E. SATELLITE TV RECEPTION COMPONENTS

The driving force behind any private cable system is the quality entertainment derived from its satellite system. The satellite programming is the bottom line. Its inclusion into an existing MATV system converts it into a satellite MATV, an SMATV, system. An engineer must design the satellite reception system to deliver a signal having both an S/N ratio and a level that match those of the off-air channels.

A satellite system consists of an outdoors assembly, the antenna, feed and LNB, designed to capture the signal, and accompanying indoor processing equipment including satellite receivers, modulators and, when necessary, decoders and stereo processors. Numerous detailed books have been written on the subject of satellite television. The intent here therefore is not to explore this area in great depth, but to discuss satellite systems only as they apply to a private cable headend.

The operation of a TVRO is conceptually simple. A satellite dish collects signals downlinked from a communication satellite and focuses them into a feedhorn. The feed directs the microwaves into a low noise amplifier (LNB, short for low noise block downconverter), the first electronically active component in the system. The LNB amplifies the signal to a level that can be processed by a satellite receiver. This receiver is tuned to one satellite transponder, usually an individual television channel, and demodulates both the audio and video inputs to produce a baseband or raw, unmodulated signal. This output is fed into a modulator that re-modulates the signal onto an AM carrier ready for mixing and distribution. Decoders are also included for those channels which are scrambled at their uplinks. Both stereo and monophonic audio outputs are available from most satellite receivers.

Satellite Antennas

With some minor exceptions, commercial satellite antennas generally have a parabolic reflective surface. These dishes are available in different sizes and are manufactured from various materials. The two most popular styles are solid and mesh. Solid dishes are usually made of fiberglass or aluminum, while mesh antennas are composed of aluminum or steel mesh.

Antenna gain is directly proportional to the size of the area which intercepts microwaves. The larger the dish diameter the higher its gain. Antenna gain is also increased as the accuracy of its reflective surface is improved. Private cable systems are designed to provide professional quality signals so an adequately sized, commercial-grade antenna should be selected (see Figures 5-52 and 5-53). Most C-band (4.7 to 5.2 GHz range of broadcasts) SMATV headends are fed by solid fiberglass or aluminum dishes in the 3.7 to 5 meter (12 to 16 foot) range. Headends fed by Ku-band signals (in the 12 GHz range) use antennas ranging from as small as 90 cm to 3 meters in diameter.

Antennas selected for private cable systems differ from those used in home reception systems in two respects. First, the antennas must be of commercial quality and they should be tested according to accepted practices to ensure that the stated gain, beamwidth and other parameters are really available for design purposes. Second, they are usually targeted onto just one satellite so that actuators which can track them across the geosynchronous arc are not required. This allows the use of stable az-

Figure 5-52. Miralite 3.7 Meter Antenna. *This C/Ku-band compatible, commercial antenna has been tested for uplink operation at frequencies up to 14.5 GHz. (Courtesy of Miralite Corporation)*

HEADEND SYSTEM DESIGN

imuth-elevation mounts. In a few situations, polar mounts like those in home TVRO installations as installed.

Feedhorns and Multi-Satellite Reception

Each C-band satellite is typically capable of simultaneously broadcasting 24 video channels and a much larger number of audio programs. Ku-band satellite transmit more variable number of channels, depending upon the channel bandwidth as well as the type of modulation technique in use. Recently introduced video compression techniques will soon allow relay of up to 8 channels per transponder. While C-band reception is explored here, Ku-band reception practices are quite similar and can be explored in more detail in some of the references.

The type of feed system chosen then determines whether 12 or 24 channels or transponders can be detected and how many individual satellites can be simultaneously received. Three types of feedhorns are used in most commercial reception systems. A conventional feedhorn connected to one LNB is capable of detecting either 12 vertically or horizontally polarized channels at one time. A probe within the feed must be physically rotated to receive the complement of 12 channels (see Figure 5-54).

The most common feed encountered in SMATV installations is the orthomode feedhorn (see Figure 5-57). This variety is connected to two LNAs or LNBs. One amplifier detects horizontally polarized signals and the other receives signals from the remaining 12 vertically polarized transponders.

In many situations, signals from more than one satellite must be simultaneously received to intercept the desired selection of available programming. Simultaneous reception from two or more satellites can be accomplished by either using one large dish having a multi-beam feed or by installing more than one antenna each of which is permanently targeted onto a different satellite. Multi-feed installations became available shortly after pro-

Figure 5-53. Microdyne 7 Meter Antenna. *The motorized polar mount on the 5 and 7 meter Microdyne antennas is designed for the off-air televisions stations as well as the SMATV and CATV markets. This apparatus can be switched between satellites in a matter of seconds. (Courtesy of Microdyne)*

Figure 5-54. Feedhorn. *This high gain polarizer is designed for use with deep reflectors having f/Ds ranging from 0.25 to 0.35. (Courtesy of Seavey Engineering Associates)*

grammers began to use more than the original Satcom I satellite to broadcast programming to cable operators. In fact, since that time signals have been uplinked to a cluster of satellites to take advantage of this technology.

In the multi-beam configuration, one or more feedhorns are offset from the central feedhorn which is located at or near the antenna focus. This then allows one or more adjacent satellites to be simultaneously detected. Either a conventional or an orthomode feed can be used at the central focus and the offset fed positions to permit 12 or 24 channel reception, respectively.

Installers have concluded after years of in-the-field experience that if a high quality antenna is used, satellites up to 4° off boresight, the main focal line, can be received with nearly equal performance (see Figures 5-56 and 5-57). However, this conclusion must be qualified. First, a multi-beam system should be used only with antennas that have diameters in the 3.2 to 10 meter range and that conform to FCC licensing requirements. These define a set of performance standards to ensure that antennas have adequately narrow beamwidths and low side lobes levels. Antennas that are smaller than 3.2 meters do not generally perform as well as necessary. For example, multi-beam systems should never be retrofitted onto an 8 foot dish. But if an installation already has a 3.2 meter or larger solid antenna that performs well on the least powerful satellite, Satcom F3 within the continental United States, chances are good that a multi-beam feed support can be retrofitted that will also produce an adequate signal when aimed a second satellite.

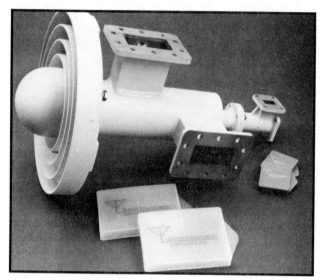

Figure 5-55. Orthomode Feedhorn. *The ESA 124-D four-port, coaxial prime focus feedhorn is designed to simultaneously receive both horizontal and vertical polarizations on C and Ku-bands without any loss comparable to that encountered when using offset configurations. Its total weight without LNBs is 5.5 pounds. (Courtesy of Seavey Engineering Associates)*

A good rule of thumb to also follow before attempting to install multi-beam assemblies it that the antenna f/D (focal length to diameter) ratio should be a minimum of 0.3 but preferably in excess of 0.33. When dishes have smaller f/D ratios, the distance between the adjacent focal points may be too small to allow easy mounting of the feed/LNB hard-

Figure 5-56. Multi-Feed Assemblies. *The components in the left photo are designed for 9 to 11 meter antennas, while those on the right are for smaller 3.7 meter dishes. (Courtesy of Superior Satellite Engineering)*

HEADEND SYSTEM DESIGN

Figure 5-57. Multi-Feed System. *This 4 meter Scientific Atlanta antenna is equipped with four LNAs attached to two orthomode feeds which can be adjusted to simultaneously detect two satellites. (Courtesy of Brent Gale)*

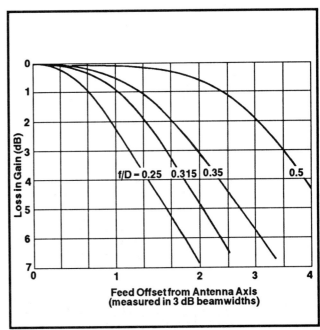

Figure 5-58. Gain Loss Due to Offset Focusing. *Gain losses that occur when offsetting either C or Ku-band feeds increase with the distance at which the feed is installed from the focal point and decrease with the system f/D ratio. In this case, the distance is measured in units of 3 dB beamwidths. Losses increase faster at Ku-band because half power beamwidths are one third compared to those at C-band.*

ware. In addition, "deep" antennas with lower f/D ratios have more "spherical aberration" at off-focus positions. This type of distortion arises when signals reflected from various positions on the antenna surface arrive at the feedhorn at slightly different times.

The amount of gain reduction when using off-center feeds has been chronicled in numerous installations (see Figure 5-58). For example, a 5 meter antenna with a 0.38 f/D and gain of 44.2 dBi has an off-boresight gain reduction of 0.3 to 0.5 dB at 3° and 1.2 to 1.5 dB at 8°. However, the same sized antenna with a f/D of 0.3 has an off-boresight signal degradation of 0.5 to 0.6 dB at 3° and 2.7 to 3.0 dB at 8° off boresight. Antenna surface accuracy also has a dramatic effect on these results. Any warpage or misalignment of petals can significantly reduce gains.

Adjacent-satellite interference may occur if the antenna is too small, deep or if the target satellite is substantially less powerful than an adjacent one. Interference is likely to be seen as degradation in cross-polarization performance. In other words, a faint picture could be observed in the background of the desired broadcast.

Signals from at most five satellites might be simultaneously received with an adequate S/N ratio by a 9 or 10 meter antenna. But generally dishes in the 4.0 to 7 meter range can detect just two or three satellite at one time. The powers of the satellites to be detected can affect these results. For example, when receiving signals from two adjacent satellites, one relatively weak and one relatively strong, the less powerful one, not the more powerful, should be targeted on axis. Or if two satellites having similar powers were being detected, it would be wise to have the feed for each one offset an equal distance from boresight.

There has been a great deal of controversy over the pros and cons of using multi-beam configurations. Nevertheless, in many cases, the resulting savings in material and space requirements have made this a favored approach.

HEADEND SYSTEM DESIGN

An alternative is to use an antenna designed specifically for multi-satellite reception. Such antennas incorporate variations of spherical and parabolic surfaces. To illustrate, the C-band Torus antenna was cut from a rectangular section of a sphere and feedhorns are mounted along a focal line. Another variety, the Simulsat antenna is a dual curvature reflector having a circular cross section along the arc of satellites and a parabolic contour at right angles. This design also causes incoming signals to be reflected to a focal line. The 5 meter version can simultaneously detect up to 35 satellites within a 70° arc and delivers a 44 dB gain (see Figure 5-59).

In North American up to seven satellites may be orbited between 125° and 137° by 1994. This will include the powerful Satcom C4 and Galaxy 5. At this point, purchasing a high quality multi-focus antenna may be a reasonable strategy for operators who own larger SMATV systems.

Low Noise Amplifiers

After the signal is collected by the feed system, it is channeled into an LNB (see Figure 5-60). This compact unit typically utilizes two or three cascaded GaAs-FET transistor stages to amplify the incoming signal. LNBs for use in reception of C-band signals usually produce IF of 950 to 1450 MHz.

The LNB is an outgrowth from low noise amplifier (LNA) technology. This device was coupled to a downconverter that lowers the frequency of the satellite signal to a range that can be efficiently transmitted with low losses via coaxial cable to indoor satellite receivers. An low noise block downconverter (LNB) combines the LNA and downconverter in one unit.

LNBs are rated in degrees Kelvin (°K). The higher the rating the more noise is produced in amplifying the extremely weak satellite transmissions. The lower the noise temperature, the lower the noise. Most commercially available LNBs fall in the range from 30 to 50°K. Parametrically cooled units rated as low as 20°K are available at higher prices.

Selecting a lower temperature LNB improves performance. However, this improvement is limited by antenna surface area. If an eight foot antenna is replaced by a ten-footer that has 56% more surface area, there will be a proportionate improvement in signal-to-noise ratio and therefore in system performance. While using a larger dish increases the gain

Figure 5-59. Simulsat 5 Meter Antenna. *This 5 meter Simulsat antenna can simultaneously detect up to 35 satellites within a 70° arc and produce a 44 dB gain.(Courtesy of Antenna Technology Corporation)*

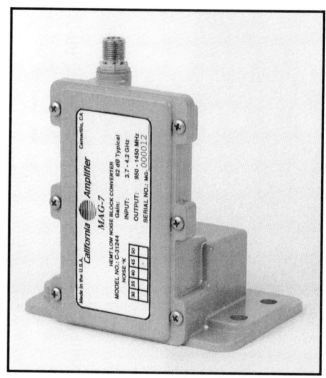

Figure 5-60. LNB. *This HEMT-technology, lightning-protected LNB has a noise temperature ranging from 25° to 50°K at 25°C and a 62 dB gain. It accepts a 3.7 to 4.2 GHz input signal and downconverts this band to 950 to 1450 MHz. (Courtesy of California Amplifier)*

of given signal, a lower temperature LNB reduces the noise level added to the signal during amplification. Therefore, the best design choice is to select a larger antenna, if possible, while using a sufficiently good quality, low temperature LNB.

It is possible to overload an LNB like any other amplifier (see Figure 5-61). However, LNBs are usually driven into compression by interfering microwave signals, not signals received from satellite.

HEADEND SYSTEM DESIGN

Figure 5-61. LNB Compression. *When signal levels entering an LNB are too strong, the amplifier is driven into compression. This means that it enters a non-linear response region in which its output is not directly related to the input signal. This usually occurs when a strong interfering signal is detected. The solution is then to relocate the antenna, use artificial screening techniques or to insert a filter between the feedhorn and LNB.*

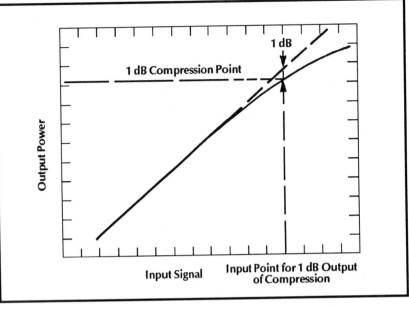

The Downconversion Step

The LNB amplifies the FM modulated C-band satellite carrier and downconverters the entire 3.7 to 4.2 GHz band to 950 to 1450 MHz.

In the pioneer days of satellite communications, an LNA first amplified the high frequency satellite signal. It was then routed via coax to an indoors satellite receiver where downconversion occurred. This technique required the use of special low-loss, costly hardline coax or waveguide. More recent systems downconverted the LNA output to an intermediate frequency at the dish so that readily available, inexpensive coaxial cables could be used to transmit the signal indoors. A single conversion system lowered the signal to the final IF in one step. A dual conversion systems required a second downconversion step in the satellite receiver to produce the final IF. Today LNAs mated with single or dual downconversion systems are occasionally used to combat severe cases of terrestrial interference. This situation is discussed later in this chapter.

Both single and dual conversion systems tune to a specific channel under command of a voltage from the satellite receiver (see Figure 5-62). The bandwidth of one transponder at a time is

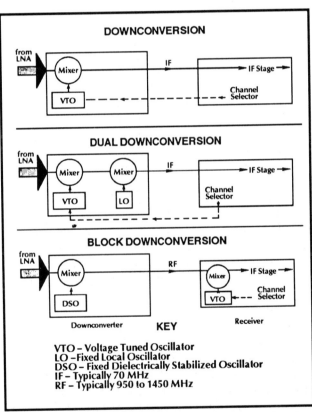

Figure 5-62. Downconversion Methods. *Single, dual and block downconversion methods all accomplish the same end result, converting the extremely high microwave signal to a lower frequency IF and selecting one of the many channels to be viewed.*

downconverted to a range centered on the intermediate or final IF and cabled indoors. Therefore, only one satellite channel is transmitted to a satellite re-

HEADEND SYSTEM DESIGN

Figure 5-63. Block Downconversion. *This system downconverts the entire satellite bandwidth to an intermediate range, typically to a range from 950 to 1450 MHz range. To accomplish this, a fixed local oscillator is located either in an LNB or a separate downconverter at the antenna. Channels are tuned by a second downconverter in the satellite receiver.*

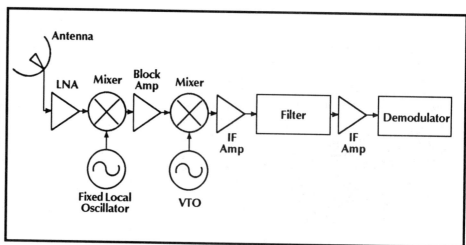

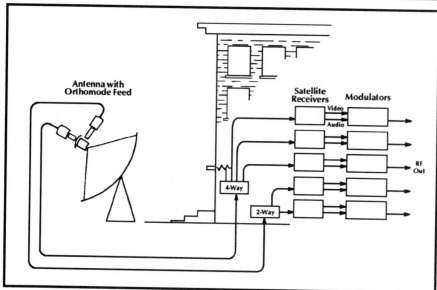

Figure 5-64. Use of LNBs. *In this diagram, signals from two LNBs on a dual feed are split into five block downconversion satellite receivers. Note that the 4-way splitter must be terminated as indicated. The labels on the modulators refer to Nexus components. (Courtesy of Nexus Engineering and Private Cable Magazine)*

ceiver at any one time. Although this design was adequate for simpler home satellite TV systems that were simultaneously tuned to just one channel, it is was cumbersome arrangement for private cable systems. In order to fulfill the requirements of SMATV and cable TV system, the block downconversion receiver was developed.

A block downconversion receiver operates in conjunction with an LNB that downconverts the entire 500 MHz satellite band to the 950 to 1450 MHz range. This block of frequencies contains signals from all 12 transponders of one polarity when a standard feedhorn is used so one or more satellite receivers can simultaneously tune to any channel in this block (see FIgures 5-63, 5-64 and 5-65).

The use of LNBs eliminates the need for a separate downconverter and its protective enclosure at the antenna and increase reliability by reducing the number of required components and connections.

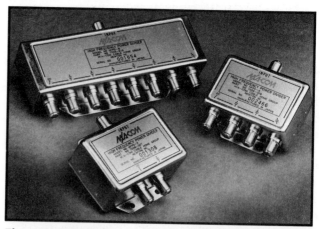

Figure 5-65. High Frequency Splitters. *These 2, 4 and 8-way splitters are used for dividing the LNB signal for input into two or more block downconversion satellite receivers. (Courtesy of M/A COM)*

HEADEND SYSTEM DESIGN

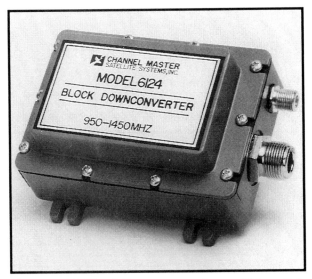

Figure 5-66. Block Downconverter. *This unit is typical of block downconverters having an N-connector input and an F-connector output. It converts C-band microwaves to the typical 950 to 1450 MHz range and is mounted either next to the LNA or just under the antenna mount.(Courtesy of Channel Master Satellite Systems, Inc.)*

The primary disadvantage of using an LNB is evident when terrestrial interference is present. It is generally not possible to insert a microwave filter in the line between the LNA and downconverter sections of an LNB. In this case an LNA and block downconverter must be installed (see Figure 5-66).

Satellite Receivers

Satellite receivers are available from numerous manufacturers and come in many different designs, shapes and sizes (see Figure 5-67). They are equipped with either tunable or fixed audio schemes. While older model receivers were designed to directly interface with descramblers, most current commercial units for use in North America have built-in descramblers and are known as IRDs – integrated receiver decoders (see Figures 5-68, 5-69, 5-70 and 5-71. These minimize the use of rack space, are easier to install and require fewer connections, and are generally more reliable than stand-alone units. The one common purpose shared by every design is demodulation of the incoming satellite signal and reproduction of the original baseband audio and video signals.

The most important specification for any receiver is its threshold, a reflection of the receiver's sensitivity to incoming signals. The lower the threshold, the

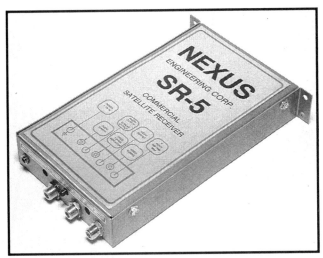

Figure 5-67. Compact Satellite Receiver. *This compact receiver designed for a Nexus Engineering headend measures 3 by 5 inches. It provides video and baseband outputs for VideoCipher decoders and an IF loop for TI filtering. (Courtesy of Nexus Engineering)*

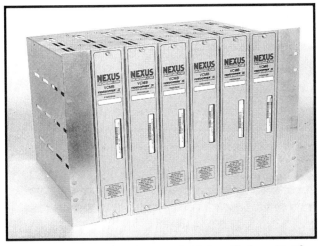

Figure 5-68. Rack Mounted VideoCipher II Plus Descrambler Modules. *This photo shows six Video-Cipher II Plus modules incorporated into one rack-mountable enclosure. These can be connected a satellite receiver such as the one shown in Figure 5-69. (Courtesy of Nexus Engineering)*

Figure 5-69. Integrated Receiver Decoder. *This is a typical racked-mountable commercial-grade IRD designed for SMATV headends. (Courtesy of Nexus Engineering)*

Figure 5-70. Compact Satellite Receiver. *The IRD-2001 from Blonder Tongue is typical of the new breed of commercial, rack-mounted, satellite receivers that have built-in VideoCipher II Plus modules. (Courtesy of Blonder Tongue).*

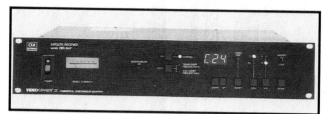

Figure 5-71. Compact Satellite Receiver. *The DX DIR-647 IRD features an on-screen display for VC II Plus diagnostics and field-changeable IF bandpass filter of 17, 22, 24, 27, 30 and 36 MHz to modify the video bandwidth. It has 24 C-band pretuned video channels with 10 kHz audio step tuning and up to 32 Ku-band channels with 1 MHz video and 10 kHz audio step tuning. (Courtesy of DX Antenna)*

higher the sensitivity (see Figure 5-72). Lowering the threshold by just one decibel equals a 26% increase in sensitivity. Without this decrease in threshold, for example, would require that the diameter of a 12 foot antenna be increased by nearly 1.5 foot to achieve the same improvement. Many receivers having thresholds of 7 to 8 dB and lower are currently available.

A variety of methods to tune both video and audio signals are commonly used. In some cases a receiver is tuned exclusively to just one transponder. Other brands are semi-agile whereby channels are selected by interchanging highly stable, frequency-fixed crystal oscillators. Today, most home receivers are completely agile and can be tuned to any C- or Ku-band channel. Tuners can use either detent or synthesized channel selection. Detent tuners have either a continuously variable device such as a potentiometer or a switch that clicks between various preset resistors to select channels. Synthesized tuners create the one-frequency mixing signal required for downconversion and tuning by using either phase-locked loop (PLL) or voltage synthesized methods. While the PLL is the most accurate approximation to the 0.005% accuracy of a single-crystal oscillator, the quartz-locked or voltage synthesized circuits which use a microprocessor to generate a

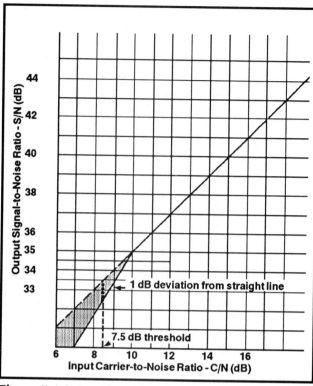

Figure 5-72. Receiver Threshold. *A satellite receiver should be provided with a sufficient carrier-to-noise ratio that exceeds its threshold. Receiver threshold is measured at the point where the departure from linearity is 1 dB.*

digital signal for each specific channel are by far the most stable variety.

The method used for audio tuning is also of concern. Many transponders transmit 4 to 6 additional channels of audio programming as subcarriers on the main video signal. Over 100 high quality FM broadcasts having a wide variety of programming menus are available. Some receivers feature continuous tuning across the entire audio band, while some are preset to a limited range of subcarrier frequencies. A few brands even require the use of a separate frequency-tuned audio board for each audio subcarrier. It is important that the equipment selected has the features necessary to tune to the desired audio programming.

The choice of satellite receiver is based on factors ranging from personal preference to technical characteristics. While style, features and price obviously enter into this decision, reliability, signal quality and serviceability should be high on the list in deciding which brand to purchase for an SMATV headend. A receiver should be capable of operating day-in and day-out for many years without interruption of service. Because problems do naturally occur, it is wise to have a spare unit on hand and to have the faulty unit serviced and back on line as quickly as possible. Choose a manufacturer who stands behind his product to help eliminate equipment headaches. Even though most well designed SMATV headends are climate controlled, receivers that have little opportunity for drifting away from the selected channel are desirable. In general, some form of synthesized tuning is suggested.

Modulators

Channel modulators, which in effect are "rebroadcast" devices, accept the audio and video baseband signals from a satellite receiver and AM modulate them onto an RF carrier having a frequency of one selected channel for input to the distribution system. This is a crucial stage in processing of a satellite signal because the resulting picture is the final product. It is important to choose a modulator that produces superior pictures with a minimum of service problems to ensure customer satisfaction.

Modulators are available in fixed and frequency agile models (see Figures 5-73 and 5-74). Fixed frequency units are channel specific and have to be ordered directly from the factory with a preset output channel. In contrast, frequency agile modulators are tunable and can be field-adjusted to any desired frequency. Fixed channel modulators are typically less expensive than frequency agile ones, although their costs are declining as the industry evolves. While frequency agile modulators reduce headend inventory needs because one modulator can generate any output channel desired, fixed modulators can be more reliable because less electronic circuitry is required than for the more complex agile brands. Some engineers have argued that once the channels that will be used to distribute signals have been determined, the output frequencies of modulators rarely need be changed so that purchasing an agile modulator is not really worth the extra expense.

A distinction is drawn between those brands of modulators with and without built-in bandpass SAW filters (surface acoustical wave filters which have a very sharp frequency cutoff). Those not incorporating good quality SAW filters may generate a "dirty" signal which can interfere with an upper or lower adjacent channel. Note that not all SAW filters are high quality devices and an SMATV operator must closely examine the specs of any components that will be purchased. It is also quite important to understand that home TVRO-type modulators do not provide either sufficient filtering or output gain to justify their use in commercial systems.

A processor can "clean up" a signal by demodulating it to baseband, amplifying the result and then filtering both the input and output signals to eliminate any unwanted sideband energy. A channel processor's output can be remodulated onto the same or a different channel from the input signal. It is important to choose crystal controlled local oscillator circuits in commercial modulators and processors in order to prevent frequency drift.

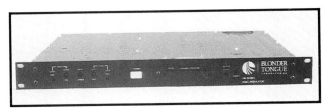

Figure 5-73. AM Series Agile Modulator. *(Courtesy of Blonder Tongue)*

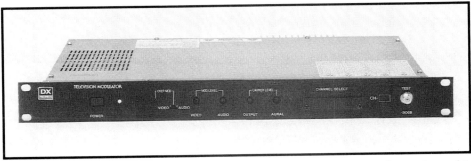

Figure 5-74. Frequency Agile CATV Modulator. *The output channel on this frequency agile modulator can be tuned by a front-panel control from channel 2 through channel YY that span the 54 to 450 MHz range. It is BTSC stereo compatible, uses PLL frequency synthesized tuning and has selectable frequency offsets of +12.5 and +25 kHz. There are separate audio/video IF loops and the output is SAW (surface acoustic wave) filtered to improve S/N ratio. Its agile ability allows an operator more flexibility in headend design and also minimizes the need for stocking excess fixed frequency modulators. (Courtesy of DX Antenna)*

HEADEND SYSTEM DESIGN

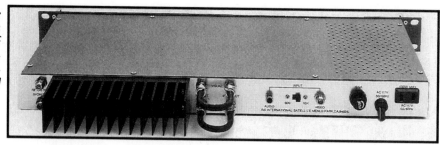

Figure 5-75. Rear View of a Television Modulator. *This photo details the video and audio inputs and IF loopthroughs. These allow the hookup of video scramblers for cable security systems.(Courtesy of International Satellite Systems)*

Most modulators have loop-through ports for both the audio and video intermediate frequencies (see Figure 5-75). These are important only when planning to use sync suppression or some other form of signal scrambling for system security. Most scrambling systems are integrated into the headend through these ports. Check that the output gain of the loop-through ports matches the requirements that the manufacturer of the scrambling system suggests.

A typical modulator has an output level ranging from 30 to 60 dBmV (31.6 millivolt to 1 volt). When feeding a small SMATV system that is designed so that the mixer output is not amplified, a modulator that generates a relatively high output must be chosen. However, if the mixing losses are high enough to warrant a higher gain, headend launch amplifier, as is usually the case, a lower powered modulator is usually more than sufficient and often desirable. Given that higher output power modulators generate substantial amounts of heat, that excellent broadband launch amplifiers are now available and that a lower power modulator has a relatively small effect on output S/N ratio, the decision is often simple.

Both audio and video as well as RF outputs on commercial-grade modulators are adjustable. The audio and video settings control the "percentage of modulation" or how much of these signals are added to the carrier. Higher video outputs make for a brighter television picture but too high a level causes a buzz in the audio and washed out pictures. Most cable-ready modulators often have the audio output preset at the correct 15 to 17 dB below the video carrier level. The RF adjustment sets the output voltage level to the distribution system.

An important modulation specification is the level of spurious output signals. These unwanted signals can interact with adjacent channels resulting in a wide variety of annoying distortions. Generally, a –55 dB or better spurious output rating relative to the video carrier level is sufficient to eliminate any visible picture distortions. In those cases where a modulator having poor output specs must be used, a bandpass filter may be a partial fix at the expense of output level and perhaps flatness of frequency response. Note that frequency agile brands tend to have slightly higher amounts of spurious outputs.

Other principle factors that impact modulator choice are reliability and amount of service that may be required. Modulator failure, in addition to causing the loss of one channel of entertainment, can create problems on adjacent channels. These are complex units that must maintain adjusted levels with a minimum of service. An inexpensive component is often a cheap unit that often unfortunately displays its true value after just a few months of operation.

Descramblers

As progressively more satellite signals are scrambled, it becomes increasingly important that headend satellite receivers are able to interface with the necessary decoders. There are two principle systems of choice in use today in North America. The most prevalent is the VideoCipher II+ system employed by HBO and most other pay-TV programmers (see Figure 5-76). The Oak Orion system is used by major programmers in Canada and by numerous corporate and special interest sports networks. Another scrambling system that is recently becoming somewhat more prevalent in the Leitch system.

In Europe a variety of systems have been and are in use. There is no defacto standard as there has been for the American home satellite market. These include Videocrypt, Payview, Irdeto and EuroCypher.

Decoders are installed in the headend between the satellite receiver and modulator. Some receivers filter out or "clamp" the part of the signal that carries the information and codes needed to operate these descramblers. Make sure that the chosen receiver

HEADEND SYSTEM DESIGN

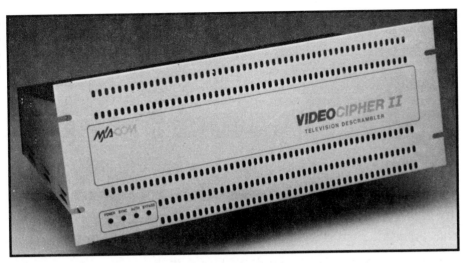

Figure 5-76. The VideoCipher II-C. *This old-style rack mounted decoder accepts only unclamped, baseband satellite signals. (Courtesy of General Instruments)*

has an unclamped video output. While IRDs that incorporate the receiver and decoder in one package have been available for the American home satellite market for a number of years, until recently most decoders for commercial applications were stand-alone units. Today commercial IRDs are available in North America. These simplify and increase the reliability of headend design.

Component Reliability

Reliability is a key factor in the private cable business. Low quality goods which are often inexpensive show their true value sooner or later. Obviously, if countless hours are spent servicing a "bargain" priced component, the savings will evaporate. An inordinate number of service calls not only wastes time but, in most cases, creates customer dissatisfaction and is often the basis for a poor reputation. Choose products from reputable suppliers. An operational headend is no place to experiment with new products and ideas. The proper place for that is in the shop on a test bench.

It is a good policy to build a small experimental headend and to thoroughly test new gear before putting it into service. Do not use home TVRO gear in a commercial environment. Most of these components are not designed for continuous duty service. They do have the specifications required to deliver high quality signals that feed an extensive distribution system and exhibit high failure rates under such conditions. If the electronic equipment is chosen correctly the first time, Saturday nights will be spent in restful peace at home instead of servicing a headend 60 miles away.

Combining and Balancing the Signal

The integration and mixing of satellite with off-air signals is a relatively straightforward process. The satellite video and audio or stereo audio outputs are fed into commercial-grade modulators whose outputs are compatible with outputs from the off-air processors. Both the levels and the S/N ratios of all satellite and off-air signals should be equalized to within ±1 dB. This requires appropriate adjustment of modulator output controls, bandpass filtering and, occasionally, the use of signal processors.

Power levels can be calculated to a fair degree of accuracy before installation. For example, if a modulator with a 40 dB gain is drawing a ±3 dBmV

HEADEND SYSTEM DESIGN

baseband signal from a satellite receiver, it will input 37 dBmV to the distribution system. If an adjacent channel is being relayed at 32 dBmV, balancing could be achieved by either inserting a 5 dB pad in the output line or by readjusting the modulator RF output level. Nevertheless, remember that no calculation will replace a signal strength meter as an essential tool for aligning headend equipment during installation.

The combining of satellite signals can be accomplished just like the process for combining off-air signals using a mixing network, a splitter or a high/low combiner (see Figures 5-77, 5-78 and 5-79). For example, an 8-way combiner could take three off-air channels and five satellite broadcasts and add them together into one output stream. Or a string of taps in series could feed modulator outputs onto one coax line for mixing with off-air channels in a 2-way combiner. Remember that insertion losses will be the same when using both devices in either forward or reverse directions.

Once all signals are combined and launched, either taps or splitters can be used depending upon the size of the final distribution system. Splitters are generally employed in smaller cable networks. Designs ranging in size from larger SMATV systems to city-wide cable networks use taps to feed signals to each customer and then use splitters to feed additional television receivers. However, a main trunk line is never tapped but employs only special CATV hybrid splitters/amplifiers so that the distribution system will be well protected from ingress interference. Then signals are tapped to individual customers from feeder lines. For example, if a hotel having two 55-room buildings were being cabled, a 2-way splitter/amplifier might be used to feed each building. Then a network of either taps or splitters could be used in each building to feed all rooms.

When complete, a properly designed and built headend fits on a compact rack (see Figure 5-80 and 5-81). All connections should be marked and neatly completed.

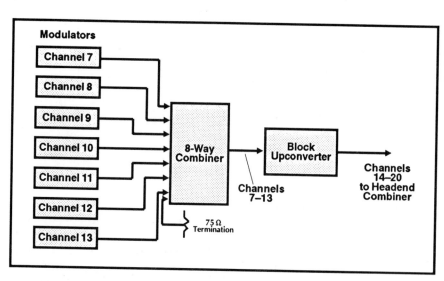

Figure 5-77. Block Upconversion Scheme. *Signals from seven standard channel modulators can be combined and upconverted all at once.*

HEADEND SYSTEM DESIGN

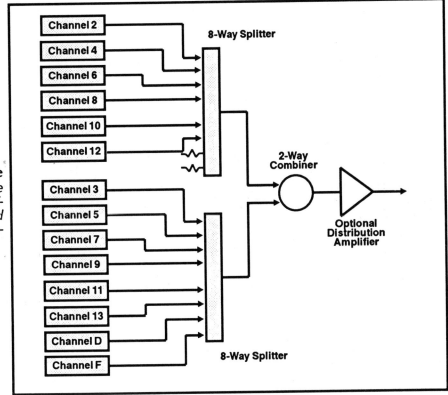

Figure 5-78. Combining Satellite Signals. *Outputs from 14 satellite modulators are combined as indicated. The 8-way splitters should have a minimum of 30 dB port-to-port isolation.*

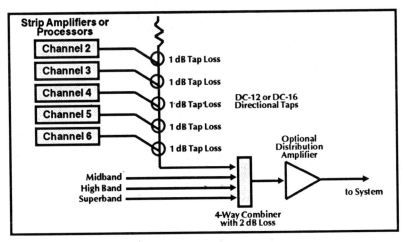

Figure 5-79. The Use of Directional Couplers. *Directional couplers provide 35 to 40 dB isolation between modulators and can be used to mix signals. These have isolation losses of 12 to 16 dB and an additional 1 dB insertion loss. This configuration results in the lower channels experiencing greater losses on their voyage to the 4-way combiner.*

HEADEND SYSTEM DESIGN

Figure 5-80. Nexus Series 2000 Headend. *This illustrates the Nexus Series 2000 36 channel headend. (Courtesy of Nexus Engineering)*

Figure 5-81. Compact SMATV Headend. *This pre-packaged 12 channel headend include six satellite receivers, six modulators and six channel processors. It requires a total of only 12 by 19 inches of space. (Courtesy of Nexus Engineering)*

F. DESIGNING THE SATELLITE HEADEND

The methods used to size satellite antennas and select LNB noise temperatures are relatively straightforward. These are outlined here but are examined in more detail in the references. The antenna/LNB system is characterized by a figure of merit known as the G/T_{sys} ratio, shorthand for the ratio of antenna gain to antenna plus LNB noise temperatures. The required G/T_{sys} can be calculated by knowing the strength of the incoming satellite signal, the receiver channel bandwidth, the elevation of the antenna and the required carrier-to-noise (C/N) ratio.

Every satellite receiver has a characteristic relationship between input C/N and output S/N ratios (see Figure 6-68). Receiver threshold is defined as the point where the linear relationship between these two parameters breaks down. At C/N ratios below threshold, picture quality suffers. In general, commercial satellite receivers have thresholds at or below 8 dB but generally should have an input C/N of at least 11 dB to achieve reasonably acceptable performance.

The link equation can be used to determine G/T. It simply states that the C/N ratio entering a satellite receiver equals the signal power leaving the satellite which is called the effective isotropic radiated power (EIRP), less free space path losses experienced on the voyage from the satellite to the receive site caused by atmospheric absorption and spreading out of the signal, plus the antenna gain less noise introduced by the antenna, LNB and other system components.

$$C/N = EIRP - Path\ Loss + G - 10\log k(T_a + T_{LNB})B$$

where T_a is the antenna noise temperature, T_{LNB} is the LNB noise temperature, G is antenna gain and EIRP is the effective isotropic radiated power from the satellite. Realistically assuming a free space path loss of 196.3 dB and a 28 MHz receiver bandwidth, this equation becomes:

$$C/N = EIRP + G - 10\log T_a + T_{LNA} - 41.68$$

The antenna noise temperature is largely a result of signals generated by the relatively warm ground entering the feed/LNB system. As a result, antenna noise temperature decreases as elevation increases:

$$T_a = 14 + 180/Elevation$$

Assuming, for example that the antenna elevation is 30 degrees, its noise temperature is easily calculated to be 20°K. If EIRP is 32.6 dBw, T_{LNB} is 80°K and the required C/N ratio is 12 dB, then:

$$12 = 32.6 + G - 10\log(20 + 80) - 41.68$$
$$= G - 29.08$$

therefore the required antenna gain equals 41.08 dB. This gain could be supplied by a 12 foot antenna of reasonably good quality. Note in the above calculations that the EIRP is expressed in dBw, decibels relative to one watt. 32.6 dBw is equal to 62.6 dBm, decibels relative to one milliwatt.

Calculating Receiver Input Levels

It is a relatively easy matter to trace the signal from the satellite antenna to the receiver input. For example, if 62.6 dBm leaves the satellite and is reduced by the free space path loss of -196.3 dB, then the difference which equals -133.7 dBm is available at the antenna. Following antenna and LNB gains of 41 and 50 dB, respectively, the signal is then increased by 91 dB to -42.7 dBm.

This signal is next fed via a 100 foot run of RG-6 coax with a 7.6 dB per 100 feet for a total loss of 7.6 dB. The net result, -50.3 dBm, is sufficient to drive typical receivers that require a minimum input of approximately -60 dBm. If a very long cable runs had been used, a line amplifier would be required.

G. A STRATEGY FOR COMBATING TI

The basic strategy in combating terrestrial interference (TI) entering via the antenna is avoidance by correct antenna placement followed by other actions when absolutely necessary. The first step in this approach is to chart and identify those sites where the potential for TI is high. Next, whether or not interference is expected, a portable antenna and a spectrum analyzer should be employed to conduct an on-site measurement for TI. This is an essential step in finding an optimal site for installing the permanent antenna. TI, like satellite signals, is very directional and can be reflected off metal objects such as a tin roof or metal siding directly into a microwave antenna. But TI is also absorbed by many materials. Therefore, even where TI levels are high, installing an antenna behind natural shields such as trees or a building may offer sufficient protection. The metal fence generally installed around satellite dishes in a SMATV system often also accomplishes the same result. Finally, if no site free of TI can be found, two options exist: to either use filters or to erect artificial screens. In the worst possible cases, a mixture of both these methods can be attempted.

Ingress interference is not included in the discussion below because it is usually easy to cure by properly grounding cables, using wall plug filters, shielding equipment interconnects and closing unnecessary gaps in metal equipment. Headends can be well isolated to completely avoid such problems.

Charting TI Problem Sites

A simple method to combat TI in advance of a site survey is by charting microwave routes on a local map. The required background information includes all present and planned locations of terrestrial 4 GHz microwave relay stations in the area as well as their transmission routes and frequencies. This information can be purchased from a professional frequency coordination firms, usually for less than $350, and needs to be compiled at most once a year. Or it can be obtained with some more work from either the FCC in the United States, the DOC in Canada or local telephone companies.

The procedure is simple. A map of the region is marked with the location of each microwave repeater station and straight lines are drawn between the sending and receiving sites. Any proposed installation which falls too close to any of these communication routes has a higher probability of being susceptible to TI. This method can be a very important complement to a site test because future communication towers are marked and potential sources of interference which might be intermittent are identified. This method supplements the natural "feel" that a competent installer begins to develop after working for a period of time in a given region.

Similar results can be obtained at a reasonable cost from professional frequency coordination companies. For a fee, these firms provide a computer printout detailing both the TI environment at an installation site and planned line-of-sight microwave sources. This includes the expected worst-case levels of interference, their operating frequencies and polarities, and their beam directions. The cost is usually around $350 but some scaled-down information can be purchased for less.

Other packages of information including a route map and an intensity map are also available for a nominal fee. The route map, a semi-transparent overlay used with a topographical map, identifies and links each relay tower by a line. The intensity map is similar but also includes the expected power of common carrier relays.

An alternative is to register the proposed reception system for a fee with the FCC. This is a wise choice for an expensive private cable installation in a large condo, apartment complex or hotel. Frequency coordination protects a system from future terrestrial relays because it is public knowledge and must be factored into the planning process of other microwave communicators. Specialized firms offer this licensing service and provide what they call "frequency protection" which warns both the licensed SMATV operator and other communicators of the potential for interference.

HEADEND SYSTEM DESIGN

Nevertheless, it is important to realize that nothing can replace a site survey. An installation which may be far from the route of a land-based microwave communication path could easily be troubled by a signal reflected off a building or metal billboard. Alternatively, a site directly between two relay towers may have just the necessary natural screening to avoid any difficulties. Any installation could also be easily susceptible to out-of-band or even ingress interference.

Methods to Screen Out Interfering Signals

TI can usually be overcome by a combination of natural and artificial screening methods. Natural screens include trees, buildings, mounds of earth or any other structures already on the site (see Figure 5-82). The best and generally least costly strategy is to make effective use of natural screens before resorting to filters or artifical screens.

All materials have varying abilities to absorb or reflect microwaves and to serve as either artificial or natural screens. Metal is by far the best reflector of microwaves. Most non-metallic structures are generally not very reflective but can absorb microwaves. For example, wood, which is a poor reflector, has limited absorbing properties. An existing wooden structure such as a garage or barn can be made into a reflector of microwaves by lining it with a metal screen mesh. This can usually be accomplished for a relatively low cost. Bricks, cinder blocks, concrete and other masonry products are also poor reflectors but reasonably good absorbers of microwaves. The moisture content of green leaves causes them to be excellent attenuators of TI. But beware of using deciduous trees as shields, because when they lose their foliage in winter so goes their value as protection against interfering microwaves.

In general, the higher an antenna is installed above the ground, the more susceptible it is to TI. The converse is also true. So placing an antenna in a depression or a valley is usually a safe bet for reducing or even eliminating interference. There is a case of an SMATV installation in a motel in California that was so plagued by TI that the least costly solution was to install the 15 foot antenna at the bottom of an empty swimming pool. A new pool was constructed as a lower cost option than curing the interference by another method!

Figure 5-82. Natural TI Screening. *Placing a dish in a protected area can shield it from wind as well as from interfering signals. Note that if deciduous trees are used, when leaves disappear in the winter, TI may return.*

Some materials are optimally designed for absorbing C-band radiation. An example is space cloth, a canvas-like fabric impregnated with carbon to make it resistive to current flow. A similar material coated with metal on one side could serve a dual purpose as an absorber in one direction and reflector in the other. Bulk absorbers can be attached to existing structures to prevent reflection of TI into a dish. Another example of such a material is a carbon-impregnated foam rubber having an array of small pointed pyramids on one side.

Metal screens like builders' "hardware cloth" are adequate reflectors. As long as the mesh openings are less than approximately one tenth of the wavelength of the microwaves, the same amount of energy is reflected as would be by a solid metal surface. Since C-band signals have wavelengths of about 3 inches, openings less than 3/10ths of an inch are adequate.

The first step in finding a screened location for a dish where TI is present is to draw a scale map of the site. The composition of all natural and pre-existing structures is noted. If the direction of the TI source is known, an ideal site may often be easily determined. Occasionally, extra man-made screens are necessary. Or lightweight screens mounted on wooden frames can be constructed. The screen height should be at least three feet above and wider than

HEADEND SYSTEM DESIGN

Figure 5-83. A Unique Site Solution. *This antenna had to be shielded from TI by a rather elaborate structure. (Courtesy of Private Cable Magazine)*

the dish. Always avoid the ping-pong effect, which causes interfering signals to be reflected into the dish instead of away from it, by tilting these screens 30 degrees or more away from the antenna. These man-made structures must be fastened down securely to prevent them from catching a strong wind and being blown into the antenna or causing other damage.

All reflecting barriers should be terminated at the edges with at least a six inch roll of the material used. This prevents creating a sharp edge from which microwaves could be diffracted into the antenna and/or feedhorn.

Occasionally terrestrial common carriers relay full video signals occupying the 30 to 36 MHz wide band from sources such as a local TV station. These will completely wipe out satellite transmissions sharing a similar frequency band. The only completely effective method of shielding against such interference is to screen out these signals with artifical or man-made barriers before they enter the feedhorn (see Figure 5-83).

Combating TI With Filters

If an interference-free site cannot be found, the next alternative is to use filter to "notch out" and suppress the interference. However, notching out the interference also reduces the available satellite bandwidth and picture quality. Shielding techniques have the advantage of not degrading the signal.

Out-of-Band TI

Light to medium to heavy power out-of-band TI can usually be cured with a bandpass filter placed at the output of the LNB. The bandpass filter is designed to reject frequencies outside the 950 to 1450 MHz band. For medium to heavy TI, a more effective and sometimes unavoidable cure is to use an LNA/block downconverter system and place a microwave bandpass filter between the LNA and block downconverter. When LNBs are used a microwave bandpass filter cannot be located before the downconverter.

If the interfering signal is high enough in power it may overdrive the LNB into compression. In this case, a special microwave bandpass filter would have to be inserted between the feedhorn and LNB to eliminate the interfering carriers.

In-Band TI

The method used to combat in-band TI depends upon the severity of the interference. For levels below –3 dB, notch filters, which remove a narrow band of signal centered on either 60 or 80 MHz, can restore nearly normal pictures (see Figure 5-84). When using receivers having an IF other than 70 MHz, specially made notch filters are required.

One notch filter can be used for each channel processed by a block downconversion satellite. First the whole 500 MHz C-band is lowered typically to 950 to 1450 MHz and then each channel is downconverted to the same final IF. A notch filter

HEADEND SYSTEM DESIGN

connected by the IF loopthrough on the receiver acts on the band of each channel that has been tuned.

As the TI level rises above −3 dB, microwave filters or traps must be used. One trap is required for each interfering frequency (see Figure 5-85). At levels above +10 dB a special type of microwave filter, a three resonator trap, is required. The complexity and cost of finding a technical solution for TI increases further as TI power grows stronger. At levels above +35 dB, very costly, specially-made traps must be placed between the feedhorn and LNB to allow it to function.

Filtering wide band interference is not possible with conventional notch filters which remove narrow bands of signal and interference. If the TI covers the full satellite broadcast bandwidth, then notching out the TI also removes the television signal. The solution to this problem other than moving the antenna or using artifical screens is phase cancellation. This method is simple on paper but more complex in practice. A sample of the TI which is taken with a second feedhorn and LNA is subtracted from the satellite signal plus TI by shifting its phase 180° and adding the signals together. This leaves just the satellite signal and eliminates the TI.

Selection of RF and Microwave Filters

Selection of the appropriate filter begins with a correct diagnosis during the site check. If possible, as is often the case, TI can be avoided. When filters are needed, the type required depends upon the source, frequency and power level of the interfering signal.

Filters can be grouped into two basic categories: notch filters, also known as traps and bandpass filters. Threshold extension filters are a subclass of bandpass filters designed for use in the IF range. Most of these can be constructed in active or passive designs. Passive filters are inexpensive and uncomplicated. They are constructed from standard electrical circuit elements and do not need ac power. Since they are not powered, they provide no gain as the signal passes through. Their insertion losses range from close to zero to substantial decreases in signal power depending upon the model.

Figure 5-84. Passive Notch Filters. *The MFC 4518-60 and 80F notch filters can be used separately or in series to reduce interfering carriers centered on a final IF of 70 MHz. (Courtesy of Microwave Filter Company)*

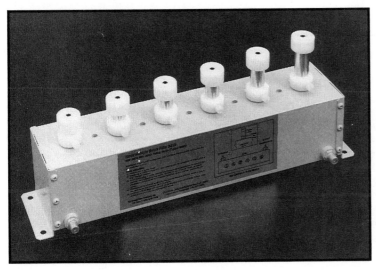

Figure 5-85. Block Downconversion Microwave Trap. *This microwave filter is designed to notch out six frequencies in the 900 to 1450 MHz band. Each notch has a bandwidth of approximately 3 MHz and can be adjusted in depth to approximately 15 dB and the center frequency can be tuned over six 80 MHz adjacent bands to cover the entire 500 MHz bandwidth. The trap is installed in the coaxial line from the LNB output but before any power dividers. (Courtesy of Microwave Filter Company)*

HEADEND SYSTEM DESIGN

Active devices need power to function and usually have an integrated circuit and either a ceramic or a SAW filter to limit the range of rejected frequencies. They can be smaller than passive units both amplifying and filtering the processed signal. Disadvantages include their need for external power and higher cost.

All notch filters are rated by the amount that the interfering signal is reduced. This corresponds to the depth of the notch. For example, a filter which was rated for −40 dB would reduce an 8 dB interfering signal to −34 dB (see Figure 5-86).

Notch filters, designed for use in the IF range, eliminate a portion of the frequency spectrum during the process of wiping out interfering carriers. The notches, which are centered on or close to either 60 or 80 MHz (or ± 10 MHz from other final IF), are characterized by their relative suppression of power and by the width or range of frequencies affected. Obviously, if a telephone carrier centered on 60 MHz and having a width of 3 MHz is attacked, the ideal filter would eliminate only this band of frequencies. Any more would cause undue loss of the desirable satellite signal.

Some notch filters are equipped with adjustments on one or all of depth, width and center frequency. Some can also be switched in and out of action so that when not required they do not degrade picture quality by unnecessarily removing a portion of the spectrum. For example, the MFC 3217LS-60 and 80 passive notch filters from Microwave Filter Company can be used separately or in series to reduce interfering carriers by about 53 dB. They can be field-tuned by using two screwdriver adjustments in conjunction with a spectrum analyzer. These filters, like most other notch filters available, can easily be inserted in the IF line to the receiver by using an F-connector input and output.

Threshold extension filters are similar to those used by some manufacturers to lower the apparent receiver threshold. They restrict the bandwidth of the signal entering a receiver and eliminate noise, some satellite signal and hopefully, in the process, the interfering carrier. For example, the E.S.P. PFG-20 and 50 active devices use a SAW filter to produce a steep-skirted frequency response and add about 4 dB of additional gain. E.S.P.'s PGF-50 provides about 53 dB carrier power reduction and a half power (-3 dB) bandwidth of only 13 MHz.

Some notch filters can have similar effects as receivers which reduce threshold by narrowing bandwidth. If the width of two filters located at 60 and 80 MHz is increased, the band between them becomes squeezed down. For example, if the bandwidth of each trap were 4 MHz then the middle bandpass region would range from 64 to 76 MHz, a bandwidth of only 12 MHz.

Bandpass filters are designed to allow only a selected band of frequencies into the downstream electronics. For example, the MFC 4352 bandpass filter passes only 3.7 to 4.2 GHz (see Figure 5-87). It directly connects into the line between the LNA and downconverter with male/female N-connectors so no jumpers are necessary. The 3966A can be adjusted to pass only one of any selected transponder frequencies in the C-band.

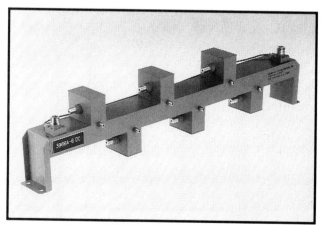

Figure 5-86. Multiple Microwave Notch Filter. This notch filter which has six resonators, each providing a 15 dB notch. These can be sync tuned to provide 15, 30 or 40 dB notches. The filter operates in the 3.7 to 4.2 GHz C-band range and is installed in the line between an LNA and downconverter via N-connectors. (Courtesy of Microwave Filter Company)

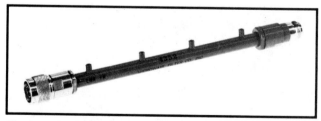

Figure 5-87. Microwave Bandpass Filters. The MFC 4352 bandpass filter is designed to pass only those frequencies in the C-band range. It directly connects into the 4 GHz line between the LNA and downconverter so that no extra cable or connectors are necessary. This filter also passes dc power. (Courtesy of Microwave Filter Company)

H. EXAMPLES OF HEADEND DESIGNS

The following systems are taken as representative examples of designs and problems which arise in headends but are by no means the only possibilities. More complex systems can easily be designed by using the fundamentals learned in studying these examples.

Example I – Typical Off-Air Processing Requirements

On-Channel Processor

Three situations are considered in this example. First, assume that channel output is being received at the output of the feedline of an off-air antenna at a relatively weak signal level of 5 dBmV. On-site measurements over a period of a few days have shown that no signal is detected on channel 10, but that channel 12 is being received from a distant transmitter at a level 40 dB lower than channel 11. To make matters even worse, channel 13 is also being detected from a nearby transmitter at a 5 dBmV level.

In this case, a good quality combination filter and strip amplifier or on-channel processor will be perfectly adequate. The curves in Figure 5-36 show that the channel 12 video carrier is reduced by approximately 15 dB so that the overall reduction is 55 dB, more than adequate. In addition, the on-channel processor reduces the channel 13 input by more than 60 dB to an acceptable level.

Heterodyne Channel Processor

Assume that the private cable system is located in a rural area equidistant from transmitters for channels 11 and 12 so that both are received at a –10 dBmV level even after rejection of channel 12 by the channel 11 narrowband antenna.

In this situation, it is not practical to use a preamp to increase the channel 11 signal strength to +10 dBmV for a number of reasons. Mast-mounted preamplifiers are not particularly reliable especially since they are installed in outdoors locations. Preamplifiers are also susceptible to being overloaded and generating picture distortions if the signal temporarily increases in power or if another interfering signal is detected. If a preamp were used, an on-channel processor still might not adequately reject the undesired signals.

The only solution here is to use a full-performance heterodyne processor to provide the necessary 55 to 60 dB adjacent channel isolation.

Notch Filter

Assume that the channel 13 signal was being detected at a substantially higher power than that of the desired channel 11. Regardless of the type of processor used, either bandpass or notch filtering would be required before the processor. The best choice is a notch filter which specifically attenuates channel 13. Its insertion loss of 1 dB will not have a noticeable effect on the amount of available signal. However, the use of a bandpass filter on a weak channel before processing could significantly lower the S/N ratio and impair reception.

Example II – A Basic Headend

In this example, off-air signals of varying levels are being collected from four different directions (see Figure 5-88). Their powers have already been corrected to represent levels relative to a dipole antenna. The desired headend output level is designed to be 40 dBmV to provide enough power for the distribution system.

Gains and attenuations are also shown in Figure 5-88. The channel 2 antenna supplies sufficient gain and is fed directly into the first mixing network. This network has built-in bandpass filters and rejects any signals from other than channel 2. This prevents, for example, channel 6 from entering the system via the channel 2 antenna which would cause phase distortions in the channel 6 picture. The signal from the channel 6 antenna is 5 dB less than the designed input requirement for the distribution amplifier. A preamp has been selected from among the options which are to increase the antenna gain, use a higher gain line amplifier or use a preamp. Here the signal level from the preamplifier is slightly too high so a 25 dB attenuator is used to lower its output to a balanced 12.5 dBmV. Channels 8 and 13 are handled in the same fashion.

HEADEND SYSTEM DESIGN

Outputs of the VHF-hi and VHF-lo channels are fed into their respective inputs on a high/low combiner. This adds increased isolation between these two segments of the total VHF band.

In general, it pays to measure and record all system operating levels. This record usually becomes a valuable aid during installation, setup and troubleshooting phases of any SMATV operation. It also is a useful cross-reference to other jobs having related designs.

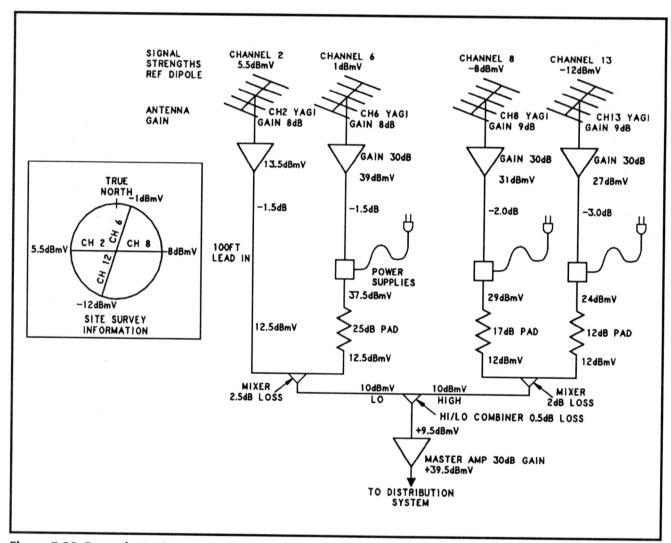

Figure 5-88. Example II. *This solution for a non-adjacent-channel MATV system is explained thoroughly in the text.*

Example III – More Elaborate Headend Design Issues

A more elaborate headend is considered here (see Figure 5-89). Ten of the possible twelve VHF channels are occupied so that a number of adjacent channels are also being used to distribute signals. This requires careful balancing of levels at the headend. It is interesting to note that since adjacent channels are never simultaneously broadcasted in one area many television sets are designed for non-adjacent channel reception and have adjacent channel traps in their IF sections to reduce the level of signals arriving from distant television transmitters.

The signal levels measured during the site survey are also shown in Figure 5-89. Since channels 2 and 4 are coming from the same direction, a VHF-low broadband antenna detects signals. A combiner operating in reverse is used to isolate these channels from each other. The recommended input levels to the headend processors are 0 to 20 dBmV. In addition, the gain of each strip amp is controlled separately and should be set so that the headend output power is as designed. Aural carrier levels are also set at the same time using a field strength meter.

Each strip amplifier has a means of mixing its output signals known as the loopthrough method. However, if all the amps were directly looped through one after another there could be serious problems with distortion. Since 0.5 dB is lost each time the signal passes through a loopthrough, after ten amplifiers the output from the first strip amp would have accumulated an additional loss of 5 dB. An even more serious concern in looping signals through from one strip amplifier to another is the interaction of adjacent channels. This is caused because the internal bandpass filters are not perfect. In this, for example, VHF channel 2 is looped through the channel 4 and 6 strip amps with approximately a 1.0 dB loss.

A dual strategy has been followed to minimize adjacent channel interactions. The VHF-hi and VHF-lo channels have been first subdivided into two non-adjacent groups of channels (channels 2,4,6 and 5 and FM, and channels 7,9,11,13 and 10) then combined in a 2-way hybrid splitter. The cross input isolation of this hybrid splitter effectively separates the adjacent channel outputs. These recombined VHF-hi and VHF-lo groups are then added together via a high/low combiner. The cross input isolation of this hybrid splitter effectively separates the adjacent channel outputs.

An omnidirectional antenna is used to detect the relatively strong FM signals. These are adjusted to equal the levels of the television audio subcarriers. The traps on the FM input line are inserted to reduce the amount of channel 6 signal entering the system through the FM processor. This signal could distort the channel 6 output after the mixing stage.

Channels 8 and 12 have been converted to channels 10 and 11. An 8 dB pad is inserted into the 12 to 11 converter input to lower the power to within the recommended 0 to 20 dBmV level. Then channels 8 and 12 are left vacant. This is required because the original signal levels are so high that ghosting could occur due to ingress into leaky connectors, into the matching transformer or actually directly into the television set. Anything else that may have been distributed on these channels could also have been subjected to intermodulation distortions.

All UHF channels originate from the same general direction and have very low powers. A UHF preamplifier is used to raise their levels to recommended converter input power. The mixing network separates them into desired channels. Converters are typically quite noisy so therefore it is wise to increase the S/N ratio at the antenna by using a high gain preamplifier. Attenuation pads could then be later used to lower the level to the range required by the converter.

HEADEND SYSTEM DESIGN

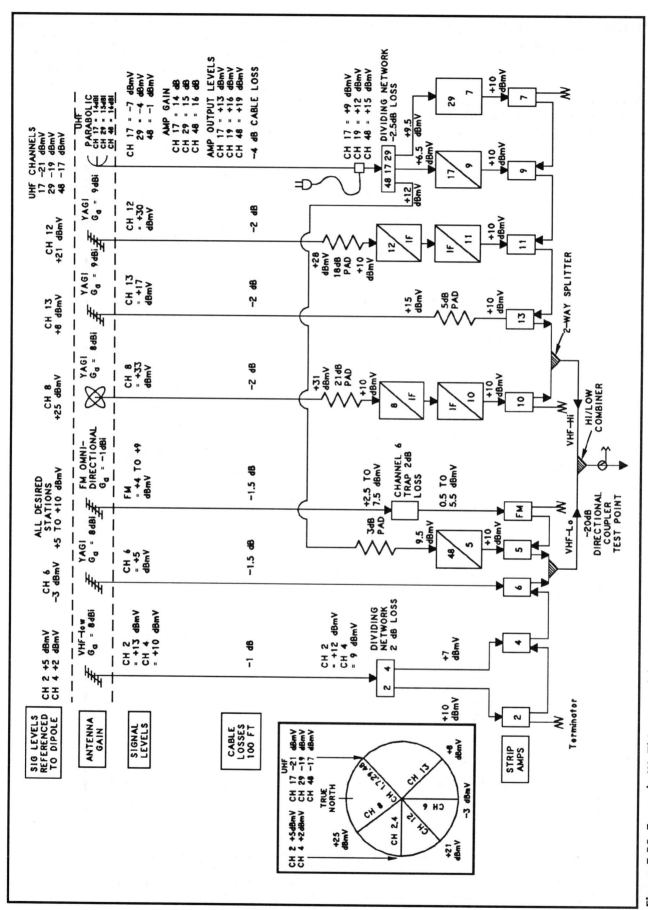

Figure 5-89. Example III. This more elaborate MATV headend is explained in the accompanying text.

HEADEND SYSTEM DESIGN

Example IV - Basic One Channel Satellite System

The basic satellite TV system consists of a single antenna and receiver whose output is combined with off-air broadcasts. A channel combiner is used to mix satellite and off-air signals (see Figure 5-90).

One line of RG-6 or RG-59 coax carries the LNB power and RF signal from the antenna to one or more satellite receivers. A VHF or UHF off-air pre-amplifier is required if the signal is relayed from a distant source or if there is a long cable run from the conventional antenna.

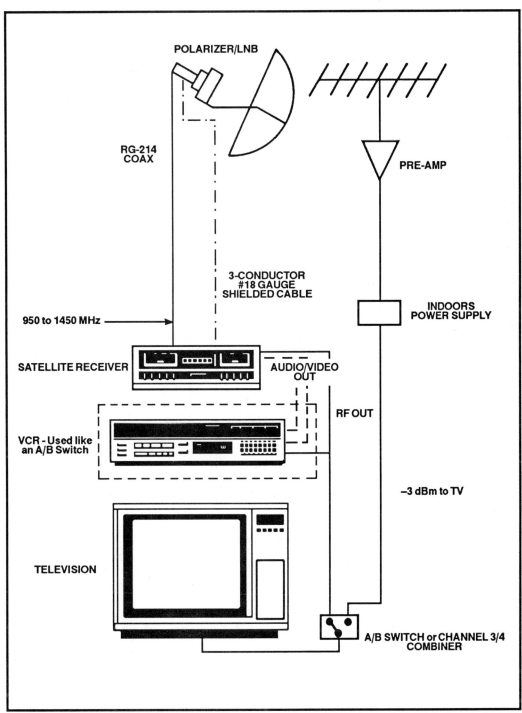

Figure 5-90. The Basic Off-Air / Satellite TV System. *The 950 to 1450 MHz input signal is fed from the LNB output through RG-6 coax to a satellite receiver. A 3-conductor, 18 gauge shielded wire carries the necessary pulses to control the polarizer.*

HEADEND SYSTEM DESIGN

Example V - 4-Channel, Single Polarity TVRO Headend

A 4-channel headend might be located at a hotel in an area where no off-air television is available. In this case, four satellite TV channels are being fed into the distribution system (see Figure 5-91). Following modulation, the channels are combined in a 4-way splitter before being added together on one output line. This system design protects adjacent channels from interfering with each other. Since no channels are adjacent in frequency, lower cost modulators without bandpass filters could probably be used. In general, this is not advised when high quality results are expected.

Example VI - 7-Channel, Dual Polarity Satellite Headend

The 7-channel headend combines four satellite channels with 3 off-air broadcasts onto one distribution network (see Figure 5-92). In this case, care has to be taken in protecting channels 8 and 9 from interfering with each other. Signal levels must be properly balanced and a good quality bandpass filter, either as an extra component or built into the channel 9 modulator, is necessary.

Example VII - Basic 8-Channel Satellite Headend

This dual-band 8-channel headend is outlined in Figure 5-93.

Example VIII - 11-Channel, C/Ku-band Satellite Headend

The 11-channel headend occupies nearly all the 12 available VHF channels. An AC-5 combiner mixes all channels together and feeds them into the distribution network (see Figure 5-94).

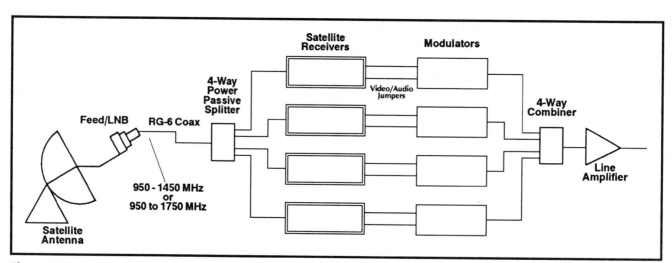

Figure 5-91. Four-Channel, Single-Polarity Headend. *The 950 to 1450 MHz LNB output is fed through RG-6 or RG-11 cable into a 4-way splitter located at the headend. The output from the splitter is fed into the satellite receiver's IF inputs. A single receiver will supply power through the power passive port on the splitter to the LNB. The audio and video receiver outputs are modulated onto channels not occupied by local off-air broadcasts. The modulated outputs are then added together using a 4-way splitter used in reverse as a combiner. The signal is coupled to a line amplifier where appropriate levels are transmitted through the distribution system.*

HEADEND SYSTEM DESIGN

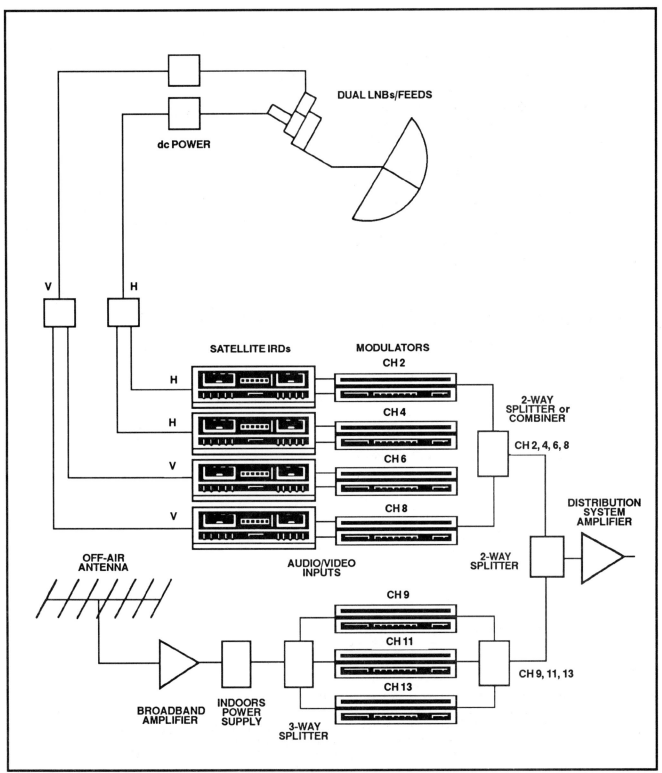

Figure 5-92. 7-Channel Headend with Off-Air Channels. *Four satellite receivers are mixed together as in the previous figure except that two horizontal and two vertical inputs are fed into the satellite receivers from two LNBs and a dual feed. A local off-air broadcast is fed into the headend from a broadband antenna. Its output enters a preamp which is mounted at the antenna. The 10 to 20 dB of gain compensates for cable losses. A power supply feeds power up the coax to the pre-amp. The amplified signal passes via a 3-way splitter into a channel processor. This device reconstitutes the signal so that it has video of equivalent quality as the satellite channels. Outputs are then fed into a 3-way splitter and subsequently via a 2-way splitter to mix with the satellite modulated channels. The combined outputs of both satellite modulators and the off-air channels are then fed through a line amp to the distribution system.*

HEADEND SYSTEM DESIGN

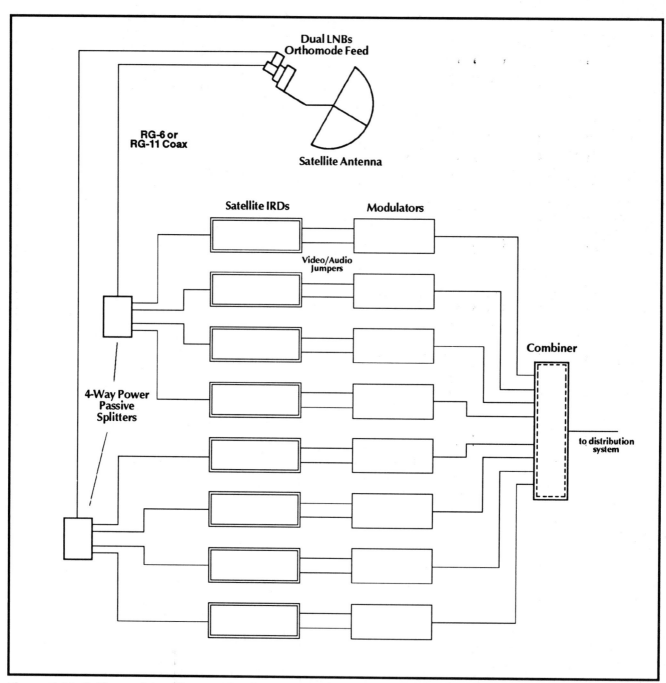

Figure 5-93. 8-Channel Satellite Headend. *Outputs from the horizontal and vertical LNBs pass via coaxial cable to the headend. The signal is then split in two 4-way splitters which produce four vertical and four horizontal outputs for relay to the satellite receiver IF inputs. A single receiver in each bank supplies power through a power passive port on each one of the 4-way splitters. The audio and video outputs from the satellite receivers are then fed into the modulator inputs for appropriate channel output selection and subsequent mixing in a channel combiner. Most combiners have a 20 dB output which combined with the modulator gain outputs can provide as much as 50 dBmV to the distribution system. All of this equipment is rack mounted and installed in a well ventilated room. Active temperature control all year round is recommended to maintain signal levels constant and to minimize frequency drift.*

HEADEND SYSTEM DESIGN

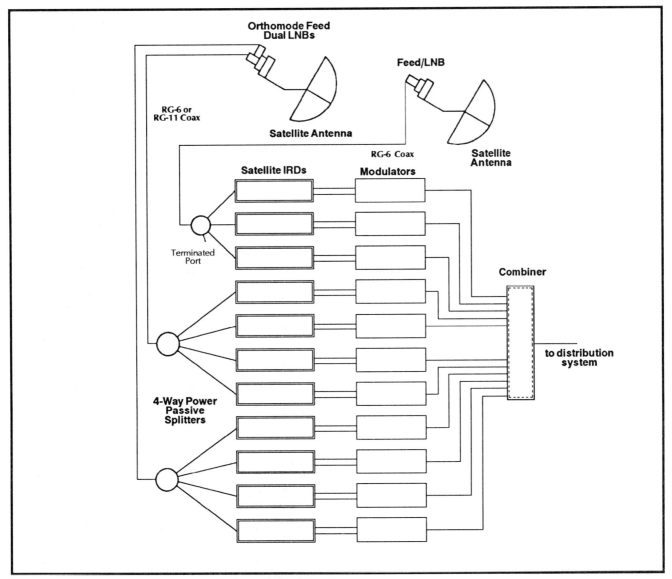

Figure 5-94. 11-Channel, 2-Antenna Headend. *The signal from the antenna on the left is mixed together in the same fashion as in the previous figure. The antenna on the right has a single polarity output selected by the feedhorn. The LNB output is fed via RG-6 or RG-11 cable into a 4-way, power-passive splitter with one port terminated. These three outputs then pass into the satellite receiver IF inputs and subsequently to modulators. All these modulated outputs are then mixed via a channel combiner with the single coaxial cable output feeding into the distribution system. Signal levels on the modulator outputs should have approximately 5 dB of tilt between the low and the high channels. In this example, the outputs which are fed into an amplifier to compensate for any high frequency roll-off are:*

Channel Number	Modulator Output (dB)	Channel Number	Modulator Output (dB)
2	20	8	23
3	20.5	9	23.5
4	21	10	24
5	21.5	11	24.5
6	22	12	25
7	22.5		

CHAPTER 6. DISTRIBUTION SYSTEM DESIGN

The goal of every private cable design is to process signals in the headend and distribute them to each subscriber's set with enough energy to produce an excellent quality picture and crisp audio. In order to accomplish this objective, the balanced signal transmitted from the headend has to compensate for insertion and isolation losses and must be sufficiently powerful to overcome losses in the cable lines and in components such as splitters and taps.

A well-designed distribution system must also achieve a number of other related objectives. (1) It must resist ingress interference from local television stations, other interfering carriers, electrical noise and distant TV stations during "skip" conditions when unusual atmospheric conditions strengthen ordinarily weak signals. (2) The system must be tight enough to prevent egress, especially if mid or superband channels which coincide with other communication service frequencies are used. (3) High isolation must be provided between subscriber ports to avoid system-wide problems with signals from isolated faulty components. It is interesting to realize that these last three objectives all relate to the choice of taps and splitters.

A. THE BASIC COMPONENTS of DISTRIBUTION SYSTEMS

The most complex configurations of distribution systems are pieced together from basic components including coaxial cables, connectors, splitters, line amplifiers, attenuation pads, terminators, dc power blocks, coaxial relays and taps.

Cables and Connectors

Coaxial cables, connectors and splitters are the conduits for carrying signals to any final destination. Each component attenuates power in a characteristic fashion. This attenuation, expressed in decibels of signal loss, must be factored in when designing a distribution system.

Coaxial cables are composed of two concentric conductors separated by an insulating dielectric layer. This assembly is sheathed in a non-conducting jacket for protection against the elements. A signal travels along the central wire and the external cylindrical conductor is grounded.

Types of Coax

A wide range of coaxial cables are available. These fall into three broad categories depending upon the construction of the dielectric sheathing material. These are hardline, conventional coax, and foam or air dielectric coax (see Figure 6-1). The central conductor of all types of coaxial cables is normally copper clad steel. This steel adds strength to

DISTRIBUTION SYSTEM DESIGN

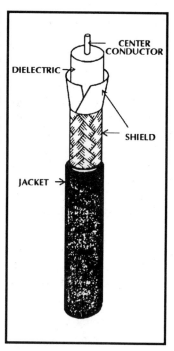

Figure 6-1. Coaxial Cable. *Coax is composed of two concentric conductors separated by an insulator called the dielectric. Signals are transmitted on the central conductor while the outer shield is designed to contain the signal and eliminate ingress interference.*

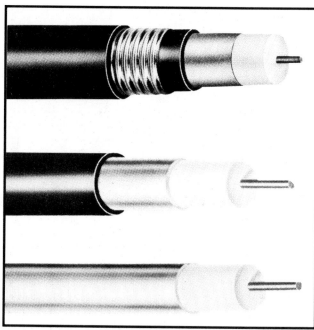

Figure 6-2. Hardline Coaxial Cables. *These three examples of coaxial cable for high frequency signal conduction are protected from interference to increasing degrees. The middle one has a polyethylene jacket. The lower coax has two metal jackets, one being a corrugated chrome plated steel armor that prevents sharp cable bends. This or any other deformation could cause impedance mismatches and signal attenuation. (Courtesy of Comm/Scope Marketing)*

prevent buckling and the copper provides a high conductivity path for the signal.

Regular coax has one or two pliable metal, grounded layers wrapped around a plastic dielectric (see Figure 6-2). Dielectrics in common use include polypropylene, a hard, translucent substance, and foamed polyethylene, a soft, white material. The foamed poly has the lowest loss and is generally the most widely used of these two materials.

The outer conductor performs a number of functions. It provides the mechanical strength when the coax is either pulled through conduit or used to span longer distances, it shields against ingress interference and it completes the RF circuit with the central conductor. There are a variety of different outer conductor configurations made from a braided copper or aluminum sheath, occasionally doubled up or replaced with a solid casing to further lower losses and ingress interference. These types include single braid, multi-layered braid, aluminized plastic tape, tape and braided wire, and tape and drain wire shields. Tape and drain shielded wires are composed of four wires helically wound on the outside of tape. This variety is not strong enough to be suitable for long cable spans, for vertical risers or for pulling through conduit. For such applications, tape and braided wire cable is a better choice.

The degree of cable braiding is rated by the percentage of dielectric surface shielded. 67% shielding is a typical value. But 100% shielded cables are recommended for SMATV installations that require good RF integrity so that the system is protected from both ingress interference and signal leakage.

The outside jacket of regular coax can either be vinyl or polyvinyl chloride (PVC). The former type is generally not used in private cable systems since it is a fire hazard. Although PVC is a safer choice, it deteriorates over time if exposed to direct sunlight. However, there are varieties of PVC cables, generally black in color, that have ultra-violet inhibitors.

Hardline is similar in construction to either of the previous two types except that it has even lower losses because it has a more rigid, metal sheath and a higher quality dielectric. The dielectric can be composed of foam or air. When the dielectric is simply air, the outer sheath and inner conductors are separated by lightweight plastic disks inserted at regular intervals.

Hardline cable is typically used in the main trunk lines of CATV or larger SMATV installations to reduce attenuation to a minimum. This type of cable is

DISTRIBUTION SYSTEM DESIGN

specified by its outside diameter. Typically 0.500 inch O.D. hardline is used in private cable installations. In addition to having reduced attenuation, these cables easily interface with many types of CATV amplifiers and passive devices which may be required in larger distribution systems. "Messengered" hardline is often used for aerial construction. The steel messenger provides the necessary additional strength for attaching amplifiers, couplers, taps and other devices to aerial cable.

Cable Performance

Whenever a signal is transmitted along a cable is it attenuated. The power loss results because every conductor has a specific resistance to the flow of electrical current. In addition, the interaction between electrical charges on the inner and outer conductors, known technically as capacitance and inductance, cause signal delays and losses. These factors that depend on the outer diameter of the central conduct, the inner diameter of the dielectric sheath and the properties of the dielectric material determine characteristic impedance (see Appendix B for details on calculating cable impedance).

Coax typically used in SMATV distribution systems is rated at 75 ohms. It is crucial to use cables having the correct characteristic impedance. If the impedance of the cable does not match that of the device it feeds, there will be substantial reflective power losses and ghosting will result.

The rate of loss over a given distance is determined by the type of cable and the frequency of the signal being transmitted (see Figure 6-3). The smaller the diameter of a cable, the greater the signal attenuation. Therefore, large diameter cables are used for long runs while smaller cables are used for shorter subscriber drops. The characteristic impedance is chosen to be 75 ohms for all cables installed. In addition, the higher the signal frequency the greater the attenuation. For example, although RG-6/U coax has 1.7 dB loss per 100 feet for signals in the VHF Channel 2 frequency range, this attenuation increases to 3.0 dB for VHF Channel 13. This 1.3 dB difference means that losses are 35% greater at Channel 13 than at Channel 2.

Cable loss varies with temperature as well as with frequency. As a rule of thumb, the loss increases by 1% for each 10°F (5.6°C) increase in temperature. This percentage loss is also somewhat less for lower frequency signals. As a result, wide variations in temperature can affect the amount of frequency slope compensation required in an extensive distribution network.

When designing a distribution system, cable runs should clearly be minimized when possible. It is also important to calculate signal losses for the highest frequency signal to be distributed. Table 6-1 lists cable loss for four different channels per 100 feet. The characteristic signal attenuations in some standard types of 75 ohm cables listed in Table 6-2 demonstrate that RG-59/U should not be used in 950 to 1450 MHz line from a downconverter to a satellite receiver because losses would be too great.

TABLE 6-1. CABLE ATTENUATION versus TV CHANNEL (dB/100 feet)

CABLE TYPE	Ch.2	Ch.6	Ch.13	Ch.50
RG 59/U	2.3	2.7	4.2	7.8
RG 6/U	1.7	1.9	3.0	5.9
RG 11/U	1.1	1.4	2.3	4.2
412	0.74	1.0	1.5	2.9
500	0.52	0.67	1.1	2.4

These values are approximate. Please consult manufacturer's specifications for actual cable used.

TABLE 6-2. CHARACTERISTICS OF COMMONLY USED COAX

Cable Type	Signal Loss (dB/100 feet) 100 MHz	1450 MHz	Impedance (ohms)
RG-59/U	3.4	11.0	75
RG-6A/U	2.7	8.7	75
RG-11/U	2.3	7.0	75

Signal losses in each particular brand of cable are dependent upon construction details such as the diameter of the central conductor and the type of braiding material. For example, Belden Alpha cables 9059 and 9803, both similar to RG-59, have different diameter central wires. As a result, they have different attenuations at 900 MHz of 35.10 and 33.46 dB per 100 meters, respectively. Specifications for the precise types of cables installed should also be obtained directly from their manufacturers.

DISTRIBUTION SYSTEM DESIGN

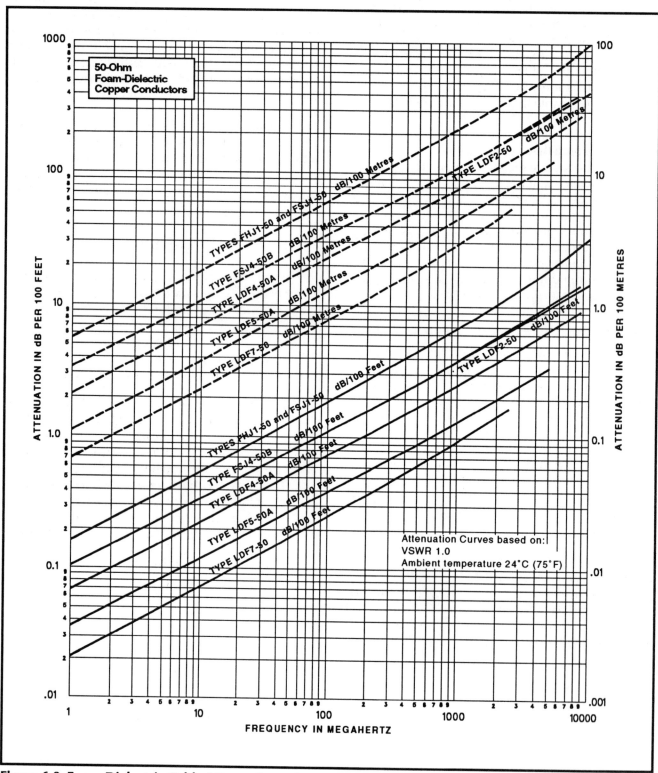

Figure 6-3. Foam-Dielectric Cable Attenuation. *Attenuation in decibels per hundred feet for various types of foam dielectric coax is graphed. (Courtesy of Andrews Antenna Corporation)*

DISTRIBUTION SYSTEM DESIGN

Installation Details

Installation errors can cause attenuation in excess of rated values. If cables are too severely bent or compressed, the resulting impedance mismatch at the kink will increase reflective losses. A suggested minimum cable turning radius is ten cable diameters. Note that of the more familiar coax used in SMATV installations, RG-11/U is the largest diameter, lowest loss cable that is still reasonably flexible.

Power losses can be substantial where connectors join cables. If these connectors are not installed correctly, impedance mismatches and losses occur. It is important to examine the inside of each connector before mating. This ensures that the center pin or conductor is not broken off and that it is extended far enough out from the connector to make secure electrical contact but not too far to short out to a chassis and damage circuitry. A connector that is improperly crimped onto a cable can also change impedances and cause signal reflections.

Water Damage to Coaxial Cables

Coax can easily be destroyed by intrusion of water, especially salt water. Also, any leaks at connectors or along the cable body can short out a signal to ground and possibly damage an LNB, active splitter, satellite receiver, distribution amplifier or any other sensitive electronic component.

Underground moisture, which inevitably comes into contact with buried cables, can cause very rapid corrosion of metal components. It is reported that tubular aluminum outer conductors have been almost completely destroyed within 90 days when in contact with water. Even small pinholes in outer jackets can allow this chain of events to occur. Poor installation or cable handling techniques or even the presence of rodents may cause such damage. For maximum reliability against rodents, a steel tape armor with over-jacketing or rigid conduit such as gray electrical PVC is suggested.

Cables have varying degrees of protection against water intrusion. For example, foamed poly coax is more susceptible to water ingress than cables having solid polyethylene dielectrics and is not recommended for use in high moisture areas unless properly sealed. Most types of cable are available with a sticky sealant inside their jackets to resist moisture. This "flooded" cable is well suited for burial in conduits.

Coaxial cables can be purchased wound on rolls up to 500 or 1000 feet in length. Different colors such as black, beige or white are available. Those colors, which blend in with surrounding walls and floors, are useful in improving installation cosmetics.

Splitters

Splitters do exactly what their name implies, they split or divide a signal into two or more branches. In certain situations, these devices may also be used in reverse to combine signals. Although splitters usually divide a signal into an even number of equal but lower level signals, most varieties having odd numbers of outputs produce one lower signal port.

Each type of splitter is rated to handle a specified range of frequencies. For example, when dividing the 950 to 1450 MHz output of an LNB, a splitter rated up to 1450 MHz (1700 in European systems) must be used or losses would be excessive. Some brands of splitters also have built-in bandpass filters so that frequencies below and sometimes above the designated range are sharply attenuated. Limiting the bandwidth in this manner can also protect cables from ingress interference. Typical insertion losses for two, three, four, eight and sixteen-way splitters are listed in Table 6-3.

DISTRIBUTION SYSTEM DESIGN

TABLE 6-3. SPLITTER LOSSES in EACH OUTPUT LEG

Type of Splitter	Loss (dB)
2-way	3.5
3-way	3.5 and 7
4-way	7
8-way	10.5
16-way	14

A two-way splitter reduces a signal to less than half its original value, a 3 dB reduction in power level. A 3-way splitter has one –3.5 dB and two –7 dB output ports; a 4-way splitter has four –7 dB ports. For example, if a 3 dBmV channel 6 television signal enters a 4-way splitter, each output leg would produce a signal of 3 dBmV reduced by 7 dB which equals –4 dBmV. This voltage level would not be enough to produce a studio quality picture because it is well below acceptable levels, so therefore noise would begin to overpower the signal. If this signal were also relayed down 300 feet of RG-59/U cable, additional losses of 8.1 dB (see Table 6-2) would result in a final signal of –12.1 dBmV at the television input. It is clear that 15 dB amplification would be required to restore a 2.9 dBmV level. In this case, a distribution amplifier would be installed before the splitter.

When splitters are used in reverse as combiners, the same losses occur as in the forward configuration. A simplified explanation can be provided using a 2-way splitter as an example. In order for the combiner to provide sufficient isolation between its input ports, it divides both signals in two. Half of each signal is available at the combiner output ports while the other half is used to cancel the signal entering the opposite port. This cancellation effectively provides the necessary isolation.

Theoretically, the losses in each port of an N-way splitter are given by:

$$\text{Loss per Port} = 10 \log N$$

Therefore, a 2-way splitter should have just 3 dB loss in each leg. In reality, splitters and combiners have internal resistive losses which account for an additional 1 to 2 dB depending upon the number of ports and the quality of their internal construction.

Distribution Amplifiers

Distribution or line amplifiers, that are inserted via F-connectors directly into a coaxial cable line, are available for boosting signals spanning either a single channel or a range of channels (see Figures 6-4 and 6-5). Less-expensive, simpler units known as "bullet amps" are powered from the dc voltage relayed down the coax. These include line amps which operate in the 900 to 1700 MHz range and are installed between the LNB output and satellite receiver input. In contrast, SMATV distribution amps are powered by ac voltage that is converted to dc power via an internal power supply.

Gain, Tilt, Bandwidth and Noise Figure

Distribution amplifiers for SMATV installations differ from off-air preamps because they are designed to deliver relatively high output levels. These amplifiers can typically attain their full output with just 20 dBmV of input. Therefore, for example, a distribution amp generating a 55 dBmV output should have a gain of at least 35 dB. These components have a specified maximum permissible input. An excessive input can cause an amplifier to overload and generate distortions. The average maximum input specification for a typical distribution amplifier is 10 to 15 dBmV. While distribution amps can have either fixed or adjustable gains, most usually have a 10 dB adjustment range. Their manufacturers specify the maximum permissible output so that gain can be reduced if input levels are too high.

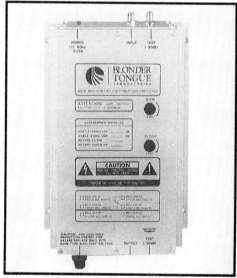

Figure 6-4. Distribution Amplifiers. *The BIDA 450 line amplifier has both gain and slope adjustment on the front casing. It is powered by 60 Hz line voltage and has an input and output test point. (Courtesy of Blonder-Tongue Laboratories, Inc.)*

DISTRIBUTION SYSTEM DESIGN

Distribution amplifiers used to generate headend output signals are also known as launch amplifiers. As a general rule, if the input signals are strong and steady, all channels are relatively well balanced in strength and the maximum headend output is less than 60 dBmV, a broadband launch amplifier is usually sufficient. In smaller systems, this component may be sufficient to power the entire distribution network. However, strip amplifiers are often used if higher headend output levels are required or if there is a particular need for good automatic gain control.

Noise figure is not of great concern in line amplifiers because, at this point in a distribution system, the input signals and the S/N ratio should be sufficiently high. Typical noise figures range up to 3 dB and 5 dB for the VHF and UHF ranges, respectively. As amplifiers are cascaded down a distribution line, the noise power can add up to the point where it might become a concern. The rule of thumb is that noise figure increases by 3 dB for every doubling in the number of line amplifiers. Fortunately, most private cable systems are not large enough for this to become a serious concern. When a potential problem exists, some simple remedies, such as choosing an amp having a low noise figure, are available. The effect of cascading distribution amplifiers is examined in Examples VI and VII in Section D below.

Many brands of distribution amplifiers designed for longer cable runs have built-in "slope" or gains that increase with frequency to compensate for tilt, the increased attenuation of signals at higher frequencies (see Figure 6-6). For example, a tilt of about 1.9 dB per 100 feet between 50 and 200 MHz (the range spanned by channels 2 and 13) is necessary when using RG-59/U coax. This is based on the difference between the 4.2 and 2.3 dB per 100 feet attenuation that RG-59/U causes at these two frequencies. For example, if the cable run were 500 feet long, the system would have to compensate for 9.5 dB of tilt over this distance, 5 times 1.9 dB. Properly adjusting or equalizing tilt minimizes the chance of adjacent channel interference.

Tilt in a distribution network is usually managed as follows (see Figure 6-7). Launch amplifier can generally be adjusted to compensate for about 3 to 6 dB of tilt. In other words the gain at higher frequencies is set 3 to 6 dB above the gain of the lower frequency channels. A value of 6 dB is used in the description that follows. After above 15 to 18 dB has been introduced as the signal travels down the cable, the tilt is re-adjusted by a device known as an equalizer or tilt compensator (see Figure 6-8). At this point the net tilt is 9 to 12 dB (15 to 18 dB less the 6 dB amplifier tilt compensation). The equalizer, which can usually compensate for about 12 dB of tilt, is then set to equalize the signal. Subsequently, its output is fed into into the next line amplifier. Then the process is repeated. Namely, the signal leaves this amplifier with another 6 dB of tilt compensation and continues on its voyage to its destination.

Figure 6-5. Distribution Amplifier. *The ASL-2000 distribution amp is power by line voltage and has gain and slope adjustments as well as a −30 dB test point. (Courtesy of Nexus Engineering)*

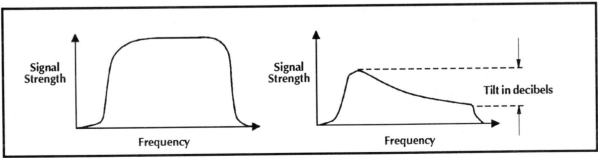

Figure 6-6. The Effect of Cable Tilt. *The illustration on the left shows that channels leaving the distribution amplifier have no tilt, i.e. they all have the same level across the frequency spectrum. After passing through a length of cable, the higher frequency channels have been attenuated more than the lower ones*

DISTRIBUTION SYSTEM DESIGN

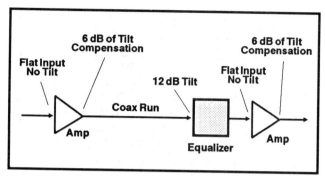

Figure 6-7. Tilt Adjustment in a Distribution System. *The accepted method to manage tilt in coaxial cables is outlined in this diagram. This first amplifier provides 6 dB of tilt compensation so the output has the level of the higher frequencies at 6 dB above that of the lower ones. After incurring 18 dB of tilt in the intervening cable run, the equalizer again sends a flat input to the next amplifier. Then the process starts again.*

Signal Distortions in Amplifiers

Distribution amplifiers and other broadband devices are never perfectly linear, especially when driven to near saturation. As a result they cause carriers of different frequencies to add and subtract together to create various types of output signal distortions. Non-linear interactions also result in the generation of harmonics of the various carrier frequencies being processed. The resulting signals are known as distortion products. Although these products can be classified into second, third, fourth and higher order products, only the lowest order distortions, second and occasionally third order products, usually have any appreciable effects.

Second order distortions occur when two signals beat together to produce sum and difference frequencies. In a system carrying VHF channels, second order beats can show up in the higher VHF channels, in the midband (channels A through I) and in the superband (channels J through W) as diagonal lines running across the television picture. For example, in the video carriers of VHF-low band channel 3 and mid-band channel E add together, the resultant equals 61.25 plus 145.25 MHz or 206.50 MHz. This sum frequency falls just 1.25 MHz above the video carrier of VHF-high band channel 12. Such second order products should be maintained less than 55 dB below the main signal level.

Second order products also include second harmonics of frequencies that enter an amplifier. For example, if the 55.25 MHz video carrier frequency of channel 2 is doubled, the resultant is 110.5 MHz, which is only 1.25 MHz above the video carrier of VHF-midband channel A-2. Better quality amplifiers that use a push-pull design can completely suppress second order harmonic products.

Composite second, third and higher order distortions, caused by the interactions among two or more frequencies. One of the more familiar varieties of such distortion that is known as composite triple beat results from the mixing of three or more video carriers to create a number of discrete beats within 20 or 30 kHz of a video carrier. The problem can rapidly worsen when more than 21 channels are present. Triple beat makes pictures look grainy and sometimes "wormy" and is often hard to distinguish from "snow" caused sometimes by a poor S/N ratio.

Cross Modulation

Cross modulation distortions, also familiar as the "windshield wiper effect," can occur when the modulated signal on one channel bleeds into the output of another channel. It is literally a crossing over of the modulation from one carrier to an adjacent one. The 15.75 kHz sideband that contains the picture sync information is generally the culprit behind this form of adjacent channel distortion. Cross mod can occur if just one channel is being distributed if the sound carrier level is too high. The result is a buzzing in the sound and "sound lines" in the picture. These and other distortion problems are examined in more detail in Section B, Chapter 10.

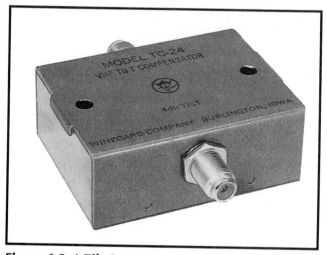

Figure 6-8. A Tilt Compensator. *This passive compensator, model TV-28 or TC-48, offers a fixed 4 dB of tilt from 54 to 216 MHz on VHF and 470 to 890 MHz on UHF by attenuating the lower more than the higher frequencies, opposite to that effect which occurs along cables. Insertion loss is less than 0.9 and 2.5 dB at VHF and UHF frequencies, respectively. (Courtesy of Winegard Company)*

DISTRIBUTION SYSTEM DESIGN

As output levels and as the number of channels being processed increase, the potential for cross mod rises. It then becomes visible as vertical or horizontal bars moving through the screen. This occurs because the TV receiver is confused with two or more sync pulses being present at one time. As cross modulation becomes progressively worse, complete pictures from one or more channels will be seen in the background. In general, similar levels of triple beat and cross modulation distortions cause similar, noticeable picture problems.

Subjective responses of television viewers surveyed suggest that if cross mod products are less than a maximum of 46 dB below the channel signal, distortions are not noticeable. However, cross modulation products were found to be visible on a blank screen when they were in excess of 51 dB below the television signal. The trade-off between signal cross modulation distortion, the number of channels processed and an amplifier's output capability is presented in Figure 6-9.

Minimizing Signal Distortions

Second and third order distortions can be minimized by maintaining the output of an amplifier within its specified operational capability. A general rule of thumb is that cross-modulation and beat distortions decrease by 2 and 3 dB, respectively, for every 1 dB drop in amplifier output. It is clear that amplifiers operate between two performance boundaries. The output should be maximized to increase S/N ratio but decreased to reduce distortions to an acceptable level.

Also, as the number of cascaded amplifiers in a distribution system is doubled, cross modulation and composite triple beat increase in impact by 6 dB. Second order beat grows by between 3 to 6 dB, and with certain designs by less than 3 dB, when amplifier numbers are doubled. Since system noise also increases for each doubling in the number of amplifiers, the motivation for minimizing the number of amplifiers and selecting top quality components becomes strong.

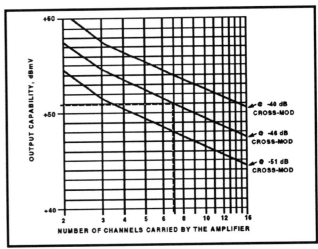

Figure 6-9. Amplifier Cross-Mod Specifications. *If the manufacturer's cross-mod specs for an amplifier's output and channel carrying capabilities are known, this graph can be used to compare performance under different operating conditions.*

Attenuation Pads

An attenuation pad, more simply known as a pad, is used to reduce the strength of an excessively powerful signal (see Figure 6-10). Pads are small, inexpensive devices which insert directly into a coax line via F-connector fittings. They are available with either fixed or variable rated attenuations. Most pads are not designed to pass dc power so they cannot be used in the line between satellite receivers and LNBs.

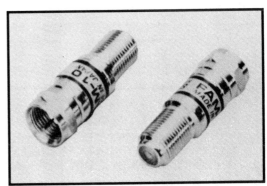

Figure 6-10. Attenuation Pads. *These pads are manufactured with one male and one female port and can easily be inserted into a coax line using F-connectors. These models have a variety of rated attenuations including 3, 6, 8, 10, 12, 16 and 20 dB with a bandwidth from zero to 1000 MHz. (Courtesy of Pico Products)*

DISTRIBUTION SYSTEM DESIGN

Terminators

Any output port on a RF distribution network must end in an appropriate device such as a television set, a drop or a terminator (see Figure 6-11). If not, interference could leak into the unconnected port or the resulting impedance mismatch could cause signals to be reflected back into the system as ghosts. A terminator is simply an encased resistor that screws onto a F-connector fitting. It has a 75 ohm matching impedance and therefore fools the cable into not being able to see an opened end or discontinuity at the unused port.

Figure 6-11. A Terminator. *Terminators must be installed on all unused splitter, taps or other ports to provide a 75 ohm impedance match. Doing so eliminates most signal reflections as well as potential ingress interference. (Courtesy of Pico Products)*

Taps

A tap directs a portion of an incoming signal to each subscriber drop while allowing most power to pass through to its output (see Figure 6-12). Taps allow a high signal level present in most distribution cable to be tapped off at a substantially lower level required by each television set. A tap can also be used in reverse as an injector where two uneven signals must be combined with high isolation provided between output ports.

A tap creates two kind of losses, insertion and isolation (see Figure 6-13). Insertion losses, which occur as a signal passes through a tap, can range from under 0.3 to over 3.7 dB and more depending on the tap value. Isolation loss is determined by the rated tap value. Each tap attenuates the signal it siphons from the line according to the value of the tap. For example, a 30 dBmV signal entering a 25 dB tap results in an output signal of 5 dBmV. In this case, the isolation loss is 25 dB. Also, for example, when a 30 dBmV signal is fed into a 23 dB tap it siphons off the difference, 7 dBmV and passes the remaining portion, less a small insertion loss of about 0.4 dB, through its output port. As seen from sample values in Table 6-4, the lower the rated tap value, the lower the isolation loss, the more signal is extracted and the higher the insertion loss.

When designing a distribution system, the insertion loss for all taps must be added together to calculate the total loss. To illustrate, if 15 taps are present and each has an average insertion loss of 0.5 dB, the total system insertion loss will be 0.5 times 15 or 7.5 dB. The total loss on any one distribution line equals the total insertion loss and the isolation value of the last tap. When selecting tap values, it is necessary to use one that will provide 0 to 10 dBmV to every TV set. This upper voltage level will not harm or overload a television and provides a customer with adequate signal levels to install VCRs or other devices.

An additional examples of tap isolation and insertion loss is presented here. A 40 dBmV signal would be reduced to 39.5 dBmV after passing through a 24 dB tap but to 37.8 dBmV following a 6 dB tap. The former tap would extract 16 dBmV of signal; the latter 34 dBmV. Note that a tap differs from a splitter

Figure 6-12. *The left photo shows a miniature tap, model 4712, that can be used as a wall terminal. The middle tap, model 4042, is for outdoor use and can be pedestal mounted or hung from an aerial cable. The third unit, model 4613, is a surface mounted, 4-outlet tap. (Courtesy of Blonder-Tongue Laboratories, Inc.)*

TABLE 6-4. TAPS and THEIR TYPICAL INSERTION LOSSES

Tap Value (dB)	Insertion Loss (dB)	Isolation Loss (dB)
41	0.3	41
38	0.3	38
35	0.3	35
32	0.3	32
29	0.3	29
26	0.4	26
23	0.4	23
20	0.8	20
17	1.7	17
14	3.7	14
8	3.7	8

Note: Exact values for insertion losses of taps should be obtained from manufacturers.

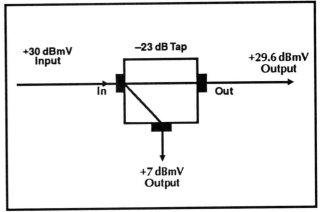

Figure 6-13. Tap Insertion and Isolation Losses. *This drawing illustrates both insertion and isolation losses. The 23 dB tap takes a 30 dBmV input, and following an isolation loss of 23 dB, delivers 7 dBmV at the customer drop. The insertion loss of 0.4 dB reduces the input signal to 29.6 dBmV which travels further down the distribution line.*

which divides a signal equally into two or more output legs. For example, a 3-way splitter, the only exception, has two equal output ports and one with twice the power. But both types of signal allocation devices are used to accomplish the same function. Taps are usually installed to pull signals off a feeder line since the throughput losses are much lower than splitters. Then splitters or taps are used at the drops to individual TV sets.

Resistive/Backmatched Taps and Directional Couplers

Three varieties of taps have been developed over the years to meet the needs of the private cable and CATV industries. The resistive tap-off, the simplest and pioneer device, consists of a resistor and usually a series capacitor. Although this type of tap is inexpensive, it suffers from impedance mismatches that cause ghosting and power attenuation, and from poor isolation to downstream reflected signals that also causes ghosting. This tap should be used only in areas where ingress and egress do not have important consequences and where only a few strong local off-air stations are being detected. Resistive taps are available in three configurations: (1) on a plate for mounting in open walls, (2) on an electrical type base that must be installed during construction and later sealed with a cover plate, and (3) an enclosed "T" tap.

A backmatched tap (BMT) incorporates a transformer to match its output port to a 75 ohm impedance. However, this type of tap still has the disadvantage of being susceptible to reflected signals from downstream, causing ghosting. BMTs are typically available in single or multi-port configurations intended for inside-wall or wall installation. A single coax then connects from the tap to a wall plate having an F-connector mounted towards the room interior.

A directional coupler (DC) uses an additional core winding to provide an extra 10 to 15 dB of isolation from upstream reflected signals. Current models provide 80 dB of RF shielding. DCs are available in 4- port, wall type or plate mounted configurations. Recently the cost of directional couplers has dropped to the point where they are as inexpensive as BMTs. Therefore, where applicable to the design in question, directional couplers should be chosen.

Installing Taps and Wall Terminals

The various tap configurations dictate the detail of their installation. Flush-mounted taps are used for wall mounting within standard electrical boxes with wall plates. Surface-mounted brands are used where cosmetic appearance is not of critical importance. Many multi-outlet taps feature this type of functional design. These either have a separate tap-off for each outlet or one tap-off followed by a splitter. Multi-outlet taps are available sealed in a weatherproofed enclosure for outdoor mounting.

DISTRIBUTION SYSTEM DESIGN

The last tap in each feeder line should be terminated. A 75 ohm resistor terminates the main throughline output of the final tap to prevent impedance mismatches and the resulting signal reflections.

Wall plates which feature a female-to-female F-connector (known as an F-81 barrel) can be used in situations where the tap cannot be installed in a wall box. In this case, a cable line is run from the tap to the wall terminal.

Special Design Requirements for Taps

Some addressable distribution systems place unusual requirements on tap-off and splitter design. dc power and low frequency addressing data must be allowed to pass through these devices while RF integrity is maintained. When purchasing these advanced systems, a number of factors should be considered. dc bypass coils must be able to handle in excess of the maximum current expected on the system. This will prevent both the saturation and burn out of the bypass coil. Also electrolytic corrosion between dissimilar metal contacts is much greater for devices carrying dc power. So these advanced taps and splitters must be examined with this possibility in mind.

Matching Transformers

A matching transformer is used to match the 75 ohm coax to the 300 ohm twin-lead input typically of older television sets or to the 300 ohm of an off-air antenna (see Figure 6-14).

A problem known as direct pick-up often results in SMATV systems whereby off-air signals are detected by television sets independently of signals being relayed by the distribution system. This can usually be traced to ingress at low quality matching transformers. While the outer conductor of a long coaxial cable can be an excellent antenna, ingress is usually prevented because of the grounded shielding. However, the junction between this coax and a poorer quality matching transformer can be a bridge that allows ingress interference. This result can be avoided if a transformer having a match of 35 dB or better is selected.

Matching transformers can be designed to transmit VHF or UHF signals or both and to pass either ac or dc power. Some models split the broadband television signal into their UHF and VHF components.

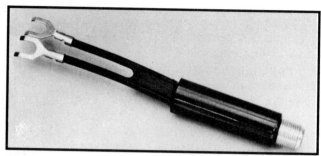

Figure 6-14. Matching Transformer. *This device is used on TV sets that require 300 ohm, twin-lead cable inputs. (Courtesy of M/A COM)*

Matching transformers are also available in either indoor or outdoor weatherproof designs.

Converters and Decoders

SMATV systems must often expand beyond the 12 channel VHF capability that is offered by most television sets. In these cases, design engineers generally first use the 9 midband channels. Once these are exhausted the 14 superband and then the higher 23 frequency hyperband channels are recruited into service in order to increase the number of programs being distributed. Higher frequency UHF band channel slots employed are rarely employed in North America (but see Section D below). Since most television tuners are not equipped to handle midband, superband and hyperband frequencies, cable-type converters must be installed at each customer's set (see Figure 6-15).

Set-top converters are available with a wide selection of features. The models begin with the most economical, basic designs. Other frills such as, programmability, IR remotes and addressability are often incorporated into the higher cost varieties in order to provide the operator with the ability to control access to pay-TV channels.

In many cases, the set-top converter becomes a combination converter/decoder. Experience has demonstrated that if more than three pay-TV channels are transmitted, it is wise to use an integrated converter/decoder or a converter in concert with an addressable tap system rather than to install a trap at each subscriber's drop box. These devices, which "trap" out individual pay-TV channels, would have to be manually turned on and off whenever a subscriber wishes to change the programming menu. (Security systems are examined in more detail later in Chapter 18).

DISTRIBUTION SYSTEM DESIGN

The set-top converter is the most desirable customer terminal because it present a distribution system with a constant 75 ohm impedance match and isolates the network from any carriers generated within the television receiver. In addition, set-tops incorporate some filtering to keep adjacent channels from interfering with each other. In contrast, wiring directly into the television receiver is the worst terminal a system can have. When the set is turned off there can be a mismatch so signals are reflected back into the system. When the set is turned on there is generally a good match to the tuned channel but not necessarily to the others. Fortunately, this situation can usually be managed with well isolated taps and splitters.

Figure 6-15. The Zenith PM2-C Compatible Addressable Decoder. *This remote controlled decoder features multi-mode scrambling that is compatible with most RF gated sync suppression systems. It is PPV and BTSC stereo compatible, has audio masking and a 84 channel, 550 MHz bandwidth. Subscriber remote control features include volume control, parental lockout, last channel recall and favorite channel scanning. (Courtesy of Zenith Cable Products*

B. CABLE TV DESIGN ISSUES RELATING to SMATV SYSTEMS

Trunk and Feed Distribution Systems

A mid-sized cable television distribution system is substantially more extensive than the typical SMATV system. Both have the function of transmitting the combined signal from the headend to subscribers via coaxial cables of varying diameters. In a standard CATV plant, this network is comprised of the main trunk and smaller feeder segments. However, in smaller SMATV systems, the distribution system most closely resembles the feeder portion of the larger plant.

It is instructive to examine some of the issues facing designers of CATV distribution system given that some larger SMATV networks have similar concerns and solutions. In both a large SMATV system or small CATV network, a separate trunk network delivers the signal to the distribution plant. It feeds signals only to distribution lines and is never connected directly to a subscriber tap.

A trunk transmits the combined RF signal at a relatively low level to minimize distortions. It also typically utilizes larger cables, having diameters of 0.750 inch and higher, than used in the feeder portion of the plant. This serves both to minimize signal losses and to reduce the number of amplifiers required. This in turn directly lowers the noise contributed by the amplifier network.

The feeder cable, usually of smaller diameter than the trunk cable, comprises most of the coax used in a distribution network. It is connected directly subscriber taps and generally carries high level RF signals. Higher voltages can be used because individual feeder segments and corresponding amplifier cascades are much shorter than those used in trunk lines. Using shorter segments serves to reduce the potential for excessive noise and distortion products. In a typical CATV system, the feed leg or distribution portion is never any longer than two amplifiers deep. However, in a small scale SMATV system, it is common to see distribution legs with three amplifiers cascaded in series. This is possible because the input to these legs typically comes di-

DISTRIBUTION SYSTEM DESIGN

rectly from the headend so the resulting distortion products are still at relatively low levels. This provides enough margin for the addition of a third or fourth line extender amplifier and the resulting distortion products.

As previously mentioned, an equalizer that is used in distribution systems to compensate for tilt or slope has the opposite effect on signal levels as frequency changes as does cable, namely it attenuates the lower more than the higher frequency signals. The intended result is to provide a near-flat input to an amplifier. Given the tilt that occurs in coaxial cables, without this component signals that were adjusted to provide an adequate input to an amplifier at lower frequencies would be too weak at higher frequencies. Alternatively, signals that had an adequate level at higher frequencies would overloaded an amplifier at lower frequencies and result unacceptable levels of distortion products.

System Design Methods

One principle objective in designing any RF distribution network is to keep individual cable runs as short as possible and therefore minimize the number of costly and service-intensive amplifiers. The two limiting factors on the number of amplifiers that can be cascaded in a CATV system are noise and distortion products. In a larger-scale CATV system, the design typically specify a mix of 25% trunk and 75% feeder lines.

The placement of amplifiers in a distribution system is determined by their specifications and the characteristics of the cable in use. The most important amplifiers parameters are their minimum and maximum operational input levels, gain, and maximum or operational output levels. An amplifier is placed at each point in the system where the signal level falls to its minimum operational input. The output level are then set up according to manufacturer's specifications.

By amplifying to just compensate for cable attenuation and passive insertion losses, the goal of unity gain is achieved. Unity gain means that the input plus cable and passive losses equal the operational output of the amplifier.

Two different distribution system design methods have been developed. The output level design method is based on the required feeder system levels. Amplifier output levels are predetermined to insure total system operation. The designer begins with the output level of the amplifier and proceeds designing down the line, installing cable and passive devices calculating signal losses until another amplifier is required to boast the signal. In the reverse design method, the designer begins at the end of the line using a specific value multitap and works back toward the headend installing amplifiers when necessary so that the RF signal equals those of the amplifier output levels. Because the reverse system uses sub-band frequencies, the attenuation loss is so low that the signal can travel much further in the reverse direction before an amplifier is needed.

One of the obstacles encountered with reverse systems are problems created with ham radio, CB and government transmissions that use the spectrum from 5-30 MHz. Depending on how the system is constructed, these could cause severe ingress and picture degradation and is also prone to ac power line interference, commonly closed voltage dots in the picture.

Distribution versus Trunk Amps

A distribution amplifier, normally known as a line extender amplifier, differs from a trunk amplifier in a number of ways. It is designed to accept much higher input levels with an average input signal of 17 to 20 dBmV. While it has a higher thermal noise figure and distortion products, it consumes less power and is less expensive. Distribution amplifiers are not designed to be cascaded. The operational output of line extenders generally ranges from 47 to 51 dBmV.

Amplifiers on the main line are commonly referred to as a trunk amps. These contribute most of the system noise because their inputs are maintained at a lower level. In contrast, most distortion products are generated by line extenders because they are operated at much higher levels than trunk amps. For these reasons, the trunk is noise limited, i.e., the inherent noise determines minimal signal levels while it is relatively transparent to distortion. However, bridger and line extender levels are designed to have higher output levels so that a maximum number of subscribers can be reached. As a result, feed lines are is distortion limited while being relatively transparent to noise.

As previously discussed, noise increases by 3 dB while distortion products increase by 6 dB whenever the number of amplifiers in a cascade are dou-

DISTRIBUTION SYSTEM DESIGN

bled. In addition, S/N ratio increases by 1 dB for each decibel that amplifier gain increases but the distortion products increase by 2 dB for every decibel of gain increase.

System Powering

Line extenders and other distribution amplifiers require a 30 or 60 Vdc supply that is powered by a 110 volt line input. Modern CATV power supplies provide voltage in a quasi square wave format. Although neither a sine or a square wave, it exhibits the best characteristics of both. Its source is a ferroresonant transformer that provides good ac regulation. The unit is current limited and therefore shockproof. Because of the transient peaks, the rectified voltage has less ripple and is easier to filter.

Power enters the coaxial cable via a power inserter that looks and operates much like a two-way splitter having about 1 dB of loss on the through leg. The power supply leg blocks RF signals from entering the power supply.

While 30 volt power supplies were used extensively in earlier cable systems, modern components require a 60 volt power supply. Most distribution amps, trunk amps and line extenders require a minimum of 40 volts. It is recommended that the minimum voltage drop across the distribution system not exceed 43 volts.

CATV Quality

The best private cable systems are those designed to meet or exceed CATV construction and performance specs. These include a bandpass of 450 MHz or better, hardline cable for all distribution runs, CATV grade equipment for all passive, active, and headend equipment, unity gain on all spacing, signal-to-noise ratio of 42 dB or better, cross modulation and distortion products at -50 dB or better, minimum subscriber outlet level of +5 dBmV at the wall plate, all signals balanced within 1 dB of each other and ac hum level not greater than 2%. (These specifications are examined in more detailed in Table 8-2).

If a system is being built in an area that is currently served by a cable network, all efforts should be made to construct it to meet the same specifications and employ equipment currently employed by the cable operator. This will allow a CATV operator to easily hook up to the SMATV system should it be purchased at a later date. Failure to meet or exceed these requirements will reduce the sale value.

C. TELEVISION INPUT SIGNAL REQUIREMENTS

The ultimate target of broadcast signals that are processed by a headend and distributed by a private cable network is the television set and audio system. Televisions function best with an input signal level of 0 to +10 dBmV, although the optimal level can vary between different brands. Nearly all modern sets have AGC circuits which compensate for excessively strong signals. This features allows inputs ranging from -20 dBmV to as high as 50 dBmV to be managed, to a point. However, it is safe practice to limit the input level to at least the 0 to 10 dBmV range.

In contrast, if the signal-to-noise power ratio is allowed to drop too low or is "in the mud," at some point in the distribution network no level of amplification will improve the situation because as much noise as signal will be amplified. In contrast, when signal power levels exceed the upper limits, attenuation pads can easily be inserted. Therefore, too much power is never really a problem but allowing signal strength to deteriorate can cause serious difficulties.

Many older televisions often have a 300 ohm, twin-screw VHF input where the familiar flat TV leads are attached. SMATV systems not using set-top cable converters transmit video signals via 75 ohm coax directly into the set. In these cases, F-connector-type inputs are required. In older sets which do not have an F-connector input, a 75- to-300 ohm transformer, known as a balun (derived from balanced- unbalanced), must be used in the transition from the coax to the twin- screw VHF terminals. The balun matches the impedance between the coaxial cable and the television receiver input to allow maximum signal power to be transferred.

DISTRIBUTION SYSTEM DESIGN

D. MIDBAND and HIGHER FREQUENCY DISTRIBUTION SYSTEMS

Many SMATV systems distribute more than 12 channels to subscribers. This requires use of either the midband, superband, hyperband or UHF channels in addition to the twelve available VHF slots. In general, the midband channels are first considered because cable, splitter, tap and other losses have similar low values as those encountered in VHF system. Midband headends are also quite similar in design and operation to those used for either low-band or high-band VHF channels. Once the 9 midband channels have been occupied, 14 and 23 superband and hyperband frequencies, respectively, are available. The 40 additional UHF slots, while rarely used in North American SMATV distribution systems, are more common in some European systems. (See Appendix E and Table 6-5 for a complete listing of all channel frequency assignments).

Off-air television stations can broadcast signals on licensed frequencies that include the VHF-low, VHF-high and UHF-bands. Other adjacent frequencies are specifically set aside for mobile radio and navigation communicators. (Note that the 88 to 108 GHz FM band contains channels that are 200 kHz wide and centered on 100 kHz spacings.) However, when signals are transmitted over shielded coaxial cables, egress interference can be minimized. So a number of channels not previously available for off-air broadcasting including mid, super and hyperbands have been allocated for use in CATV and SMATV systems (see Figure 6-16 and Table 6-5).

Mid, Super and Hyperband Distribution

Distributing broadcast signals in the mid, super and hyperbands has its advantages and disadvantages. The most obvious benefit is that 46 extra channels are available. This number is usually more than enough for any SMATV system. In addition, no modifications in headend or distribution designs or special components are necessary when using the midband channels. However, as the frequency increases into the mid and superbands, losses become greater and design techniques as described in the next section are required.

The major disadvantage in using these bands is the need for either CATV-type converters or cable TV-ready televisions. Most standard sets are only equipped to tune in VHF and UHF channels. Using converters is an extra expense and can reduce a system's reliability. In addition, converters do not necessarily interface with the existing remote control capabilities built into standard televisions.

Midband distribution systems can also suffer from interference caused by second-order distortions generated in older line amplifiers. Such spurious signals can be created when the sum frequencies of low-band VHF channels are amplified. For example, the channel 5 plus channel 6 video frequency equals 160.5 MHz which falls within midband channel G (converter channel 20). When channel G is used, interference caused by beats between the sum signal and the channel G signal can cause picture distortions. In a new SMATV installation, such interference can generally be avoided by careful design. In such cases, an alternative is to simply replace all older amps with new push-pull amplifier technology in which second order products are not a concern.

UHF Distribution

When distributing UHF channels, two factors must be considered. First, UHF tuners are not as sensitive to input signals as VHF tuners. UHF input signals to televisions should be a minimum of +5 dBmV in order to achieve professional quality pictures. Second, losses in coaxial cables, taps, splitters and other passive devices are substantially higher. Therefore, signals must either be amplified more often during distribution or be launched from the headend with correspondingly more power than normally the case in VHF distribution (Figure 6-17).

For example, the losses in one type of RG-59/U foam-dielectric cable at VHF channel 2, VHF channel 13 and UHF channel 14 are 1.8, 3.6 and 5.5 dB per 100 feet, respectively. Therefore, if a 1000 foot cable run were being used in the distribution system, the signal would have to be launched 37 dB higher on UHF 14 than VHF 2. At the UHF frequen-

DISTRIBUTION SYSTEM DESIGN

TABLE 6-5. NORTH AMERICAN CATV FREQUENCY ALLOCATIONS

Channel Number/ Letter	Bandwidth (MHz)	Carrier Frequencies (MHz) Video	Color	Audio
Sub-Band				
74/T-7	5.75-11.75	7.00	10.58	11.50
75/T-8	11.75-17.75	13.00	16.58	17.50
76/T/-9	17.75-23.75	19.00	22.58	23.50
77/T-10	23.75-29.75	25.00	28.58	29.50
78/T-11	29.75-35.75	31.00	34.58	35.50
79/T-12	35.75-41.75	37.00	40.58	41.50
80/T-13	41.75-47.75	43.00	46.58	47.50
VHF-Low Band				
2/2	54-60	55.25	58.83	59.75
3/3	60-66	61.25	64.83	65.75
4/4	66-72	67.25	70.83	71.75
5/5	76-82	77.25	80.83	81.75
6/6	82-88	83.25	86.83	87.75
FM Band				
FM-1	88-94	89.25	92.83	93.75
FM-2	94-100	95.25	98.83	99.75
FM-3	100-106	101.25	104.83	105.75
VHF-Mid Band				
69/A-5	90-96	91.25	94.83	95.75
70/A-4	96-102	97.25	100.83	101.75
71/A-3	102-108	103.25	106.83	107.75
72/A-2	108-114	109.25	112.83	113.75
73/A-1	114-120	115.25	118.83	119.75
14/A	120-126	121.25	124.83	125.75
15/B	126-132	127.25	130.83	131.75
16/C	132-138	133.25	136.83	137.75
17/D	138-144	139.25	142.83	143.75
18/E	144-150	145.25	148.83	149.75
19/F	150-156	151.25	154.83	155.75
20/G	156-162	157.25	160.83	161.75
21/H	162-168	163.25	166.83	167.75
22/I	168-174	169.25	172.83	173.75
VHF-High Band				
7/7	174-180	175.25	178.83	179.75
8/8	180-186	181.25	184.83	185.75
9/9	186-192	187.25	190.83	191.75
10/10	192-198	193.25	196.83	197.75
11/11	198-204	199.25	202.83	203.75
12/12	204-210	205.25	208.83	209.75
13/13	210-216	211.25	214.83	215.75
VHF-Super Band				
23/J	216-222	217.25	220.83	221.75
24/K	222-228	223.25	226.83	227.75
25/L	228-234	229.25	232.83	233.75
26/M	234-240	235.25	238.83	239.75
27/N	240-246	241.25	244.83	245.75
28/O	246-252	247.25	250.83	251.75
29/P	252-258	253.25	256.83	257.75
30/Q	258-264	259.25	262.83	263.75
31/R	264-270	265.25	268.83	269.75
32/S	270-276	271.25	274.83	275.75
33/T	276-282	277.25	280.83	281.75
34/U	282-288	283.25	286.83	287.75
35/V	288-294	289.25	292.83	293.75
36/W	294-300	295.25	298.83	299.75
Hyperband				
37/AA	300-306	301.25	304.83	305.75
38/BB	306-312	307.25	310.83	311.75
39/CC	312-318	313.25	316.83	317.75
40/DD	318-224	319.25	322.83	323.75
41/EE	324-330	325.25	328.83	329.75
42/FF	330-336	333.25	334.83	335.75
43/GG	336-342	337.25	340.83	341.75
44/HH	342-348	343.25	346.83	347.75
45/II	348-354	349.25	352.83	353.75
46/JJ	354-360	355.25	358.83	359.75
47/KK	360-366	361.25	364.83	365.75
48/LL	366-372	367.25	370.83	371.75
49/MM	372-378	373.25	376.83	377.75
50/NN	378-384	379.25	382.83	383.75
51/OO	384-390	385.25	388.83	389.75
52/PP	390-396	391.25	384.83	395.75
53/QQ	396-402	397.25	400.83	401.75
54/RR	402-408	403.25	406.83	407.75
55/SS	408-414	409.25	412.83	413.75
56/TT	414-420	415.25	418.83	419.75
57/UU	420-426	421.25	424.83	425.75
58/VV	426-432	427.25	430.83	431.75
59/WW	432-438	433.25	436.83	437.75
60/XX	438-444	439.25	442.83	443.75
61/YY	444-450	445.25	448.83	449.75
62/ZZ	450-456	451.25	454.83	455.75
63/AAA	456-462	457.25	460.83	461.75
64/BBB	462-468	463.25	466.83	467.75
65/CCC	468-474	469.25	472.83	473.75
66/DDD	474-480	475.25	478.83	479.75
67/EEE	480-486	481.25	484.83	485.75
68/FFF	486-492	487.25	490.83	491.75

DISTRIBUTION SYSTEM DESIGN

TABLE 6-5 continued... NORTH AMERICAN CATV FREQUENCY ALLOCATIONS

Channel Number	Bandwidth (MHz)	Carrier Frequencies (MHz) Video	Color	Audio	Channel Number	Bandwidth (MHz)	Carrier Frequencies (MHz) Video	Color	Audio
14	470-476	471.25	474.83	475.75	49	680-686	681.25	684.83	685.75
15	476-482	477.25	480.83	481.75	50	686-692	687.25	690.83	691.75
16	482-488	483.25	486.83	487.75	51	692-698	693.25	696.83	697.75
17	488-484	489.25	592.83	593.75	52	698-704	699.25	702.83	703.75
18	494-500	495.25	498.83	499.75	53	704-710	705.25	708.83	709.75
19	500-506	501.25	504.83	505.75	54	710-716	711.25	714.83	715.75
20	506-512	507.25	510.83	511.75	55	716-722	717.25	720.83	721.75
21	512-518	513.25	516.83	517.75	56	722-728	723.25	726.83	727.75
22	518-524	519.25	522.83	523.75	57	728-734	729.25	732.83	733.75
23	524-530	525.25	528.83	529.75	58	734-740	735.25	738.83	739.75
24	530-536	531.25	534.83	535.75	59	740-746	741.25	744.83	745.75
25	536-542	537.25	540.83	541.75	60	746-752	747.25	750.83	751.75
26	542-548	543.25	546.83	547.75	61	752-758	753.25	756.83	757.75
27	548-554	549.25	552.83	553.75	62	758-764	759.25	762.83	763.75
28	554-560	555.25	558.83	559.75	63	764-770	765.25	768.83	769.75
29	560-566	561.25	564.83	565.75	64	770-776	771.25	774.83	775.75
30	566-572	567.25	570.83	571.75	65	776-782	777.25	780.83	781.75
31	572-578	573.25	576.83	577.75	66	782-788	783.25	786.83	787.75
32	578-584	579.25	582.83	583.75	67	788-794	789.25	792.83	793.75
33	584-590	585.25	588.83	589.75	68	794-800	795.25	798.83	799.75
34	590-596	591.25	594.83	595.75	69	800-806	801.25	804.83	805.75
35	596-602	597.25	600.83	601.75	70	806-812	807.25	810.83	811.75
36	602-608	603.25	606.83	607.75	71	812-818	813.25	816.83	817.75
37	608-614	609.25	612.83	613.75	72	818-824	819.25	822.83	823.75
38	614-620	615.25	618.83	619.75	73	824-830	825.25	828.83	829.75
39	620-626	621.25	624.83	625.75	74	830-836	831.25	834.83	835.75
40	626-632	627.25	630.83	631.75	75	836-842	837.25	840.83	841.75
41	632-638	633.25	636.83	637.75	76	842-848	843.25	846.83	847.75
42	638-644	639.25	642.83	643.75	77	848-854	849.25	852.83	853.75
43	644-650	645.25	648.83	649.75	78	854-860	855.25	858.83	859.75
44	650-656	651.25	654.83	655.75	79	860-866	861.25	864.83	865.75
45	656-662	657.25	660.83	661.75	80	866-872	867.25	870.83	871.75
46	662-668	663.25	666.83	667.75	81	872-878	873.25	876.83	877.75
47	668-674	669.25	672.83	673.75	82	878-884	879.25	882.83	883.75
48	674-680	675.25	678.83	579.75	83	884-890	885.25	888.83	889.75

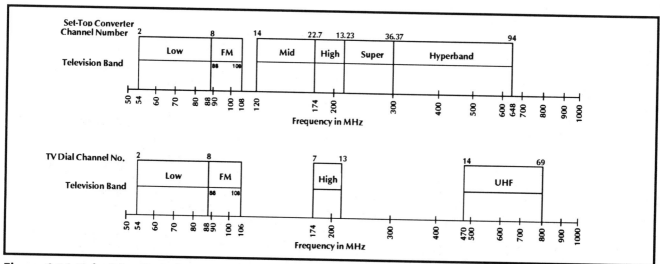

Figure 6-16. Television Broadcast Frequency Allocations. *The bottom diagram illustrates the allowed off-air frequencies. However, many more channels in the mid, super and hyperbands can transmitted over cable networks. Of course, cable systems also have access to UHF channels.*

DISTRIBUTION SYSTEM DESIGN

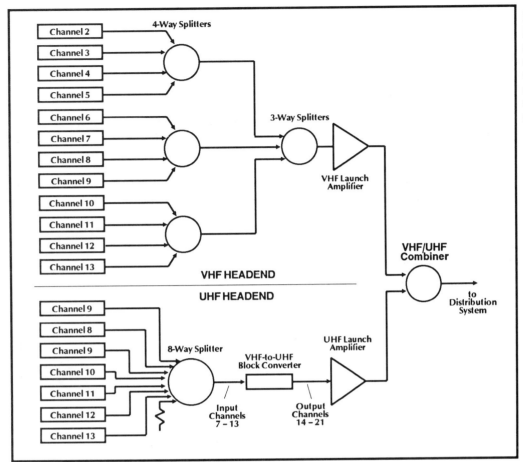

Figure 6-17. UHF/VHF Headend. *This combination UHF/VHF headend uses crystal-controlled, block upconverters to feed UHF channels into a UHF launch amplifier*

cies the losses as well as the cable tilt would be even higher. While such tilt and losses would be unacceptable in CATV networks, they can be accommodated in smaller private cable systems.

Splitters and directional couplers that manage both VHF and UHF frequencies are available. However, as would be expected, losses are slightly higher in the UHF band. For example, the attenuation in a 2-way splitter increases from 3.5 dB at VHF to 4.0 at UHF. A 12 dB directional coupler has 12 dB loss at VHF but 13 dB at UHF. These losses must be factored into all design calculations. In the above example, if 10 directional couplers had been used, the headend would have had to generate an additional 10 times 1.0 dBmV which equals 10 dBmV of signal.

Three more important design changes must be considered in those rare cases where UHF distribution systems are being installed. First, most line amplifiers are specified for operation up to a maximum of 250 to 300 MHz. However, amplifiers are available that function at over 500 MHz. The simplest design would be to combine outputs from two separate UHF and VHF launch amps. Second, UHF modulators tend to be quite expensive. However, some manufacturers offer crystal-controlled VHF-to-UHF block upconverters that can be used to replace modulators. Third, a simple matching transformer that generates separate UHF and VHF outputs can be used at television sets. This device adds an additional loss of 1.0 and 2.5 dB attenuation for the VHF and UHF signals (see Figure 6-18).

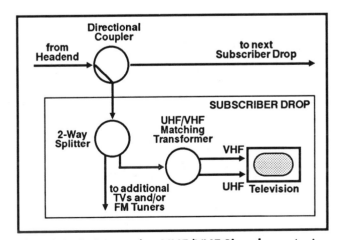

Figure 6-18. Separating UHF/VHF Signals. *A simple UHF/VHF matching transformer can be used to separate signals in both frequency bands for delivery to a television receiver.*

DISTRIBUTION SYSTEM DESIGN

Most private cable engineers generally argue against using UHF distribution systems because of the higher losses and the need for higher power, specialized line amplifiers, splitters and other passive components. They claim that more than enough frequency space is available in the other bands and that a UHF system will generally have both higher initial costs and maintenance expenses.

Satellite IF Distribution

In some situations it is economical to directly distribute the LNB frequency modulated IF output of a satellite reception system. This signal, that falls in the 950 to 1450 MHz range in North America and 950 to 1700 MHz range in Europe, contains twelve or more channels. Each customer would be supplied with a satellite receiver in order to tune to these channels. If channels of both polarities are to be received, an orthomode feed and two LNBs would be hooked into a two-cable distribution network.

IF signal distribution is feasible when a limited number of customers in a multi-unit environment wish to view satellite television transmitted from a single satellite. For example, in Europe each high powered Ku-band Astra satellite can potentially relay up to 48 channels. One 2 to 3 foot antenna can then serve all customers, each of whom uses a block downconversion satellite receiver in place of the standard cable converter. This arrangement avoids the need for multiple, often unsightly antennas.

IF distribution systems have a number of inherent advantages as well as disadvantages. Some of these are listed below. Note that most of the design considerations discussed in the previous section on UHF distribution apply to IF systems.

- When the number of subscriber drops is small, typically less than 50 units, costs can be dramatically lower than in systems where a full-scale SMATV headend must be installed. For example, in a 10-unit building only ten home-grade satellite receivers are required to receive 31 channels now available on Astra. In contrast, a standard SMATV headend would need 31 commercial quality receivers to achieve the same result.
- IF systems suffer the inherent disadvantage of being required to manage relatively high coaxial cable losses. Attenuation in the satellite IF band can be over 5 times that in the VHF television band.
- Connector, splitter and other passive losses are also somewhat greater in the higher frequency IF range.
- While all the standard rules of channel load, distortion and system loss calculations apply, IF systems have the advantage of distributing FM signals. This modulation format provides an inherently immunity to noise compared to AM systems, in large part because more bandwidth is used for each channel transmitted.
- This system is the perfect bulk billing situation. Subscribers lease or purchase a satellite receiver, pay a monthly connection cost and are billed directly by programmers who charge for their service.

Two typical customer drops are shown in Figure 6-19. Amplifiers, splitters, taps and other components in this system must be rated for operation over the entire range in use. Thus, for example, if off-air VHF-low signals are being combined with an LNB output in the 950 to 1450 range, equipment must be rated for the entire 50 to 1450 MHz range.

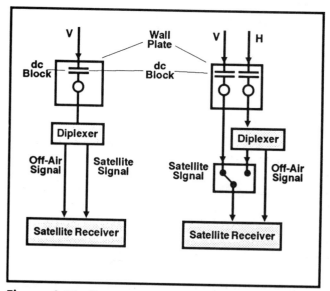

Figure 6-19. Customer Drop Configurations in IF Distribution System. *This diagram illustrates two types of customer drops that can be used on the output of an IF distribution system. The set-up on the left receives satellite channels on one polarity. A diplexer separates the off-air from satellite signals. In the configuration on the right a diplexer also separates satellite from off-air signals. Then a H/V switch controlled by the satellite receiver switches upon command between horizontal and vertical polarity signals.*

DISTRIBUTION SYSTEM DESIGN

E. SAMPLE SYSTEM DESIGNS and CALCULATIONS

A number of hypothetical situations are studied in this section beginning with rather simple examples but ending with real-world complex problems. The principle task in these examples is to calculate all the losses in the distribution systems so that the signal level required at the headend can be determined, to ensure that adequate power reaches all televisions. Tables 6-1, 6-2, 6-3 and 6-4, that give representative values for cable, splitter and tap losses, are used throughout these calculations.

Before designing an actual SMATV distribution system, numerous questions must be addressed. The number and frequency of channels to be carried must first be determined. All mechanical features of the system must be planned. These include the location of antennas and the headend, the layout and distances of cable runs, the placement of drops and outlets, as well as the location and number of splitters, taps, amplifiers and other components. In some cases, a system may be designed to be CATV compatible in preparation for a future connection to the local cable TV network or for an outright buy-out.

Example I – The Basic System

The first project is a small system having two distribution legs (see Figure 6-20). RG-59 coax cable with a loss of 4.5 dB per hundred feet has been chosen for the distribution system. Line A uses a total of 60 feet of coax. This type of cable has 4.2 dB of loss per 100 feet at channel 13 frequencies. Therefore, the total attenuation for line A is 60/100 times 4.2 dB which equals 2.5 dB. Similarly, line B has a total length of 180 feet and therefore an attenuation of 7.6 dB. There is only one splitter in the system located at the headend output. This two-way device splitter generates a loss of 3.5 dB in each leg.

In general, it is common to use a median value of 0.7 dB to determine tap insertion losses when the tap values have not been exactly specified. This number is an average value of insertion losses representing those for a 17 dB tap. Therefore, on line A there are 3 taps with a total insertion loss of 2.1 dB. Line B has 7 taps with a total insertion loss of 3.5 dB. If the precise values of these taps were known, a

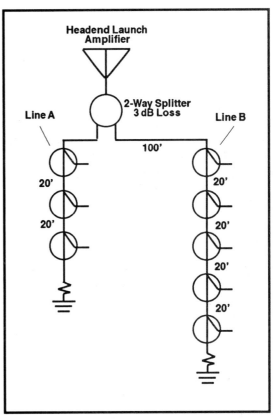

Figure 6-20. Example I and II. *This diagram is explained in the text of Examples I and II.*

closer estimate of total tap attenuation in each line could easily be calculated.

The last line tap must clearly be the lowest value tap in the system because, at this point, the signal level is at its lowest point. This device is assumed to be an 8 dB tap, although there are even lower value taps available from some manufacturers. On the basis of this assumption, the last tap on the line has an isolation value of 8 dB.

The total system losses are now easily calculated:

LOSSES	LINE A	LINE B
Cable loss	2.5 dB	7.6 dB
Splitter loss	3.5 dB	3.5 dB
Insertion loss	2.1 dB	3.5 dB
Isolation loss	8.0 dB	8.0 dB
Total loss	16.1 dB	22.6 dB

DISTRIBUTION SYSTEM DESIGN

In this case, the losses for line B are greater than those for line A. This clearly suggests that when figuring system losses, the total attenuation for the longest line on the system should be first calculated as the worst possible case. If enough signal is available to reach the end of the line with the greatest attenuation, than sufficient power will be present for all the other lines.

The total loss on line B is 22.6 dB. Therefore, at least 22.6 dBm must originate from the headend to insure that the last tap receives at least 0 dBm of signal. If a 10 dBm output were desired at the last tap, the headend output would have to be increased by 10 dB to 32.6 dBm. Most systems designers figure in an extra signal margin of 6 dB in selecting the headend output to cover any minor design variances. Remember, it is easier to deal with too much signal than not enough.

Example II – Calculating Tap Values

Most SMATV systems incorporate a headend having a known output. This typically ranges from 45 to 60 dBmV depending on the variety of equipment selected and construction practices followed. For purposes of this discussion, the same example as above is used to calculate tap values. The assumption is that 10 dBmV is required at each tap and the headend output is 45 dBmV.

Tap values can be calculated by adding in the losses from tap to tap and by choosing the tap value that yields the desired output (see Figure 6-20). The 45 dBmV headend signal first enters a two-way splitter with a resulting loss of 3.5 dB. The 41.5 dBmV output then travels down 20 feet of cable on line A. This contributes an extra 0.9 dB of attenuation. Therefore, 41.5 dBmV less 0.9 dB or 40.6 dBmV is available to the first tap-off on line A. By using a 29 dB tap, 40.6 dBmV less the isolation value of 29 dB which equals 11.6 dBmV is available its outlet. The 29 dB tap has an insertion loss of 0.3 dB, so that 40.3 dBmV exits from the port on the first tap.

The 40.3 dBmV from the port on tap one then passes through another 20 feet of cable with a resulting loss of 0.9 dB so that 39.4 dBmV enters the second tap. Using another 29 dB tap-off yields a tap output of 10.4 dBmV to the television set. Following this tap insertion loss of 0.3 dB an output of 39.1 dBmV is available.

The 39.1 dBmV signal then passes through an additional 20 feet of cable, again with a loss of 0.9 dB, resulting in a 38.2 dBmV input into tap three. The next 26 dB tap yields a 12.2 dBmV output to the television. This tap has an insertion value of 0.4 dB, resulting in an output of 37.8 dBmV. The end of this cable line is terminated here. However, there is still sufficient signal left to power a longer length of cable and taps.

By using this same procedure, all tap values for line B can similarly be calculated. Remember the design constraint is that a minimum of 10 dBmV must be available at the output of each tap. If an exact solution is not possible, the tap that gives the value closest to, but not under, 10 dBmV output should be used. As an exercise, these calculations can be performed and answers checked against the solutions in Table 6-6.

TABLE 6-6. SOLUTIONS to EXAMPLE II TAP VALUE PROBLEM

TAP #	LINE A INPUT LEVEL (dBmV)	TAP VALUE (dB)	OUTPUT VALUE (dBmV)
1	37.0	26	11.0
2	35.7	23	12.7
3	34.4	23	11.4
4	33.1	20	13.1
5	31.4	20	11.4

TAP #	LINE B INPUT LEVEL (dBmV)	TAP VALUE (dB)	OUTPUT VALUE (dBmV)
1	31.5	20	11.5
2	26.2	14	12.2
3	43.5	32	11.5
4	38.7	26	12.7
5	30.2	20	10.2

DISTRIBUTION SYSTEM DESIGN

Example III – Distribution Amplifiers and Gain

The previous example involved a very small system with all power provided by output from the headend. In many cases, the system is large enough to require more signal than the headend can generate. Then amplifiers may be required to boost the signal for effective distribution.

This system has relatively long cable runs between taps (see Figure 6-21). As a result, after the signal has passed through the second tap, there is insufficient power to pass through a third tap and still provide the required 10 dBmV output level. The signal level at this point is decreased by the 31.5 dB cable attenuation in 600 feet of coax. Therefore, the signal must be amplified at this point, the input of tap 3. The line amp contributes 30 dB gain to its 13.5 dBmV input signal yielding a 43.5 dBmV signal at the input to tap 3. Note that the input to this amplifier is within the typical 15 dBmV specification so that overloading is avoided.

Example IV – 32-Unit High-Rise Apartment Building

This sample system is designed for an 8-story high-rise having four rooms on each floor. The layout and cable footage are illustrated in Figure 6-22. The assumptions are:

- Although a minimum signal level of 0 dBmV is required for snow-free pictures, a full 10 dBmV is made available to provide for a margin of safety if, for example, a customer wishes to split the signal for VCRs or other in-home components.

- RG-59 foam dielectric coax having an attenuation at channel 13 of 4.0 per 100 feet is selected. This type of cable is considered to have the best cost efficiency for such medium sized VHF distribution systems.

- The amount of signal loss between the customer's TV and the wall tap depends upon the length of cable and whether or not a matching transformer is required. The worst case is assumed. A transformer loss of 0.5 dB and 10 feet of cable loss of 0.4 dB for a total of 0.9 between the wall tap and TV.

- This system uses surface mount wall taps that fit into standard 2 x 4 inch wall boxes. Taps values, available from their manufacture are:

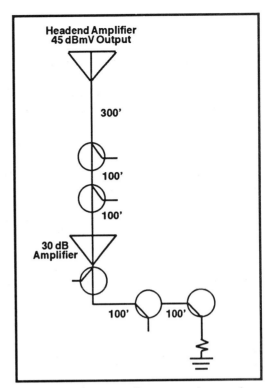

Figure 6-21. Example III. *See the text for a description of this example.*

	Model Number		
	DC10	DC14	DC19
VHF Isolation (dB)	10	14	19
VHF Insertion (dB)	1.5	0.6	0.5

There is enough information in hand to design the distribution system. The technique of working backwards from the outlet having the greatest total loss is used here. In this case, all four vertical riser paths are equivalent. In general, the most efficient design strategy is to follow the line having the highest total attenuation.

The last tap used has the least amount of isolation or tap loss. This tap must have its throughline output port terminated with a 75 ohm terminator, a close-ended F-connectors having an internal 75 ohm resistor. Taps having the throughline already terminated are usually designated with a "T" suffix.

Each television must receive at least 10 dBmV plus the 0.9 dB lost between the TV and wall tap. Therefore, the amount of signal required at the last tap equals 10.9 dBmV plus the lowest tap isolation value of 10 dB or a total of 20.9 dBmV. The output signal required at the previous tap is 20.9 dBmV plus

DISTRIBUTION SYSTEM DESIGN

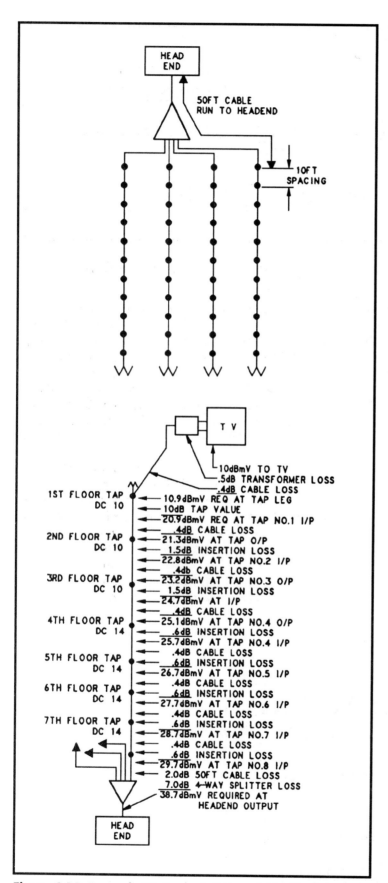

Figure 6-22. Example IV. *Example IV is explained in the accompanying text.*

the 0.4 dB cable loss in the short run between the two taps. The signal entering this 10 dB tap equals the previous sum, 21.3 dBmV, plus its insertion loss of 1.5 dB. The signal entering the second of last tap is then 22.8 dBmV. Following the 10 dB isolation loss, the second of last drop 12.8 dB.

These calculations are followed in this fashion all the way to the top of the riser. This figure is drawn upside-down down to facilitate the process. Notice that when the signal level out of the next tap becomes higher than the sum of room signal losses plus the level required at the TV, this is the sign that the next higher tap value should be used. For example, if a DC10 and not a DC14 tap had been used on the fourth floor, 15.7 dBmV signal output would be too great for the television to handle and could result in overdriving and distortion. However, by using a DC14 tap having a 0.6 dB insertion loss, the signal level becomes an acceptable 11.7 dBmV. Note that it is generally better to be a little above design requirements than a little below.

The output level required at the headend to adequately power all televisions can be found by continuing this process and then adding in the splitter loss of 7 dB and attenuation in the final cable run. The result is that 38.7 dBmV is required at the distribution amplifier output. It is not necessary to add in losses in any of the other three risers. Each riser is identical here but, in any case, other routes are independent of each other.

These calculations were completed for signal losses at channel 13 frequencies. However, all coaxial cables exhibit tilt. Losses for RG-59 foam of 2 dB per 100 feet at channel 2 are less at the 4 dB per 100 feet for channel 13 than at the other end of the VHF spectrum. Since the total cable length is 130 feet, losses throughout the length of the cable of 2.6 dB (1.3 times 2) at channel 2 are 2.6 dB less than the 5.2 dB (1.3 times 4) attenuation at channel 13. The last television will therefore receive 2.6 dB more on channel 2 than the 10 dBmV present on channel 13. This excess signal does not really present a problem since 12.6 dBmV is not enough to overload a television set. The differences between channels 2 and 13 signal strengths progressively decrease for those outlets closer to the headend.

DISTRIBUTION SYSTEM DESIGN

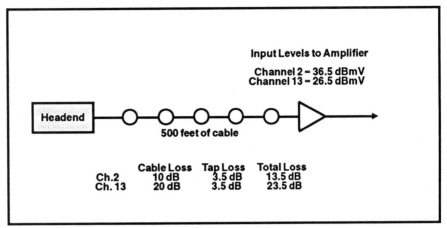

Figure 6-23. Example V. *The details of example V are explained in the text.*

Example V – Distribution Amplifiers and Tilt

In this example a headend is feeding a simple distribution system that has one line amplifier (see Figure 6-23). Signal levels at the headend for both channels 2 and 13 are 50 dBmV. Five hundred feet of cable and 5 tap-offs lie between the headend and the amplifier. The tap insertion losses are 0.7 dB for a total of 3.5 dB at both frequencies. But cable losses of 10 and 20 dB for channels 2 and 13, respectively, do vary with frequency.

Simple subtraction shows that strengths of the channel 2 and 13 signals at the amplifier input are 13.5 and 23.5 dB. Therefore, there is 10 dB of tilt. Note that this is the same value as the differences in cable attenuation because insertion losses of passive taps are not frequency dependent.

The common strategy for managing tilt is to set the output of the launch amp by 3 to 6 dB higher on the higher frequency channels prior to insertion into the distribution system. When the signal arrives at the subsequent amp, an equalizer is used to create a flat level into to this amp. Again the second amplifier's output is tilted by 3 to 6 dB.

Example VI – Noise in Cascaded Line Amps - Part I

This simple example is designed to demonstrate how the signal-to-noise (S/N) ratio is effected by cascading distribution amplifiers (see Figure 6-24). Some of the concepts here have been explored earlier in the Section C of Chapter 5. Note that this section is somewhat more detailed than others and can be skipped on the basis of its conclusion that cascading amplifiers generally have little effect on S/N ratio in small to moderately sized private cable systems.

The headend generates an output of 50 dBmV with a S/N ratio of 40 dB. The noise power leaving the headend is therefore 50 dBmV less 40 dB which equals 10 dBmV. The first amplifier is spaced after 30 dB of cable attenuation so that it has an input signal of 20 dBmV. The original 50 dBmV is then restored by its 30 dB gain. The situation is apparently very simple. But what happens to the noise level and S/N ratio? Assume that the amplifier has a noise figure of 16 dB which is quite typical.

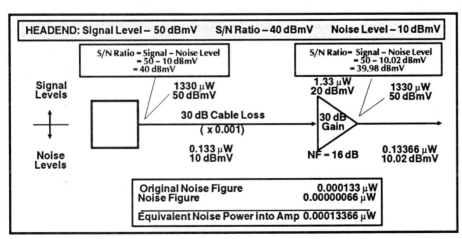

Figure 6-24. Example V. *The precise calculations to determine S/N ratio are outlined in the text.*

DISTRIBUTION SYSTEM DESIGN

The most important element to understand in the following calculations is that the noise entering the first line amp adds to that generated internally in the amplifier. Decibels are added or subtracted to show relative changes in signal power or voltage levels, i.e. multiplying or dividing levels. However, this notation cannot be used when powers are added together. Therefore, the actual noise powers in microwatts must first be determined.

A simple calculation using Ohm's Law shows that when a voltage of 1 mV is applied to a 75 ohm resistor, the power generated is 0.0133 microwatts. Therefore, the noise level in dBmV is expressed relative to the power generated at a voltage of 1 mV in the following equation:

Noise in dBmV = 10 log (Power in µvolts/0.0133)

or by turning this equation inside out:

Noise power in µwatts = $0.0133 \times 10 e^{\text{Noise level}/10}$

Now the noise power at the input to the line amplifier can easily be found. The noise power at the output of the headend is 10 dBmV which translates to the following value in microwatts:

Noise power = $0.0133 * 10 e^{10 \text{dBmV}/10}$
= 0.133 microwatts

The 30 dB of cable loss also reduces the noise power by 30 dB or a factor of 0.001. Therefore, the noise power entering the amplifier is 0.000133.

Next the noise power added by the amplifier must be calculated. The equivalent noise figure of the line amp is given by –59 dBmV plus 16 dB or –43 dBmV. This number, equals the –59 dBmV input noise of a standard TV channel plus the amplifier noise figure. Now the noise added by the line amp can be calculated:

Noise power added by amp = $0.0133 \times 10 e^{-43/10}$
= 6.6×10^{-8} µwatts

The total noise power at the amplifier input is the sum of the input noise power plus the noise power added by the amplifier:

Input noise power = 0.000133 + 0.00000066
= 1.3366×10^{-5} µwatts

This is amplified by 30 dB or multiplied by a factor of 1000 to a final value of 0.13366 microwatts. Therefore the noise level at the output of the amp is:

Noise level in dBmV = 10 log (0.13366/0.0133)
= 10.02 dBmV

The S/N ratio at the amplifier output is then the 50 dBmV signal less the noise level of 10.02 dBmV. This equals 39.98 dB which is only slightly below that of the S/N ratio output at the headend of 40 dB. The conclusion is clear. Once an adequate S/N ratio has been set by the headend, the cascading of distribution amplifiers will not greatly affect its value.

Example VII - Distortion in Cascaded Distribution Amplifiers

Given the distortion specifications of an amplifier, both the number of channels that can be managed and the output level can be chosen. For example, if a broadband amplifier lists an output capability of 54 dBmV with a cross mod distortion of –46 dB when 7 channels are distributed, what would the output have to be in order to carry 14 channels at the same level of distortion? What would the output level have to be to carry this increased number of channels at –50 dB cross mod?

As the number of channels carried increases, so does the distortion if the output level is kept constant. However, if the output level is reduced, the same amount of distortion can result when distributing a greater number of channels. The rule of thumb states that if the number of channels is doubled, the output power must be reduced by 3 dB to maintain the same cross mod level. Therefore, if the output is reduced to 51 dBmV, then 14 channels can be distributed with cross mod products at –46 dB relative to the carrier.

The cross modulation also decreases by 2 dB for every 1 dB increase in amplifier gain. Therefore, to reduce the distortion to –50 dB, the gain has to decrease by 2 dB. The output signal also drops by this amount to become 49 dBmV. A similar rule holds for composite triple beat which decreases (increases) 3 dB for every 1 dB drop (increase) in gain.

Example VIII – Distortion in Cascaded Line Amps - Part II

In this example, a medium sized system that is typical of a trailer park or condominium complex is examined. This system at 300 pads is large enough to require CATV-type distribution amplifiers. Cable is laid via underground direct burial cables with taps and amplifiers mounted in pedestals. The headend has been designed to generate an output level of 50 dBmV at an S/N ratio of 50 dB into four equal distribution legs, each operating at an S/N ratio of 46 dB.

A full 6 dBmV is delivered to each outlet with an S/N ratio of 46 dB.

A simple formula can be used to determine the required input signal to a series of cascaded amplifiers having equal noise figures and inputs so that the desired S/N ratio is attained:

$$\text{Signal In} = S/N - 59 + NF + 10 \log N$$

where Signal In is the required input to the each amplifier, NF is the noise figure and N is the number of amplifiers. So if these two cascaded amplifiers have noise figures of 16 dB, in order to achieve the desired S/N ratio of 46 dB, the required input signal is given by:

$$\text{Signal In} = 46 - 59 + 16 + 10 \log 2$$
$$= 6 \text{ dBmV}$$

This is the lowest input level that will produce the required 46 dB S/N ratio. If the input power is increased, the S/N ratio will increase but so will the amount of signal distortion. If these particular amplifiers had gains of 25 dB and output capabilities of 45 dBmV carrying 12 channels and producing -57 dB cross mod, the 6 dB input level would be sufficient to produce a 31 dBmV output having reasonably low levels of distortion and an acceptably higher S/N ratio. If the input signals were increased, for example, by 5 dB to 11 dBmV, the S/N ratio would increase to 50 dB and the output would still be an acceptable 36 dBmV.

Note that this formula is only correct when all amplifiers are operating at the same input level and are subjected to the same amount of input cabling attenuation. If the spacing were uneven, the power method as outlined earlier in Example VI would have to be used.

A simple method is also available to calculate the S/N ratio reaching the drops given the headend output S/N ratio and the S/N ratio in each leg. First, the difference between these two S/N ratios is taken. Then the 4 dB point on the x-axis of the graph in Figure 6-25 is used to find the intersection point on the data reference line. This corresponds to 1.5 dB. This value is then subtracted from the distribution leg S/N ratio of 46 dB to arrive at the final 44.5 dB S/N ratio reaching the drops.

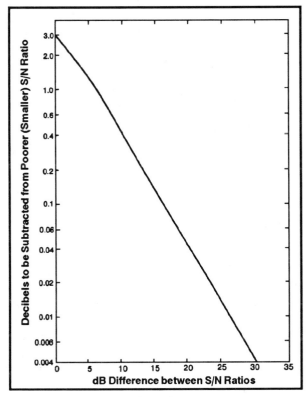
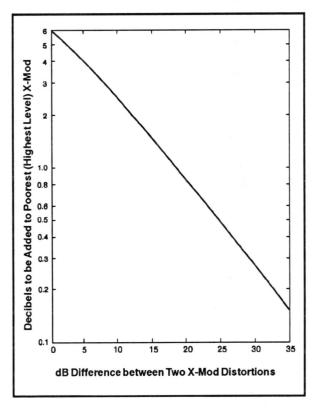

Figure 6-25. Combining Unequal S/N Ratios and Cross Modulation Distortions. *The use of these graphs in calculating the combined effects of unequal S/N ratios and levels of cross modulation distortions is explained more thoroughly in the text of Example VIII.*

DISTRIBUTION SYSTEM DESIGN

F. 18 GHz MICROWAVE LINE-OF-SIGHT TRANSMISSION SYSTEMS

There is often a need to distribute the broadband signal from a central headend to a remote location in the operation of both SMATV and wireless systems. This may occur when, for example, an SMATV operator secures a contract to serve a building complex that is located in the same area as an existing system and it would be most economical to avoid building a second headend. In such cases, a number of technical options are available. One method is to hard wire buildings via coax cable. In the United States this may involve obtaining a CATV franchise if the cable route crosses city rights-of-way. This approach is usually impractical if the two systems are separated by more than a city block or two. An alternative low cost option is the use of microwave or laser links. The use of laser links for relatively short hops is explored in the following section.

Applications

Microwave and laser links can transmit broadband signals from one central headend to other locations within a 10 to 15 mile radius. They allow an operator to serve multiple systems with a minimum investment in headend components while reducing operational and maintenance costs. They also provide a method for circumventing franchise restrictions that prevent an SMATV distribution system from crossing public easements, highway and railroad lines.

Microwave relays are used extensively in wireless systems to relay signals form various educational facilities to the transmit site. These studio transmission links (STLs) are often used by operators who lease excess ITFS capacity from educational broadcasters. Most of these facilities have production studios that are located some distance from the transmit site. STL links that operate in the 12 GHz CARS frequency band are the most common link (see Table 6-7). These systems typically use single channel FM modulation to achieve superior S/N ratio and greater range, up to 15 miles or more. FM modulated systems can successfully detect much lower signal levels. Since most ITFS interconnects relay just one to a maximum of four channels, the additional cost of the single channel CARS band amps are not excessive as would be the case if many channels were linked like in a broadband 18 GHz system.

18 GHZ microwave relays can also be used to relay signals from a remote satellite receive antenna location to a central SMATV headend or MMDS transmission site. This may be necessary, for example, if the headend or wireless transmitter is located on top of a building or another location where terrestrial interference is severe, or if antenna and/or equipment space is limited at the site. While 18 GHz links are less common than CARS band systems, they are the only form of line-of-sight relays other than laser links that can be used to connect SMATV headends in the United States.

TABLE 6-7. CARS BAND FREQUENCY ALLOCATIONS

CARS Group	Channel Range Number	Channel Range Frequency	Bandwidth (MHz)
A	A1-A20	12700.0 – 13200.0	25.0
B	B1-B19	12712.5 – 13187.5	25.0
C	C1-C42	12700.5 – 12946.5	5.0
D	D1-D42	12759.7 – 13005.7	6.0
E	E1-E42	12952.5 – 13198.5	6.0

DISTRIBUTION SYSTEM DESIGN

History

Modern private cable microwave relay systems have evolved from those employed by the CATV industry. Cable companies have long used microwave systems operating in the 10 to 23 GHz range to deliver broadband signals to outlying areas within a major cable franchise market. In very large system, microwave is also employed in a "super trunking" mode to replace long trunk lines that are operating at or near their maximum amplifier cascade specification. In this mode, a central headend site broadcasts the signals to numerous hub sites where they are fed into the front end of a local distribution plant.

These systems operate in both FM and AM modes, with the AM mode being favored by current operators. Originally, these systems were composed of expensive single channel amplifiers combined through mixers and waveguide into a multichannel mode. Equipment incorporating klystron tubes was also used to achieve a multi-channel transmission format. Both these types of designs were expensive to construct and operate.

In the early 1980s advances in GaAsFET technology enabled designers to construct broadband amplified CARS band transmitters and receivers that could be enclosed in small housings and mounted directly behind transmit and receive antennas. This evolution in electronic components brought the cost of these systems well within the reach of smaller private cable operators.

In February of 1991, the FCC approved the use of the 18 GHz spectrum by private cable operators for the first time. It allocated over 70 channels of spectrum for the re-transmission of video, audio and data carriers. As a result of this ruling, SMATV operators now have a method to serve multiple locations with a single headend site and thus profitably expand their businesses.

Licensing in the United States

The Federal Communications Commission (FCC) is the regulatory body that licenses all microwave links in the United States. The two basic types of microwave line-of-sight systems, CARS band and 18 GHz, are regulated by different divisions within the FCC. Each has its own set of rules and regulations for issuing licenses.

CARS Band Licenses

The issue of CARS band licenses by the CARS Branch of the FCC is governed by Part 78 of the FCC Rules and Regulations. They are available only to franchised cable television operators and licensed wireless operators. Private cable operators may not use this band for line-of-sight transmission needs.

To apply for a CARS band license, an operator must first have a frequency coordination report prepared by an engineer. This report is included with the CARS band application, Form 327, and a filing fee form 155. The paperwork for this application is much more detailed than that for an 18 GHz license and failure to properly complete these forms is a major source of problems in the application process.

After the application is filed with the FCC, it is typically placed on public notice within two to three weeks of receipt at the commission. It remains on public notice for 30 days to allow for filing of petitions to deny the application. This might occur if another microwave system operator feels that the new service would interfere with one already in service. If the application clears public notice, the commission grants a construction permit to build the system, typically within two to four weeks. Although the entire process usually takes about 90 days from start to finish, the FCC may take a substantially longer time depending upon the current backlog of applications.

CARS band licenses are issued for a period of five years. Once the permit is issued, the operator must install the link within a year or the license automatically expires. After the system is completed, the operator must file a notice of completion with the FCC.

18 GHz Licenses

18 GHz microwave systems are licensed by the Microwave Division of the FCC Private Radio Bureau and their issuance is regulated by Part 94 and Part 21 of the FCC Rules and Regulations. The application is filed on form 402 for Private Operational-Fixed Microwave Service. Although this form is less detailed than Form 327, it still requires a frequency coordination report, filing form 155, and a statement of eligibility. The 18 GHz band is open territory to all communicators engaged in private commercial activity.

DISTRIBUTION SYSTEM DESIGN

After filing the application, the approval process take from 90 to 120 days. The application is placed on public notice within two to three weeks of receipt by the FCC and remains there for 30 days. Following issuance of the permit, the operator has one year to build the system at which time the FCC sends the operator a construction response form to verify completion of the system. This is also a five year license with the same provision for expiration if it is not installed within one year.

The application process for any FCC license is a very exacting exercise in properly filing all required paperwork. Any error in the process will most likely result in an out-of-hand rejection by the commission. The use of reputable legal counsel who specializes in FCC law as well as a competent engineer can be crucial in minimizing errors and subsequent difficulties.

Figure 6-26. 1. STL Microwave Link. *This STL link is being used in the Corpus Christie wireless cable system to receive broadcasts for re-transmission as ITFS educational television (Photo Steve Berkoff)*

Technical Operation

Microwave links generally employ broadband amplifiers that translate input signals in the 50 to 550 MHz band to those in the 18 GHz range. The transmitter accepts a standard balanced broadband signal in the +15 to +30 dBmV range. Although such a signal can be obtained at any point within a typical SMATV distribution network, the preferred placement of a microwave link is as close to the headend as possible where signal distortions are minimal.

Microwave signals are transmitted and received via parabolic reflectors ranging in size from 2 to 6 feet in diameter (see Figure 6-26). Most 18 GHz systems use a mast mounted, broadband amplifier to reduce costs and minimize the need for installation space (see Figures 6-27 and 6-28). However, systems that must transmit signals over a relatively greater range still employ single channel per carrier amplifiers that are rack-mounted in an indoor environment to achieve maximum output levels (see Figure 6-29). Such systems route signals via waveguides and are substantially more expensive than their broadband counterparts.

The output from the receive antenna and associated electronics is a downconverted signal in the 50 to 550 MHz range at a +15 to +30 dBmV level which can be used to drive a typical distribution plant. The

Figure 6-27. 18 GHz Transmitter. *The Nexus PC Linc is a 18 GHz transmitter. (Courtesy of Nexus Engineering)*

transmitter and receive electronics are phased locked via one or more pilot carriers. This serves to eliminate possible frequency drift between the transmitter and receiver and eliminate co-channel reception problems that otherwise could result.

DISTRIBUTION SYSTEM DESIGN

Figure 6-28. AML Transmitter. *This enclosure can be fitted with a broadband outdoor transmitter, repeater or receiver. The 72 channel transmitter is a block upconverter that accepts an input signal of 22 dBmV in the 54 to 490 MHz range and produces output in the 18.14 to 18.58 GHz range. It has a pilot automatic level control to maintain a constant output. The output C/N ratio varies from 61 to 53 dB as the number of channels transmitted increases from 12 to 72. The range is approximately 1.5 miles. The repeater is a broadband linear microwave amplifier used to re-transmit microwave signals without changing their frequency. The receiver downconverts the multichannel microwave block of signals to VHF frequencies in the 54 to 490 MHz range for input to distribution networks. The factory-set output 56 db C/N ratio can be increased by trading off for increased distortion products. (Courtesy of Hughes, Microwave Products Division)*

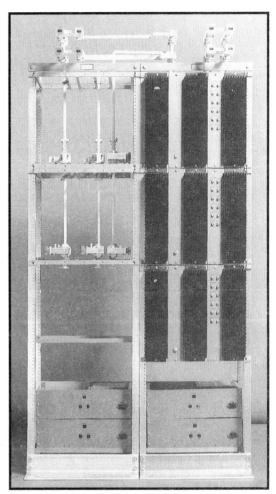

Figure 6-29. Rack Mounted Transmitter. *This rack carries nine single channel per carrier 18 GHz transmitters. Up to 24 channels can be mounted in a single rack. Power output levels range from +24 to +33 dBm per channel. Four power supplies drive the transmitters and output are channel through the waveguides near the top of the rack. (Courtesy of AML Marketing, Inc.)*

Technical Design Consideration

There are numerous technical considerations that affect the design and installation of a microwave link. Line-of-sight clearance, Fresnel zone clearance, path loss, rain fade and multi-path problems are major considerations in system design and siting. (Although some of these design concerns are similar to those encountered in relaying wireless cable TV signals as outlined in Chapter 13, these effects are more pronounced at the higher frequencies discussed here).

All microwave line-of-sight systems require an unobstructed view between the transmit and receive antennas since microwave signals travel in relatively straight lines and are strongly absorbed by intervening materials. A visual verification of the transmission path with binoculars or other sighting methods is therefore a crucial first step in any installation.

In addition to a clear line of sight, the path must be well clear of objects because the wave-like microwaves can be diffracted by nearby objects. This is known as the Fresnel zone effect (see Figure 6-30). To avoid propagation problems, the boresight of the transmitted beam must be greater than approximately 0.6 times the radius of the first Fresnel zone. The Fresnel zone clearance, in most 18 GHz applications, must be more than approximately 30 feet.

DISTRIBUTION SYSTEM DESIGN

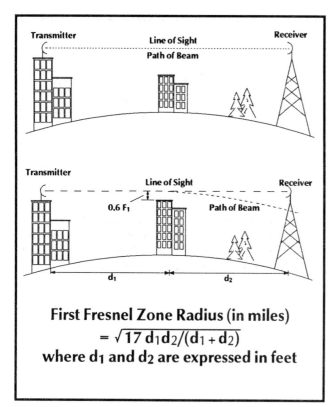

Figure 6-30. The Effect of Diffraction on a Line-of-Sight Path. *In order to have effective transmission of signals, no objects should be closer than about 30 feet from the line-of-sight path. Otherwise the microwaves may be diffracted, namely the signal is partially deflected and weakened in power. This phenomena is known as the Fresnel effect. The rule is that the path should be at least 0.6 times the first Fresnel zone radius to avoid appreciable attenuation in signal power. (Courtesy of Nexus Engineering)*

Path loss stems from two sources. First any microwave signal spreads out and weakens as it travels towards it target. Second, intervening materials such as oxygen and water vapor absorb and reduces signal power. This attenuation is directly increases both with path length and rainfall rates.

Water vapor is the principle absorber of microwave in the atmosphere. Therefore, heavy rainfall can result in substantial attenuation, cause the C/N ratio to fall below 35 dB and thus produce unacceptable pictures at the receiving end (Figure 6-31). The system availability factor, expressed either as the number of hours per year the C/N ratio falls below 35 dB or the percentage of time that the C/N ratio remains above 35 dB, should be no greater than 8 hours per year or its equivalent, a 99.9% reliability factor, respectively. A microwave engineer can correct for the effects of rainfall by examining historical data in the area in question and then by designing an adequate fade margin or excess power into the system. The larger the fade margin, the higher the system reliability.

In addition to ordinary path losses, microwave signals can be reflected by intervening terrain such as hills, buildings and water towers. In addition, the right combination of temperature and moisture content in the air can also actually cause the signal to be reflected along a path through clear air. These events create the condition known as multi-path (see Figure 6-32). When the reflected signals arrive at the receive antenna, any phase differences from the direct signal can result in increases or decreases in resultant power levels.

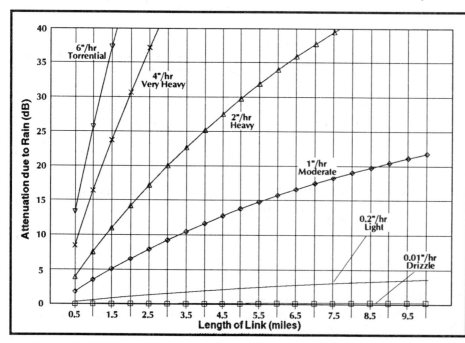

Figure 6-31. Signal Attenuation versus Length of Link and Rainfall Rates. *Signal attenuation increases both as path length and rainfall rates increase. (Courtesy of Nexus Engineering)*

DISTRIBUTION SYSTEM DESIGN

Figure 6-32. Signal Reflections and Multipath.
A microwave signal may reach the receive antenna along paths of different lengths. As a result, signals arrive with variable phase relations to the main beam and add or subtract depending upon the exact phase difference. (Courtesy of Nexus Engineering)

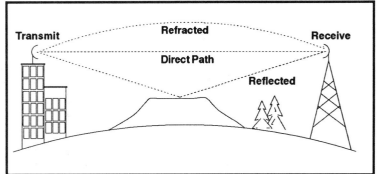

Path losses and signal reflections are further compounded since a parabolic antenna has a very narrow beamwidth at frequencies as high as 18 GHz. As a result even a minor misalignment of either the receive or transmit antennas can easily result in a 2 to 3 dB decrease in gain (see Figure 6-33). This can be caused by wind loading that causes an oscillatory movement of the antenna. This outcome can be avoided by using a smaller, larger beamwidth antenna or a radome cover, both at the expense of antenna gain.

The signal power reaching the receiver input depends upon all of these factors as well as gains and losses in antennas, amplifiers and interconnects. An example of such "link calculations" is presented in Figure 6-34 below.

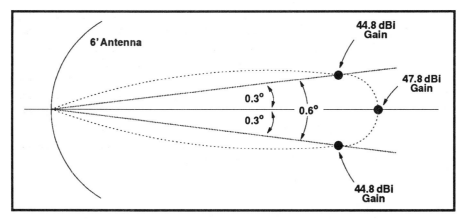

Figure 6-33. The Effect of Antenna Misalignment. *This figure illustrates how antenna misalignment can result in a decrease in transmitted or received power. Here a pointing error of 0.3 degrees results in a substantial drop in gain from 47.8 to 44.8 dBi, a 3 dB reduction. These figures are based on the fact that a 6-foot antenna has a beamwidth of 0.6 degrees at 18 GHz. (Courtesy of Nexus Engineering)*

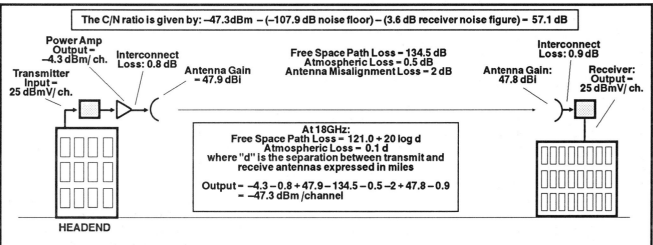

Figure 6-34. STL Link Calculation. *This illustrates the factors that enter into an STL link calculation. Among these factors the most variable are atmospheric losses and antenna misalignment losses.*

149

DISTRIBUTION SYSTEM DESIGN

The Financial Advantage

While microwave links can provide solutions and overcome a number of technical constraints, the economic advantage is the principle motivation for their use. Such a link can allow an operator to expand the service area, reduce initial capital requirements, generate additional revenue streams and minimize maintenance costs.

Modern SMATV headend cannot economically serve buildings having less than 200 to 250 units. A simple 36-channel headend can cost between $40 to $75 thousand or more to purchase and install, a practical expense only in buildings with large numbers of potential customers. However, by using a microwave link, customers in a relatively small building can be economically served. Microwave links range in price from $40,000 for a single link, high power system for long haul applications to as little as $10,000 for a multiple link system operating at short hops over distances of one to four miles. This considerable reduction in headend investment allows servicing smaller-scale buildings to provide a good return on capital.

A microwave link can also be used to expand the number of services provided to an existing customer base. Most 18 GHz systems now are capable of transmitting up to 72 channels in addition to a pilot carrier signal. One set of receivers and modulators at one headend location can therefore be used to feed additional services to existing headends.

By employing just one headend site that is linked to multiple distribution systems, an operator can invest in state-of-the-art technology that would not otherwise be justified on a stand-alone basis. This parallels a major CATV company investing in top quality components to serve ten of thousands of subscribers. A private cable operator would thus be able to amortize the investment over a larger customer base. This additional investment in quality equipment can also include inclusion of addressable scrambling technology. These systems that allow greater revenue control, have lower construction costs at customer drops and give an operator the ability to offer expanded pay-per-view services (see Chapter 18 for more details).

In addition, the use of a centralized headend can allow an operator to create additional revenue streams via insertion of local advertising. To this end, an advertising insertion system is installed at the main headend and time is sold to local merchants. This strategy can provide substantial additional revenues for a relatively small initial investment.

Finally, the use of a centralized headend can result in substantial reductions in overall maintenance costs. For example, ten headends each offering 24 channels represents in excess of 500 separate electronic components to be installed and maintained as well as 24 separate headend buildings to be constructed, climate controlled and secured. Each location must be serviced at least once a month to effect tasks such as balancing and aligning equipment. Such as system would also require authorization of as many as 240 decoders. Replacing these headends with ten microwave links eliminates the need for redundant headend and satellite reception antenna space and reduces maintenance to a minimum. Most modern 18 GHz radios are exterior mounted at the receive antenna and require little routine maintenance.

G. LASER LINKS

If a link is required to simply cross a street or transmit signals across a similar relatively short distance, the application process to obtain a license for a microwave link may be too time consuming and the cost may be too high. The alternative low cost technology, other than a direct cable run, is a laser link, also known as an infrared link. Such equipment can reliably relay signals from 1,000 to 3,000 feet at a cost ranging from $8,000 to $10,000. No license is required.

Laser links modulate the broadband RF signal onto light waves in the 780 to 830 nanometer wavelength range. The modulated beam is then collimated and transmitted to an optical receiver at any other site within view. These links are similar to fiber-optic cables except that no physical connections are required. As with microwave links, the input to an infrared link should fall in the +15 to +30 dBmV range; the output is within the same level. A laser system can similarly be used to replace distribution or trunk amplifiers (see Figures 6-35 and 6-36).

DISTRIBUTION SYSTEM DESIGN

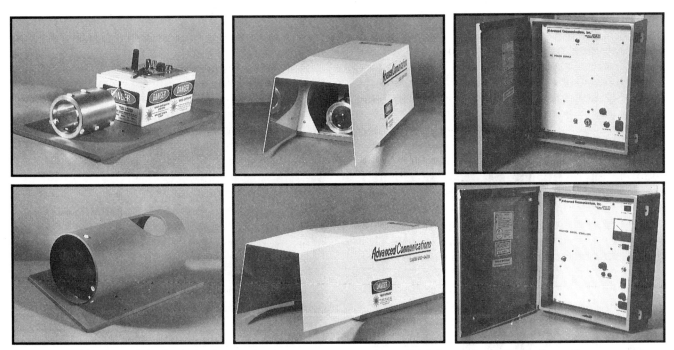

Figure 6-35. A Laser Transmission System. *The ACS1001 and ACS200 laser transmit and receive systems are illustrated here. The top three photos are, from left to right, the transmitter without its enclosure, the assembled transmitter, and the transmitter power supply and modulator. The bottom three photos are the receiver without its enclosure, the assembled receiver, and the receiver power supply and demodulator. (Courtesy of Advanced Communications Systems)*

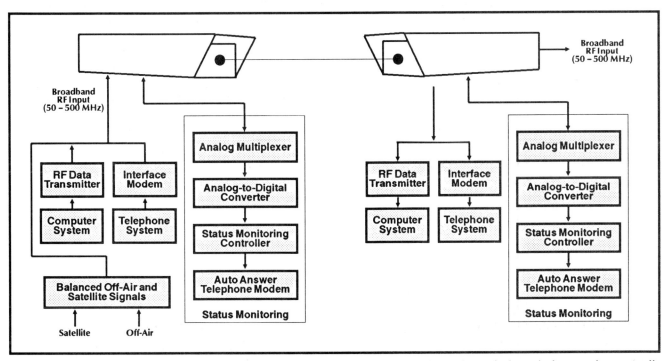

Figure 6-36. An Infrared Line-of-Sight System. *Both the transmitter and receiver of a laser link are schematically illustrated here. First, the off-air and satellite signals are processed, combined and balanced in the headend. Data or voice communications can be interfaced via a data transmitter or a modem, respectively. All the signals in a 500 MHz bandwidth then modulate the output of a light emitting diode (LED) and are collimated for transmission. At the receiver, the beam is focused onto a detector, either a photocell, optodiode or photo multiplier, where it is reconverted into an electrical signal. At this point the S/N ratio is quite low, so the signal must be amplified by a low noise preamp before being demodulated. The status monitoring capability of this particular unit allows control from a remote processing center.*

DISTRIBUTION SYSTEM DESIGN

Although pioneer infrared systems were limited to an output of less than 1 milliwatt of power, modern transmitters typically generate between 10 and 25 milliwatts. These devices feature AGC and temperature control circuitry, highly stable mounts as well as sophisticated RF processing and modulation designs.

Infrared line-of-sight systems have a number of advantages as well as some disadvantages. At infrared frequencies of approximately 500,000 GHz, there is more than enough bandwidth to carry millions of television channels. However, most commercially available systems use their available power to transmit on the order of six channels. Electromagnetic interference is not a concern in optical links. In addition, like 18 GHz links, no cables need to be routed between sites, so installation is a matter of hooking a prepackaged system to line power and then aiming the transmitter towards the receiver.

Infrared links are low power systems that have limited transmission ranges if signals are to be received at C/N ratios in excess of 48 dB with reasonably small amounts of intermodulation products. While advanced components may transmit signals as far as 3,500 feet on a clear day, atmospheric disturbances such as rain, snow, smog or fog can attenuate the signal and often limit the effective range to under a thousand feet. System reliability can be greatly enhanced by using shorter path lengths. The need for short paths can be overcome to a certain extent by increasing output power. However, as the transmission distance increases, other limiting factors become important. These include the resolving power of the optics as well as the structure and accuracy of the transmitter and receiver mounts. A light beam tends to spread out as it travels through space. An excellent optical system can only partially counteract this natural behavior. The narrow beamwidths of an infrared link dictate that it must be mounted rock-steady and kept in perfect alignment to function properly. Consequently spotting scopes are routinely used to install laser equipment. The requirement for stability can pose a problem when lasers are installed on tall buildings that may vibrate or oscillate in high winds or move throughout the day in response to their thermal contraction and expansion.

In summary, although some manufacturers specify transmission path lengths in excess of half a mile, the best use of laser links is for short hops on the order of 1,000 feet to cross roads and other short distances between distribution systems. In those applications that require greater distances, a microwave system can provide higher reliability and improved operational flexibility.

CHAPTER 7. PROJECT BIDDING

Once all the background data have been collected and the system design has been completed, the time has come to estimate costs and to carefully prepare a bid. The underlying cost information is crucial when building a private cable system either as a contractor or as an owner/operator. Accurate cost forecasts can be prepared only with a realistic design in hand and with a solid working knowledge of the bid process.

A. BID PREPARATION

The key to the bid process is the ability to accurately "take off" labor and materials needs from system plans. This take-off is an essential part of the construction planning effort and leads to a complete materials list as well as an estimate of labor needs. It is important to pay great attention to detail when preparing take-offs and then to double check the results. A poorly prepared take-off is at best an educated guess and at worst can cause material and labor shortages resulting in costly construction delays.

Most of the data for this process itself is generated from plans of one kind or another. Site and floor plans provided by an architect are an important input, so learning how to read blueprints is an essential skill required for bid preparation. Most architectural plans are drawn to scale. Common scales use 1/4" to represent 1 foot for most structural details, and 1" is equal to 100 feet for plot plans and street maps. Architectural rulers, known as scales, have these conversions engraved on their faces for easy use in reading plans. It is useful to purchase such a ruler and a couple of sets of different plans in order to practice measuring distances and identify objects on the drawings. This procedure is straightforward, so learning all that is necessary requires just a few hours of practice.

After the project has been evaluated on paper and a preliminary design and bid developed, it is time for a trip back to the site. Although the plans may appear complete, there may be changes and hidden obstacles at the job site that have not been previously observed. A careful on-site comparison of the proposed SMATV plans will assure the contractor that nothing has been overlooked. This is especially important if the system is being installed in a newly constructed project. During the building phase of any complex, changes are often made that are not reflected in the plans. Serious problems can arise if the SMATV contractor is not aware of these alterations.

Once the system design and plans are completed, a materials list can be prepared. This procedure is simply a matter of working through the plan one step at a time. The number of directional couplers, splitters, amplifiers and various headend components need to be counted and the total footage of cable and fittings calculated. All construction materials including an overage factor to eliminate any shortages should be estimated. Small items and hardware such as screws, anchors, fitting adapters and PVC fittings can be included in a miscellaneous category that reflects an average cost per job for these items. Table 7-1 outlines a typical materials list for a two hundred unit building complex.

PROJECT BIDDING

TABLE 7-1. 200-UNIT SMATV SYSTEM TYPICAL MATERIALS LIST

HEADEND

1	Satellite dish with mount
1	Dual feed
4	Low noise amplifiers
4	Block downconverters
1	VHF broadband antenna
1	Antenna mount pole and hardware
1	VHF separator
4	High frequency satellite signal splitters
5	VHF band pass filters for channels 2,4,5,7,9
5	VHF strip amps for channels 2,4,5,7,9
7	Satellite receivers
7	Frequency agile modulators
1	8-way mixer
1	VHF hi/lo combiner
1	2-way splitter
1	Equipment rack
1	1000' roll of RG 6/U headend cable
200	Fittings
lot	Screws, adapters, staples, hardware & concrete

DISTRIBUTION SYSTEM

2,300'	0.500 hardline cable
2,300'	2" PVC
25	4-way taps
15	Lock boxes
2	Directional couplers
60	0.500 fittings
25	PVC sweep and couplers
lot	Miscellaneous hardware

POST WIRE EQUIPMENT

23,000'	RG 6/U coax cable
400	RG 6/U fittings
200	Wall plates
2,000	Channel molding
200	F-81 fittings
400	Screws and anchors
lot	Miscellaneous hardware

In general, labor costs can be the most difficult to estimate. Manpower skills as well as initiative and drive vary from person to person. To successfully estimate the amount of work a person can accomplish within a specific time frame is an exacting science that requires a great amount of experience. One useful solution to this difficulty is to assign a piecework structure and pay rate for every job item on a project. A rate can be set for each outlet installed and for each foot of cable laid in the ground. This can be an effective strategy because even highly technical work such as splicing passive and active splitters as well as racking headends can be broken down into piecework items.

The work load can be subdivided into piecework items and a pay rate can be established by obtaining bids from local subcontractors. When using an in-house work force, overhead costs and profits from the subcontractor's bids can also be estimated. This method is a good starting point for determining a fair labor rate. It is important to realize that some projects require more work to perform the same task than would be the case on another seemingly similar job. This is especially true of post wire construction which may be quite variable. The difficulty of each project should be evaluated to permit an approximate adjustment of rates. Piecework estimating permits an accurate judgment of labor needs as an input to bid preparation.

It is important to know if union labor will be required on a job. If a bid were submitted based on the assumption that the project is non-union and it is later determined that union labor is required, the increase in costs might eliminate all profits. In addition, if the SMATV contractor does not officially operate a union shop, he may experience some difficulty in procuring union labor. Even for a project built as an owner/operator system, labor unions can cause difficulties if the private cable installation occurs during the construction phases of the complex. When it is a union job, the SMATV operator's non-union employees may not be welcomed on the site. This can be a touchy point that needs to be clarified with the general contractor and developer before any crews are allowed on site.

On jobs that require union labor, a call to the local electricians or communications workers' unions can provide the contractor with all the rate information necessary to prepare a bid. The contractor must also realize that union jobs have specific requirements that dictate the exact nature of the work that each class of employee can perform as well as the rate that is paid for that work. These rules necessitate careful work load planning and scheduling. The common complaint and often the result is that union jobs always seem to take longer to complete. None of these factors should be overlooked when preparing a bid for a project that requires union labor.

The last item of bid preparation is determining the profit margin. When building an SMATV system as a contractor, not as a owner/operator, profit is added to reflect the expense of time and expertise. An owner/operator requires a good estimate of a system's market value as an input to both planning and company financial statements. It is poor business sense not to earn a healthy profit for hard work performed and years of experience.

The standard method to determine profit mark-up is to add material and labor costs and then to multiply the sum by a profit factor. This then yields a total system cost that includes the well-earned profit. For example, if material and labor cost are $8,500, the revenue of $13,600 is calculated by multiplying total costs by 1.6. This yields a profit of $5,100 which equals the revenues less total costs:

Material Costs	$5,550
Labor Costs	3,000
Total Costs	8,500
x 1.6 multiplier	x 1.6
Total Bid	13,600
less Total Costs	8,500
Total Profit	$5,100

Once accurate costs have been estimated, the key to this process is determining the multiplier. This number determines a company's competitive position and whether or not the bid will successfully lead to signing the contract. The best strategy is a healthy balance. If the multiplier is too high most bids will be failures; if it is too low, profits will disappear.

It is wise practice to understand why this multiplier should vary with the ratio of material to labor on each job. Labor intensive jobs require a greater mark up than contracts that are mostly for materials in order to account for the risk involved. For example, workers can be unreliable, can get sick or may not perform up to expectations due to inherent difficulties at the site. The formula below can be used as a basis for defining this relationship. As labor percentage increases, the multiplier in turn increases. Of course, this numerical relationship is only a suggestion and is subject to interpretation. The relative rates vary between different regions and this formula should be adjusted accordingly.

Margin Factor = Ratio of Labor to Total Job Cost + 1.3

For example, if material and labor costs are $50,000 and $20,000, respectively, then labor is 2/7ths of the $70,000 total expected job costs. Then the margin factor is calculated as:

$$\text{Margin Factor} = 2/7 + 1.3$$
$$= 1.59$$

and the total system price should be bid as 1.59 x $70,000 which is equal to $111,000. This has a built-in profit of $41,000.

B. THE BID DOCUMENT

Bid preparation is just as important a step as completing the final contract. In many cases, it forms at least a major part if not all of the substance of a contract. When the bid document is accepted it is an obligation to perform all duties as proposed. It is imperative to outline every aspect of the project in detail and to list all the tasks that will be performed under the terms of this contract. Chapter 2 has presented more information on preparation of contracts.

The bid should include the number of satellite and off-air channels that will be distributed. Issues such as financing, project timetables, guarantees and servicing agreements should also be addressed. In addition to the total job price, the bid should outline the possibility of additions or exclusions such as licenses, taxes, fees and permits. The type and make of various components can also be included and perhaps enhanced with other information such as company brochures and "cut sheets" describing their technical specifications. The written bid reflects the policies and professionalism of the SMATV company. If it is professionally composed, it can be the difference between winning or losing bids.

The presentation of a bid legally constitutes an offer to engage in a contract. The acceptance of the bid then converts it into a legally binding contract. In another words, if a client notices a mistake in the proposal that is to his benefit, his acceptance of the bid can legally bind the contractor to perform the

PROJECT BIDDING

job, even if substantial losses would be incurred! The legal ramifications of offer and acceptance rules require that the bid be extremely carefully prepared. Everything must be double checked. Of course, experience is the key for developing a feel for estimating project costs based on inputs such as total number of channels to be offered and units to be cabled.

If a bid seems too low, figures must be rechecked. On extremely large projects, it is wise to hire the services of a designer or contractor to aid in preparing the plans as well as the bid. The quoted price is the one the client is obligated to pay. It is foolish to submit a bid that cannot be completed with near 100% certainty.

C. NEGOTIATIONS

There are often many opportunities to negotiate after submitting a bid. This is especially true when the client/SMATV company relationship has been previously established to the satisfaction of both parties. For example, a rate may be proposed that is not as competitive as desired by the customer or perhaps there may be a question as to the need for a specific design item. Whatever the case, the customer may have already selected the SMATV company and will negotiate for those specific items that need attention before the bid can be accepted. This time is an excellent opportunity to steer the contract through careful negotiations and additional planning.

Negotiation is a joy for some and a necessary evil for others. Those with egos that prevent them from conceding any contract point will soon find that they are not being asked to bid on the more lucrative jobs. Each negotiation point should be carefully reviewed. For example, is the charge too much for a specific item or does the client just perceive it that way? If the price is too high, it may be necessary to lower it. If the customer is misinformed, he or she may need education as to the difficulty and cost of the particular job or item. Sometimes a knowledgeable client may know a vendor that can supply items at a lower cost. When working with this vendor, it may be an advantage to amend the bid based on the alternative vendor's pricing for materials. Sometimes the bid can be reduced if the client suggests an insightful design change or a substitution that results in lower costs. It is important, however, to be sure that acceptance of the proposed changes will not jeopardize the integrity of the project. All bid changes should be documented in writing and a new contract prepared once all points have been negotiated.

The time usually comes when a client informs the SMATV expert that he has found a company that will do the same job at a lower price. This is usually but not always a negotiation ploy and the time has come for using refined sales skills. One strategy is to ask the client to review the competitor's track record and previous jobs while, at the same time, simultaneously conducting an investigation of their reputation. The client should be resold by showing him some more samples of work, either photographs or preferably on-site visits, and list of satisfied customers. If possible, the competitor's proposal can be reviewed in order to make an item for item comparison. Often this will uncover discrepancies in quality or design and perhaps substitution of inferior products. The bottom line is a mixture of sales technique combined with a healthy dose of reality. Usually the more professional contractor having a price in the competitive ball-park and a stable reputation wins the contract. Private cable customers are usually sophisticated buyers not necessarily solely motivated by price. However, if too many bids are lost, the time has come to evaluate the SMATV company's pricing structure and operational procedures.

A successful negotiator must have and convey a sense of caring to his client by being responsive to all concerns and addressing each one in turn. The client must know that the project will be completed on schedule and in a professional manner. A useful tact to assume is to find a few points important to the client that can be conceded. But negotiating is a give and take process that ultimately should allow both parties to win by successfully transferring a good product in one direction and profit in the other.

CHAPTER 8. CONSTRUCTION and INSTALLATION

The tasks of site surveying the property, designing a private cable system and bidding and negotiating a contract all lead to the final construction phase. The coordination of the manpower and materials needed in construction can make the contract process seem easy in comparison. As in bid preparation, planning is again the key to successfully completing a project. The data assembled during bid preparation serves as useful input to planning all construction needs. Using this background information to the fullest extent can minimize or avoid many potential problems.

The most important factors involved in building an SMATV system are examined below. As is common in any construction project, certain basic skills are required to complete the job. The goal of this chapter is to outline these skills to aid in evaluating how to be proficient in private cable construction. When lacking in some or all of these and related areas, alternate sources for construction help are discussed. Remember, there are as many ways to build a system, as there are methods to skin a cat.

A. THE PERMIT PROCESS

Before beginning any construction process, building permits must be secured. This is especially important in commercial projects such as apartments, condos and hotels. Requirements vary between cities, states, provinces and countries. In many areas, only certain components, such as antenna foundations, require permits to proceed with construction. It is important to thoroughly check all city requirements and to obtain all necessary permits before the start of construction.

In almost all cases, plans and documentation are required as input to obtain a permit. In addition, most cities must receive a set of load and stress calculations which are certified by a licensed structural engineer. The task of preparing these documents is usually a long and arduous process. Therefore, unless the in-house skills of a good draftsman and engineer are readily available, this work is best left to an independent architect and civil engineer. Depending on the details involved, the fees for such services usually range from around $500 to $2,000. This is money well invested in meeting schedule requirements because most city permit departments will return improperly prepared documents, often resulting in lengthy delays.

Other readily available sources of plans and drawings include the building owner and his architect as well as those on file with the city planning department. A careful search of these resources can often uncover very detailed drawings that can usually be purchased for just the cost of duplication. These documents can reduce the amount of time an in-house architect spends on the job and can save a considerable portion of his or her fee.

CONSTRUCTIONS & INSTALLATION

Most municipalities require engineering data regarding load and stress factors affecting the satellite antenna. This is especially important when attaching a dish to an existing roof structure. An 8 to 15 foot parabolic antenna has a large surface area and can be an excellent sail and generate tremendous forces especially when mounted in windy locations above ground. Most antenna manufacturers can provide such background operational data upon request. This data then can be forwarded to the architect or engineer involved in the preparation of the plan so that he can verify that the antenna installation will comply with all local engineering requirements.

Before obtaining permits, many municipalities will also require granting of zoning and planning approval. This process can often try the patience of a saint. This is especially the case because most planning committees are not familiar with the private cable industry. Many city officials and legislators have developed a negative impression of satellite antennas as a result of the rapid and often controversial growth of the home satellite industry. This attitude is further compounded because some irresponsible TVRO dealers have installed neighborhood eyesores. In some communities, legislators have attempted and succeeded in enforcing an outright ban on satellite dishes.

Zoning and planning committees must be assured that a well designed and well planned private cable project will not detract from the beauty of the community. If possible, the satellite antenna should be installed out of sight so as not to be visible from the street. Intelligent use of landscaping can enhance the site appearance. Zoning and planning committees respond favorably to instructive drawings of the proposed site and to photographs of other attractive projects. Photos of existing installations as well as letters of recommendation from satisfied complex owners can go a long way towards alleviating the concerns of the planning committee. During the presentation, all objections must be positively addressed. A committee will develop a respect for an SMATV expert who presents himself as a concerned businessperson who is receptive to suggestions from its members.

The permit process revolves around compliance with city codes. Become acquainted with all the provisions that apply directly and indirectly to the private cable industry. In general, there are few rules that are specifically addressed to the satellite industry, so interpretations applying to an SMATV project must therefore be derived from more general codes. Policy statements regarding these interpretations are often available upon request. Some codes are related to certain construction practices and methods and mandate that plans for any type of project comply with these practices. In those cases where variances are required, obtain written approval prior to construction. Knowledge of and compliance with all applicable codes will help avoid costly corrections of non-standard work.

Applications for permits can sometimes take quite a while to process. This can cause a domino effect because long delays in approval can create problems in meeting contractual obligations. It is important that permit processing time is included in construction scheduling and that permit applications are carefully prepared. Be certain that all the required drawings and the correct number of prints as specified by the building department are obtained. If these require engineering certification, check that each print is stamped by the civil engineer. Be sure to include all necessary fees for the permit. Most fees are based on a percentage of the value of the system. It is important to be fair in these estimates because building departments generally are aware of the temptation to save money by lowering the value of projected work.

If applications are improperly prepared, chances are good that additional data will be requested and that the entire package will have to be re-submitted. This will certainly result in more delays. Developing a reputation of thoroughness and professionalism with the local building department will go a long way to speeding the entire process of obtaining permits.

Every city will have its own method and procedures to process the permit application. It is a good practice to become aware of all the different steps and departments that the application must pass through before the permit is issued. If the contractor learns which personnel manage these tasks, the permit process can then be accurately tracked and unexpected delays can be minimized. The development of personal contacts within the various planning departments can very effectively expedite future applications.

CONSTRUCTIONS & INSTALLATION

B. SUBCONTRACTING the CONSTRUCTION

A qualified subcontractor can be an essential aid for the small SMATV operator/builder. Having the skills and know-how to successfully complete a 4-channel, 30-unit SMATV system is commendable but may not easily lend itself to installing and operating a 20-channel private cable system in a 600-unit luxury apartment. A subcontractor who is experienced in multi-unit construction can complete a project on time and in a professional manner. Using the services of a subcontractor can eliminate the need to maintain a large equipment inventory and can cut out substantial parts of overhead expenses including the need for a large weekly payroll. In addition, all construction costs can be fixed within the terms of a subcontractor's contract before the job ever starts. Any defects in workmanship are the responsibility of the subcontractor. This factor alone may save countless dollars if unexpected problems are encountered. These reasons have motivated many experienced private cable operators to rely solely on qualified subcontractors for all their construction needs.

The process of finding a good subcontractor can be similar to finding a talented, reliable auto mechanic. In any industry there are reputable craftsmen as well as crooks. Although most SMATV subcontractors have developed some experience in cable TV construction, both the private cable and CATV industries are rough and tumble environments having high attrition rates. Fortunately, the skills developed in CATV contracting are identical to those needed for private cable construction. The methods for finding and dealing with a cable TV subcontractor can be important survival skills.

Finding a Subcontractor

Finding a good subcontractor does not necessarily have to be as difficult as it may at first appear. This statement is supported by all the successful cable systems in operation. In most cases, these have been built by subcontractors who are usually seeking additional work. The key is knowing where to find them. The tried and tested method to do so is by referral. Friends or trade acquaintances in the industry will readily recommend firms. The local cable company can also usually supply a list of reputable subcontractors. On a number of occasions, one of the authors has been occasionally contracted to build projects outside of his home area. The construction manager of the local cable companies in these areas most often was able to recommend one or two good local construction firms. This method was quite successful and a letter of of appreciation was always sent to the construction manager and his supervisor.

Armed with the names of a few contractors, the interview process can begin. The key factors to examine are the contractor's reliability and the quality of his work. It is important to see a sample of this work. Most subcontractors are glad to show photographs of their favorite projects or to arrange a site visit. Review the photos and inspect the site personally. The difference in appearance between the current site conditions and the original glossy photos indicate the quality of workmanship. If possible, projects that are two to three years old should be examined to see how the work has stood the test of time. Look for any sagging and failure in the conduits and joints. Check the condition of any landscape restoration required following underground construction. Ask the site manager how responsive the contractor has been in repairing any construction defects. It is essential to ask questions because the quality of the subcontractor's work reflects upon the party who hãs the principle contract, the SMATV operator/builder.

In addition to checking construction quality, investigate the subcontractor's reliability. A high quality job is not worth the price if it is not completed on time. Have the prospective subcontractor provide a list of references and check them out thoroughly. Many questions need to be answered. Was the project completed on schedule? Were there any additional charges not covered in the original bid? Did the subcontractor's crew act in a professional and courteous manner on the job site? The subcontractor and his crew represent all parties on the project. His performance becomes a direct reflection on the overall professionalism of the project. If he is constantly late in meeting schedules, the client will have reason to doubt the abilities of all concerned. If the subcontractor or his crew are rude to the project staff and residents, chances are good that future jobs will be jeopardized. Always choose a contractor who is courteous, prompt and professional.

CONSTRUCTIONS & INSTALLATION

Subcontractor Costs and Fees

The most important factor in selecting a subcontractor is naturally the cost required to complete the construction. Once the task of selecting a few subcontractors who meet the necessary standards of quality has been completed, the next step is to choose the one who charges the most reasonable rates. Most contractors are quite experienced in dealing with franchise cable operations, organizations which have notoriously tight purse strings. An intelligent policy to follow is to understand how the CATV companies operate in order to develop a strategy for negotiating a fair rate schedule with their subcontractors.

Almost all franchise cable operations pay for construction and installation on a "piece work" basis. Every job item is categorized and has it own rate established. These are standard line items common to most CATV contracting, although the specifics in furnishing certain materials such as conduit, concrete slurry and asphalt capping materials may vary. Among the standard cost items are:

- Cable trenching (per foot)
- Asphalt and concrete cutting (per foot)
- Under-driveway boring (per foot)
- Multi-unit post-wire (per unit or outlet)
- Multi-unit pre-wire (per unit or outlet)
- Installation and splicing of active and passive electronics (per piece)
- Firing up and balancing of system (per foot of total system length)
- Installation of customer drops and service (per drop or outlet)

The strategy is to determine which of these and other items are appropriate to the project in question and then to prepare a job sheet that includes all the necessary labor activities. This can be presented to the prospective subcontractors who then will prepare a rate sheet. Their relative competitiveness can easily be determined by comparing their responses on an item by item basis. In addition, many contractors are willing to work on a turn-key basis. They will provide a list of all materials required as well as their labor and warranty terms for the project. The fact that they will manage the materials inventory helps to eliminate costly warehousing as well as waste of materials. Always avoid subcontractors who work only on a time and material (T & M) basis. Any subcontractor who operators only on T & M and who cannot prepare an accurate bid for a specific project, is a contractor who is unsure of his own professionalism. He will most likely create unnecessary and numerous headaches on the job.

Many subcontractors have to wait a substantial amount of time before receiving payment after completing work under contract with franchised CATV operations. This creates an unnecessary financial burden that can force a contractor to submit a high bid. A private cable operator who can pay within 30 days of subcontractor billing might be able to lower bidded costs and save some money. If such an arrangement is negotiated with a subcontractor, delivering payment on schedule can ensure a successful, long-term relationship. Most subcontractors will be more flexible in negotiations if the possibility of a long-term relationship that will span many projects exists. A commitment which is based on mutual satisfaction may result in lower line item pricing.

Negotiations are the key to determining the final price. They begin after labor requirements are prepared and the prospective subcontractor has submitted a bid. Most businesspeople expect to encounter some negotiations and generally find these an aid to establishing a good working rapport. Never accept a first offer as final. Use information from a competitive bid but never quote it directly. Both parties should be also willing to concede a few points. But remember the truth in the old adage that "you only get what you pay for." The best price must be obtained without sacrificing quality or service.

Legal Considerations

After the prospective contractor has provided convincing evidence about his quality and pricing, it is important to ensure that he is legally qualified to complete the work. There are numerous requirements for licenses, insurance protection and construction bonds in order to operate as a contractor. All these items must be verified.

In most provinces and states a contractor's license is a basic requirement to perform work on commercial projects. Check all local and state regulations regarding these requirements. Verify that the license number of the contractor is valid and that it has the appropriate classification. Checking for any violations or suspensions of the license can indicate weaknesses in the record of a subcontractor's past performance. In addition to a contractors license, verify the status of his business license and tax permits. Be sure that he is current in meeting tax liabilities and payroll commitments. The party that hires his work force may assume some responsibility for meeting these obligations. A subcontractor who does not take care of his business responsibility could soon be bankrupt, leaving his contractual partner holding the bag.

A subcontractor's insurance status must also be verified. The two key items are workman's compensation and liability. Remember, principle liability clearly rests with the SMATV builder who has signed the original contract to perform the work required on the project. If a workman with inadequate insurance is injured during construction and is not properly covered, liability for his medical bills and support may shift hands. If anyone else is injured or if property damage is incurred, a lawsuit could be filed that may force the SMATV builder into bankruptcy. Make sure that the subcontractor has purchased adequate insurance. A $1 to 3 million umbrella policy is considered standard coverage in these days of fast-draw attorneys. It is wise to be named as an additional insured party on the contractor's insurance policies and to keep in touch with his agent to be sure that no policy changes are slipped in under the wire when problems arise. This insures that if the contractor fails to make premium payments, both parties will be notified of any impending cancellations. Proper insurance is crucial. Under no circumstances should a project be undertaken if the insurance status is in doubt.

One final area of legal concern is the matter of liens. In most states and provinces, if a subcontractor fails to pay a materials bill or meet a payroll obligation, both parties can be held responsible for debts incurred during construction. These debts can become actual liens on the property title. Needless to say, if a lien is filed against a client because of something the subcontractor did or failed to do, the private cable firm will most likely be thrown off the project. This possible outcome can be prevented in advance with a lien release signed by all subcontractors who will work on the job. By requiring lien releases for all laborers and materialmen, both the SMATV firm and the client are well protected.

A related legal concern is the rights of the subcontractor in those situations where the system is built specifically for sale to the project owner. If certain well-defined procedures are not followed during the construction phase, the contractor can lose his right to lien a project if a payment dispute does later arise. The procedures include the timely filing of preliminary notices and notices of intent to the owner. These documents inform the building owner that the contractor is performing work for pay on the project and that any future disputes might result in liens being placed against the property title. A complete understanding of local and state or provincial regulations will prevent the contractor and his subcontractor from overlooking these items and possibly from losing the right to recourse if a dispute arises.

Subcontractor Management

Once the final selection of the subcontractor is completed, the task begins in earnest. Throughout this process, the key to a successful relationship is planning and communication. If everybody knows what needs to be accomplished as well as how and when to implement the project, construction will run smoothly. It is wise to devote as much time as possible to planning the operation with the subcontractor.

The first planning document is the written agreement with the subcontractor. It should spell out in detail the amount and kind of work to be performed as well as the labor rates for each item. Provisions should also be included for unforeseen straight labor rates for any job not covered in a schedule. Include details regarding billing and payment schedules. A good method is to provide progress payments of a predetermined amount when pre-assigned phases of the project are completed. Be sure to retain payment of a specified amount, typically 10%, until 30 days after completion of the project. This will help insure that any problems which arise will be corrected.

It is important to prepare a game plan leading to completion of the project. List all critical operations and the expected completion dates of each. Track these on a large chart onto which daily progress data can be entered. If delays result from material or manpower shortages or weather conditions, these should be factored into the plan as the construction progresses. By tracking daily progress, manpower and material requirements can be updated when necessary to guarantee that the project is completed on schedule. There are many good project management computer programs on the market that utilize PERT and GANT charts and other useful management tools (see Figure 8-1). If an in-house computer is available, an investment in one of such programs would be a good idea. The main idea here

CONSTRUCTIONS & INSTALLATION

is to utilize good planning and management skills to maximize efforts and minimize costs.

Set standards of conduct for the subcontractor and his crew. The expected code of conduct should be written down and then signed by the subcontractor and his employees. This will eliminate the possibility for later confusion about what was required of the subcontractor and whether or not the employees had understood their roles at the site. These people are the builder's representatives to the client and to all potential subscribers. Courtesy, professionalism and decency must be required of all workers. It is a prudent policy to obtain photocopies of all workers' drivers licenses so that if a a theft or other problems such as harassment occur this information can be given to police. Although the subcontractor should be well aware of the performance goals, the actual management of his crew is under his direction. If any complaints are received, he must be notified and should immediately take corrective action. If a worker receives different orders from the subcontractor and private cable manager, the result will be confusion and low productivity. While clear communication is essential, the subcontractor must also understand his role as a subordinate.

At the conclusion of each stage, all tasks must be inspected to verify that all the work was done correctly. Construction is a real world activity. In almost all cases, expect variations from the original plan. A reliable contractor will provide an "as-built" drawing of the completed project. This includes any deviations from the original plan, as well as actual system performance levels as measured at each tap outlet. These must be inspected thoroughly with the subcontractor present. It is essential to fully understand what has been done and where any potential service problems might exist. If any items need correcting, prepare a written list with a time limit for completion. Do not release the final payment until all repairs are made and the project has been accepted and signed off by the property owner. Be sure to verify that all negotiated warranties are in effect and that all material and labor lien releases have been signed. This verification is a solid protection against any portion of the contractor's debts.

These discussions should make it clear that the use of subcontracted labor can provide many benefits for the private cable operator. By utilizing the subcontractor's equipment and manpower as well as his existing business structure, a substantial amount of overhead can be saved. In addition, his or her experience in cable construction can result in projects that are completed on time and within budget. The guidelines presented here can be instrumental in locating a good subcontractor and in developing a business relationship that will prove financially beneficial to both parties.

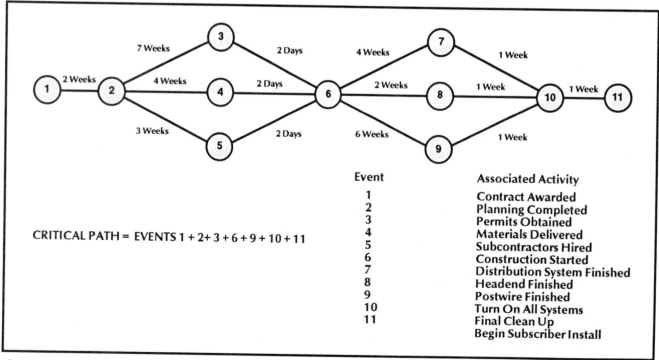

Figure 8-1. PERT Chart. *A PERT chart is a useful tool in scheduling a complex sequential or parallel tasks. In this example, the total time to completion has been estimated to be 17 weeks and 2 days.*

CONSTRUCTIONS & INSTALLATION

C. IN-HOUSE CONSTRUCTION

As a private cable business grows, the time may come when some or all construction activities can be managed in-house. This situation may gradually evolve with the first step being the establishment of an in-house group to handle small or difficult jobs. Or perhaps only those installations which are owned and operated and which have directly managed subscribers would be tackled. In any case, it is necessary to proceed to organize and staff a construction crew. The experience gained from planning and working with subcontractors is the perfect preparation for this stage.

The equipment and manpower requirements depend upon the work to be performed. Obviously it takes a much larger investment to cut up streets and run underground cable than it does to install a drop to a customer's apartment. A clear evaluation must determine how extensive these construction activities will be. Evaluate previous construction contracts and develop a list of those activities that were common to all these jobs. Examine the amount of work performed on each job and determine those areas where it might be more profitable to retain the services of a subcontractor for specialty items. Start with those activities that require the least investment in tools and equipment and work into the more complex jobs.

The Basic Approach

The complete installation of an SMATV system involves expertise drawn from many disciplines. For example, concrete work is required for mounting satellite dishes or towers for off-air antennas, metal work is needed for custom mounting brackets, electrical work must be performed on the electronic components in the headend and distribution system, and carpentry work is required for headend buildings and finishing touches throughout the complex. Of course, any phase of the entire job can be subcontracted. But for those tasks completed in-house, numerous specialized tools are required in addition to the standard collection. These include items such as a crimping tool, fish tape, cable lubricants, butyl rubber sealing tape, liquid neoprene, cold-weather electrical tape, a cable staple gun, a multi-meter, a field strength meter and a spectrum analyzer.

The electrical circuits should be installed by a licensed electrical contractor, whether the expertise is obtained in-house or from a hired independent. The law usually requires that most, if not all, electrical work be completed by a licensed professional. If a fire or other unexpected problem occurs, a licensed electrician would bear the brunt of any necessary legal protection. Note that although most headends require less than 250 watts of power, when the needs of all the installation tools, such as soldering irons, are added together, substantial amounts of power may be needed at various times during installation.

The most natural area to begin in-house construction is the building of headends. Most private cable operators have first entered the SMATV business with some technical knowledge and experience. Construction of the headend is the most technical of all private cable tasks. Managing this phase in-house can reduce costs and increase system reliability. In contrast, underground or aerial cabling is perhaps the most difficult phase of any installation. The equipment is relatively expensive and the task requires a great deal of coordination with the local government as well as with water, telephone and electric companies. Therefore, it best to have at least the first few jobs subcontracted to an experienced professional.

Pre-Installation Testing

It is a wise practice to measure the performance of electronic components before they are installed. This procedure serves two functions. First, it avoids the need to conduct extensive troubleshooting on a new SMATV system if a defective or substandard component had inadvertently been installed. Second, it allows the performance of components such as amplifiers to be measured and catalogued so that periodic maintenance checks can reveal any departures from normal behavior before breakdowns do occur.

New amplifiers should be bench tested and aligned prior to being installed. A sweep generator and oscilloscope can be easily used to verify that the response is flat across the 40 to 250 MHz or other applicable frequency range. This trace should be

CONSTRUCTIONS & INSTALLATION

copied and kept on file for later use. Attenuators can be inserted to match the expected total losses and tilt in the final system, so that a technician may ascertain that the amplifier can function up to specs in a situation similar to its final operating conditions.

The levels of hum modulation should also be measured and catalogued. Hum bars caused by 60 or 120 cycle per second ripples in power supplies can be an annoying system difficulty. Power supply ripples measured before installation can be compared with manufacturer's specs to determine if any potential problems exist.

Headend Construction

If at all possible, the headend should be installed by those technicians who will be directly responsible for its operation and maintenance. This allows them to know exactly how it was designed and assembled, how it operates and to be aware of any potential design or component problems. This feeling for its operation can be important when troubles actually occur. A subcontractor may install a shoddy headend and then disappear, but a competent technician who has to operate and live with it day-by-day will be sure that it is installed correctly from the beginning.

Headend construction is a reasonably straightforward process. The equipment is installed in a rack and is wired following a clear set of plans. The quality of this wiring determines the quality of the headend. Poor connections are the number one cause of headend and system problems. It is important to learn to make tight and effective fittings and to pay attention to each and every one installed (see Figure 8-2). An average headend may include 200 or more fittings. If each one is installed as if it was the most important element in the system, many future troubles can be avoided.

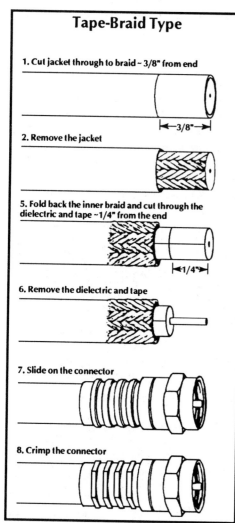
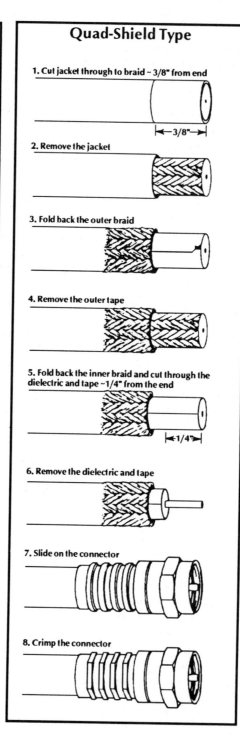

Figure 8-2. Connector Installation.
It is crucial that connectors are properly and securely attached to coaxial cables. A correct method to install non-sealed connectors to both quad-shield and tape-braid coax is illustrated here. (Courtesy of Times Fiber Communications, Inc.)

CONSTRUCTIONS & INSTALLATION

When installing F-connectors the following suggestions may eliminate service problems:

- Do not score or puncture the exposed center conductor, especially when using copper clad aluminum conductors.
- Remove the cut-off burr at the end of the center conductor.
- Be sure that the dielectric material has been cleanly removed from the exposed length of the center conductor. Remove any clinging material before inserting it into the female connector.
- Be certain that the aluminum grounding sheath is not burred, flared or distorted. A properly prepared sheath should slide easily and freely into the F-connector sleeve.
- Crimp the connector using the proper tool. An unevenly crimped ring can result in cable distortions and impedance changes.

A common installation mistake is the improper use of coaxial cable. The highest quality cable available rated with a 100% shield should be used in headends. Quad-type shielding is best. Coax must not be bent too severely because a short turning radius can kink the cable, cause distortion in the nonconductive dielectric, slightly change impedance at this point and result in signal reflections and losses. The radius of this bend should be at least 10 times the diameter of the cable. Many companies now use an all-copper quad shielded cable for headend construction. This variety is easier to bend than aluminum and if the copper conductor should oxidize after years of use, it will still retain a reasonably high conductivity. When an aluminum cable oxidizes and changes to a non-conductive metal, attenuation increases and the S/N ratio drops.

All cables should be labeled with tape or tags at both ends for easy identification. Input and output cables should be segregated into separate groups and these bundles should be tied together with wire wraps. Do not use tape to wrap up cables because it ages and will soften and get gummy or break away in a relatively short time. Then, a sticky mess will have to be handled if and when cables need to be serviced. The manner in which cabling is handled can make the difference between a well-organized wire harness and a rat's nest.

All components must be housed in an appropriate headend building. These electronic devices are a substantial investment and must be well protected. Choose an area that is both reasonably close to the satellite antenna and to a central point in the distribution system. In some cases, it may be necessary to build a special mini-building or to buy one of the prefabricated enclosures especially marketed for this application.

All equipment should be installed in a 19-inch rack having safe locks and, if necessary, a more sophisticated security system might be incorporated into the headend building. In general, racks are designed so that the off-air and satellite signals work their ways from the top to the bottom to allow for easier installation of the necessary cables (see Figure 8-3).

It is very important to maintain a reasonably stable temperature in the headend in order to prevent drift and unnecessarily rapid aging of the electronic components. For example, modulator gains can increase by over 10 dB when temperatures drop from those of a hot summer day to a cold winter night. The headend must usually be air-conditioned to reduce dust accumulation, control humidity and to prevent overheating. If an excessive amount of water vapor accumulates, electrical components could degrade and have a shorter than normal lifetime. Those rack-mounted devices that draw appreciable amounts of current must also be well ventilated. Good headend maintenance will result in years of trouble free service.

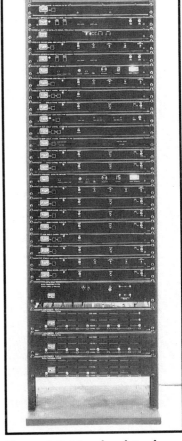

Figure 8-3. Headend Rack.
(Courtesy Pico Macom)

CONSTRUCTIONS & INSTALLATION

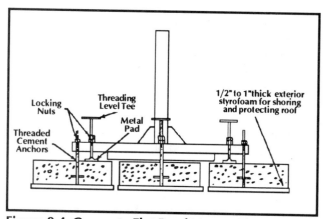 

Figure 8-4. Concrete Flat Roof Mount. *This mount is bolted onto four concrete slabs which are spread over a 10 square foot area. "Tea handle" screws are used to allow leveling of a polar mount. No leveling is necessary for az-el type mounts. (Courtesy of Upper Midwest Satellite Supply)*

Satellite Antenna Installation

It is important to use the services of an experienced TVRO technician to install a satellite antenna. This is a precision device having very close tolerances that must be met to obtain optimal performance. It is wise to follow the manufacturer's directions to the letter when assembling and installing the antenna. The resulting signal quality depends on a number of factors including accurately centering the feed and correctly adjusting its position at the focal point. This "feed peaking" is critical to obtaining a high strength signal.

The relatively large antennas used in SMATV systems are generally mounted on a pole or an azimuth/elevation tripod either of which is secured to a concrete pad. This structure must be strong and stable. If the mount will be installed well above ground level or on the roof of a building, its is crucial to have the designs approved by a certified structural engineer (see Figure 8-4). Note that a 3.7 meter parabolic antenna can generate a wind load of over 10,000 pounds in a severe storm.

Once the antenna is installed and the feedhorn and LNB are mounted, this dish must be aimed at the target satellite. While most home satellite reception dishes are installed to track across the entire belt of satellites, SMATV antennas generally are fixed on one target at a time. Therefore, while the alignment process is quite similar to that followed in home TVRO installations, it is more forgiving of errors. If the dish must be re-aimed at some point in the future, fine adjustments can be made with the antenna targeted at just one satellite. All the fine details describing installation and aiming of antennas are described in references listed at the end of this book.

The simplest method to aim the antenna begins with a list of satellite coordinates specific for the site latitude and longitude. In North America most SMATV antennas are usually targeted onto Galaxy I, Galaxy III, Galaxy V or Satcom F1. The alignment procedure begins by connecting a spectrum analyzer, and/or a satellite receiver and TV into the output of the LNB. Then the antenna is slowly moved back and forth past the targeted satellite until even the smallest signal registers on the equipment. The identity of the satellite can be determined by viewing different transponders and then by comparing the programming to that listed in a satellite TV guide.

After this initial step is accomplished, the signal level must be peaked on the analyzer or meter by making fine adjustments to antenna position. Note that many receivers have a connection point for VOMs to measure tuning or AGC voltages. When installing a dual-beam system, the second and subsequent feeds should also be adjusted to a position where the maximum signal strength is detected from all LNBs. In this case, follow the manufacturer's instructions provided with the multi-feed mounting structure. When LNAs are installed, the downconverters should be mounted tightly to the side of the feed or preferably in weatherproof enclosure on the mounting pole just below the dish. Finally run the quad shielded cables to the headend where any any high frequency splitters that are needed can be installed.

Off-Air Antenna Installation

The prime concern when installing off-air antennas is to insure that they will not topple over or rotate in their supporting pole under wind, snow or other loads. The most secure assembly is a tower that provides greater stability and longer service than a pole or guy-wire mount. While using a tower also makes servicing antennas and preamplifiers easier, poles are less expensive and easier to install. However, whatever type of mount is chosen, it must be correctly installed (see Figure 8-5).

In general, all masting should be at least 1.5 inch, thick-wall, galvanized conduit. The tops should be capped and the bottoms left open to allow water to drain. All mounting hardware should be stainless or galvanized steel. Cadmium plated hardware will not stand the test of time in most environments. A rusted fastener is both a safety hazard in a tower and poor advertising for a professional outfit.

Install towers according to their manufacturer's instructions. The local supplier should know the weight of each antenna in order to calculate wind and, if applicable, ice loading so that an adequate tower can be selected. The installation procedure usually involves setting a base section in a pit and filling it with concrete. The concrete must set for at least 24 hours or more to properly cure and to gain sufficient strength before other sections are installed. (This same consideration also applies when constructing satellite pads). When mounting a tower on a roof top, be absolutely sure that the lag bolts are screwed into the joists and not just into the plywood sheath. If the roof joists were not used for anchoring, the plywood could be ripped off from the roof in a high wind. One way to verify that each bolt is properly secured is to drill a small pilot hole for each one to make sure that a joist is in its path. The same procedure should also be followed for anchoring all the down-guy-wire hooks. If the test hole had been drilled in the wrong location, if obviously must be waterproofed before the job is completed.

When installing a pole mount on a roof, follow similar procedures as for a tower (see Figure 8-6). Insure that the bolts for the base and down guys are each securely screwed into roof joists. In order to waterproof the roof, each pilot hole should be filled with roof sealant prior to bolting down the mounts, and then all mounts and hooks should be well covered with a liberal application of sealant after installation. This is a crucial step in preventing water leaks and corrosion to the supporting bolts and wooden structure. Use turnbuckles on all down-guys to allow the position of the antenna and the tension on each guy wire to be adjusted. The turnbuckles should be installed so that the adjusting screws are turned out at least 3/4 of the way and then tightened after installation. Wrap all guy wires at least twice through the turnbuckle and hook eyes to prevent them from pulling out. It is a good practice to avoid locations near electrical connections or devices such as those encountered in the vicinity of outdoor air conditioners. If the conductive tower or antenna does topple and contact an electrical "hot-point" serious damage may result.

Well conceived planning is the key to properly installing a pole mount. For example, once the pole is raised into the air, it would be an annoyance to have to pull it down because the installation of the third row of guy wires had been overlooked. Most pole mounts have at least one telescoping section to

Figure 8-5. Typical SMATV Antenna Array. *This array features a large, broadband VHF antenna which receives signals from local transmitters located 15 miles away. The UHF antenna detects programming from local stations at a distance of 40 miles.*

CONSTRUCTIONS & INSTALLATION

allow the antenna to be extended up to an additional 50 feet in length. Install all the equipment needed before, not after, raising the pole. Once all the guy wires have been arranged in bundles to prevent them from becoming tangled, the antenna can be mounted to the pole top and secured with a bolt or cotter pin through both the pole and antenna.

Next raise the pole in its mount and attach the first set of guy wires. At this point, a free standing section of usually 10 feet is in place with the remainder of the telescoping sections on the interior. Then the other sections can be raised. When doing so, it is useful to have two other technicians on hand to steady the pole with guy wires as they are extended. These guys wires cannot be fastened until all the sections have been raised into place. After each section is fully extended, install a cotter pin or bolt through its body. This prevents the individual section on the pole from turning except when all others do so. After all sections are up, tighten all guy wires and adjust the turnbuckles to secure the pole in a vertical orientation.

When preamplifiers are being mounted on the antenna mast, they should be connected within easy reach for servicing but as close to the antenna as possible to minimize downlead losses.

After the antenna is installed, it must be aligned with the broadcast transmitter. To do so, connect a signal meter to the downlead and tune in the desired channel. While turning the antenna or pole, watch for the position where the peak level reading occurs on the meter. Do the same alignment procedure for each antenna installed. After peaking, the antenna must be firmly fastened to its mount. This alignment should be periodically checked after the system is up and running. When installing multiple antennas on a mount, it is necessary to observe some important rules regarding their spacing. Antennas which are too close to one another will cause reflections that can result in ghosting or S/N ratio decreases due to phase problems in the received signal. Antennas must be separated by at least half a wavelength as measured at the lowest frequency channel being detected. Wavelengths for VHF channels are listed in Table 8-1 (also see Appendix E for a more complete list). Antennas should also be installed at least one wavelength above the roof to minimize their effects on performance.

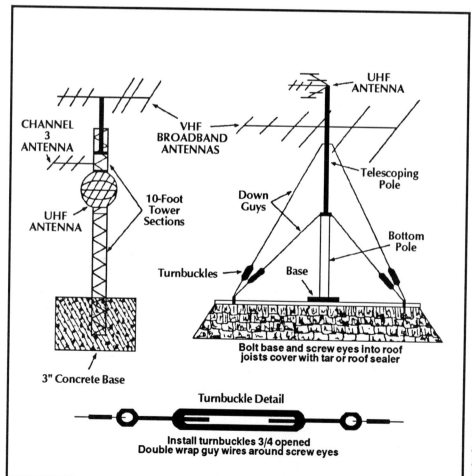

Figure 8-6. Installing an Off-Air Antenna. *The details of mounting and securing an off-air antenna are illustrated here.*

CONSTRUCTIONS & INSTALLATION

TABLE 8-1. WAVELENGTHS for VHF TELEVISION CHANNELS

CHANNEL	WAVELENGTH (Inches)
2	205.0
3	186.0
4	170.0
5	148.0
6	138.0
7	66.5
8	64.5
9	62.2
10	60.5
11	58.5
12	57.0
13	55.2

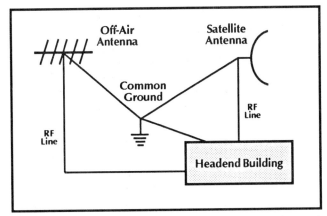

Figure 8-7. Grounding SMATV Antennas. *The typical grounding system layout for both satellite and off-air antennas is illustrated here. Adequate grounding provides both noise voltage reduction as well as transient protection. Satellite antennas, must be properly grounded, even when mounted on a pole in the ground.*

Once the off-air antennas and necessary preamps have been installed, all the downleads can be routed into the headend. All these cables should be wire-tied to the pole or tower. It is a wise policy, whenever possible, to run all lines in conduit to protect them from weather damage and vandalism. Complete the job by weatherproofing all entrance holes to prevent water from entering the headend and apply silicon sealant on all connections at the antenna. When antennas are installed correctly, they should last at least 5 to 10 years under normal operating conditions.

Grounding and Surge Protection

Adequate grounding is necessary to protect an SMATV system from a number of negative outcomes. Grounding can reduce the hazards of electric shock, can limit over-voltages and thus protect equipment, and can provide a rapid, non-destructive path to shunt away lightning strikes.

The off-air antenna is usually the highest structure in the vicinity and is a prime target for lightning hits. It should be grounded by connecting a lighting rod, usually constructed of copper or copper-clad steel, via a woven, bare copper conductor directly to a 0.5 inch by 8 foot long copper-clad steel ground rod. The satellite antenna should also have a lightning rod attached to its highest point which is connected by a number 0 cable to a ground rod. All points in the electrical system, both headend and distribution sections, should be protected by being securely grounded together (see Figure 8-7).

Surge protectors must also be used to protect electronic components from damaging voltage spikes. Some higher quality brands can handle surges of up to 50,000 volts and are able to respond in microseconds.

Distribution System Construction

Constructing distribution systems is more complicated than installing headends because it involves a wider range of tasks and talents. Cable lines can follow both underground and above-ground paths or a combination of these methods. Each has its own particular set of equipment requirements and techniques. For example, underground construction requires the use of special trenchers and boring tools, while pre- and post-wiring requires ladders, large drills with long drill bits used in conjunction with concrete hammer drills. Whatever the design, the in-house installation crews must be experienced and capable. For example, handling hardline cable requires special skills to prevent kinking and damage. Also, in general, constructing an SMATV system requires the disruption and alteration of a client's property. This further mandates that the necessary skills and knowledge are in hand before any of the following procedures are attempted.

CONSTRUCTIONS & INSTALLATION

Above-Ground Signal Distribution

The simplest distribution systems are those having a minimum amount of underground construction. It is much easier to route wiring through attics and between buildings than it is to open up trenches and streets. A good strategy in an SMATV design is to always try to minimize underground construction. The savings can be quite substantial.

The most common cable routes are in attics or basements. It is fortunate when a building happens to have attic or basement access. Cable can then be run from each apartment via the attic or basement and across and out to an externally mounted lockbox. When postwiring, namely wiring an existing system, cables may have to be run through stacking closets or on exterior walls before being fed back into an attic or basement. In contrast, when wiring a building under construction, it is a simple task to drill through the studs and bring the wires together in the attic or basement. Having an open structure to wire before the drywall has been installed certainly makes running cables much easier. In any case, an attic or basement can generally provide the easiest route to install and distribute signals.

Some buildings are constructed on slab foundations or have no attics. These usually require external wiring to reach every apartment. However, if such a structure has stacked closets, cables can enter at the bottom or top units and the rest of the distribution system can be internally installed. If stacked closets cannot be used, then the wiring must be routed on the exterior walls. Always try to find a location for cable runs that is as unobtrusive as possible. It is best to avoid the front of a unit and instead look for routes next to rain gutters or downspouts. When running interior or exterior feeds, use conduit or molding to hide and secure cables.

Routing the RF signal from building to building can be accomplished in many cases without resorting to underground construction. If the buildings are reasonably close to each other, cables can be looped between roofs with minimal visual impact. Self-supporting, messengered cable, coaxial lines having an additional cable for strength that also travels along the route, can be strung between buildings and secured to hooks. The advantage of having the bulk of the distribution system running between the top of each building and perhaps through the attics is security from vandalism. But it is important to check local city ordinances for any restrictions regarding aerial cables or wires and to be very careful if other wires, especially power lines, are present.

Messengered cables are generally supported by an 1/8 inch galvanized wire, although runs in excess of 100 feet use 1/4 inch wire. This special cable should be tensioned between the antenna mast and point of entry to the headend building and between any other two points of use. The coax should be wrapped around the messenger in a spiral of one turn every two feet. For larger diameter, low loss aluminum sheathed cables, a thicker 3/16 inch messenger is required. In this case, instead of wrapping the coax around the messenger, a 0.045 inch stainless steel lashing wire is used to attach the coax to the messenger. Although lashing machines that accomplish this task are commercially available, most shorter runs can easily be installed by hand.

Although running aerial cables from building to building may be an easier installation task, most operators today prefer to bury all of distribution cables.

Underground Signal Distribution

There will always come a time when it is necessary to run a cable line in the ground to bridge the gap between buildings. This situation may arise because city codes or aesthetic considerations prevent use of aerial drops. Or the design of a large project may demand that hardline coax be installed in underground conduits. Whatever the case, installing an underground distribution system involves more than just digging a ditch. Skill, care and common sense are required to prevent serious mishaps. Attention must be given to protecting cables from the elements as well as from accidental damage.

The first thing to know before opening a trench is the location of all other utilities in the ground. Indiscriminately tearing open the earth without any idea of the underground structure is inviting disaster. A broken gas line can explode with tremendous force and level whole buildings, severed power lines can quickly electrocute a man, and damaged phone lines, aside from being an embarrassment, are extremely expensive to repair. It is imperative to locate all utility routes by contacting all the local utility companies before beginning excavation in order to avoid such outcomes. Their personnel will visit the site at no charge to mark the exact location of their lines. In addition, some cities have companies that specialize in tracing utility lines. If these professionals mark an area as clear but problems later arise, they pay to fix it. Not calling for "locates" and hitting a line causes a loss of credibility as well as money.

CONSTRUCTIONS & INSTALLATION

After mapping the utility routes, excavation can begin. When using power equipment such as a trencher or a concrete saw, it is important to be familiar with its operation (see Figure 8-8). Practice in an area that is free from utilities and that can easily be refilled. Excavating raw earth without any surface landscaping is a simple task. However, if there is sod or grass present, lay a sheet of plastic next to the proposed trench line. After using a flat shovel or a commercially available powered sod cutter to gently cut sections of the top layer of grass and dirt, lay them on the plastic. When the trench is completed and refilled, this sod can be easily replaced. Minimum required trenching depths vary from 12 to 20 inches depending on local codes. Check local ordinances in order to comply with specifications. In those cases where the trench has not been cut sufficiently deep, the local inspector has the option to insist that the cable be pulled out and all the work be redone.

Another popular method of routing cables underground is to directly bury them using a vibratory flow machine. This device makes a slit in the ground to the required depth. The cable which is encased in an armor casing or a flexible PVC conduit is inserted into the ground immediately after the plow has made its cut. While this method has the advantage of reducing installation time, once the burial is completed in most cases the cable cannot be easily removed. Servicing a severed line may also require retrenching an entire section.

Once the trench is excavated, the cable can be installed in the conduits. Although some brands of "direct burial" cables are designed to be buried without any extra protection in the ground, the best policy is still to use PVC conduits for additional protection against water, minerals and small animals like rodents which can cause damage. Using conduit also makes replacement of bad or damaged cables or insertion of additional cables possible even after the burial has been completed.

Schedule 40 conduit with sufficiently large internal diameter is normally used to allow cables to be easily threaded into and out from the lines, although some cities will permit the use of thin-wall, communications conduit. Conduit is extremely fragile and must be handled with care. All sections and sweeps should be glued together to prevent water ingress. It is important not to use any right angle fittings for corners or bends because large radius sweeps are necessary to allow the cable to be easily pulled through the conduit. Use duct tape on the exposed ends to prevent any dirt from entering its interior prior to in-

Figure 8-8. Power Trencher. *This machine is capable of digging a 1, 2 or 3 inch wide trench from 8 to 12 inches deep at a rate of 20 to 25 feet per minute. (Courtesy T.H. Riley Manufacturing)*

stalling coax. After all the conduit is in place, fill in the trench with the removed earth, replace any sod or grass that has been previously removed, and compact and rake it so that the original condition is approximated. Then water the entire trench, clean up any remaining earth and sweep the area clean.

The next step is running the cable through the conduit interior. This is accomplished by installing a length of pull rope which is made of polypropylene designed specifically for this purpose. Fasten this rope to a lightweight plastic cone or ball slightly smaller than the conduit diameter. This is known as a mouse. Insert the mouse into the conduit and deliver a blast of compressed air to propel it to the other end. This not only inserts the pull rope in place, it insures that the conduit is open and clear.

The coax can be connected to the pull rope with a device known as a "Chinese finger". This is a long wire basket much like a child's finger trap. Have one person pull the cable while the other pushes it into the conduit. Using an adequate amount of cable pulling lubricant makes the job easier. Be very careful not to kink the outer shield or to skin off the cover on the cable. Installing cables, especially hardline, requires patience and skill to avoid damage.

Underground cables are often routed to distribution pedestals (see Figure 8-9). These units are typi-

CONSTRUCTIONS & INSTALLATION

cally installed halfway between buildings or at other convenient points where the amplifiers, couplers and other interface equipment can be installed for easy access.

System Activation

Once the headend has been installed and the distribution plant is in place, the time has come to activate the system. In a large franchise operation this step is considered to be a major operation. However, in a smaller scale SMATV system it may be as simple as powering on the headend. As is always the case, a little experience and practice can help in simplifying the operation.

All voltage levels at the headend and throughout the distribution network must first be adjusted and then these levels should be recorded. Both FCC recommended system parameters and "good engineering practice" parameters at the subscriber drop are listed in Table 8-2. Although these are recommendations that were designed to apply to top-quality cable systems, they are useful guidelines for those who wish to adhere to stringent practices and deliver excellent video and audio signals to customers.

A field strength meter (FSM) is an indispensable tool in activating a system as well as during installation and troubleshooting (see Chapter 10 for more details). Most professional units can be operated from line or battery power and feature a video or earphone output. The unit used during system activation must be capable of monitoring all channel frequencies that are being distributed.

The first step in activating a headend is to "burn in" all the electronic equipment. These components should be run for 24 hours to achieve temperature stability. Tune the satellite receivers to their chosen transponders and, if using frequency agile modulators, set them to their assigned output channels.

Figure 8-9. Exterior Distribution Pedestal. *This unit contains a Jerrold directional coupler and a line extender amplifier having a 31 to 35 dB gain. The input is set at 15 dBmV and the AGC output is 48 to 50 dBmV. It is part of a 2000 unit system. The CWI pedestal is installed halfway between two buildings.*

CONSTRUCTIONS & INSTALLATION

The FSM is then used to align the off-air antennas while checking for ghosts with a portable, color television. If signal levels are stronger or weaker than were indicated during the site survey or if there is any unexpected interference, the headend design should be consulted to create a simple fix, if possible. These problems should be solved before proceeding with the next step. In those cases where a preamp has been installed on a particular antenna and the level is too high, a simple solution may be to pad the preamp input.

Next, the off-air input mixers should be balanced in order to deliver an even output signal across the entire bandwidth. Set the RF output levels on the modulators and strip amplifiers to the desired output level, usually 45 to 55 dBmV, and to within 1 or 2 dBmV of each other. When the strip amps feature AGCs, follow their manufacturer's procedures in setting levels.

If a broadband amplifier is used on off-air channels, balancing may be somewhat more difficult. In this case, the signals must be equalized using the input levels from the antennas. If the channels are not adjacent in frequency, the levels should be set within a maximum of 6 to 7 dBmV of each other. If they are adjacent, they should be adjusted within a maximum of 1 dBmV of each other. The output of the amplifier should operate within its designed level and 2 to 3 dB below its maximum output level. Use the manual gain control while observing its output on a field strength meter.

In general, the audio signal levels should be set at 13 to 15 dB but not more than 17 dB below the video carriers. Most strip amplifiers and modulators have aural carrier reducers to accomplish this task, although the success of this step depends, to a certain extent, on the aural levels of any non-adjustable gear. When necessary, separate aural carrier traps can be used to accomplish this task.

The video and audio modulation levels must be set to match those of the strip amps. The same settings should be used for off-air and satellite signal modulators and processors. The headend can then finally be connected to the distribution plant.

It is wise practice to test a distribution system with an ohmmeter for short circuits and non-terminated lines before powering on. If distribution amplifiers are used, a short could cause equipment damage. The characteristics of the various passive devices in the feed lines should be understood before testing. Usually taps show an open circuit between the input and ground and pass dc from input to output. If this is the case, connect the meter between the center conductor and ground at the input to each feeder line at the headend. If there are no

TABLE 8-2. FCC REQUIRED PERFORMANCE PARAMETERS at SUBSCRIBER TAP and GOOD ENGINEERING PRACTICE VALUES

Signal Parameter	FCC Recommended Levels	Good Engineering Practice
Minimum Video Signal Level	0 dBmV (1000 µV)	3 dBmV (1400 µV)
Maximum Video Signal Level	below overload	10 dBmV (3140 µV)
Max. Difference between Adjacent Video Carriers	3 dB	1 dB
Max. Difference between any Video Carriers	12 dB	7 dB
Minimum A/V Carrier Ratio	−13 dB	−13 dB
Maximum A/V Carrier Ratio	−17 dB	−17 dB
Long-Term Signal Variations	12 dB	10 dB
Maximum FM Station Level	not specified	−7 dBmV
Minimum FM Station Level	not specified	−20 dBmV
Maximum Difference between FM Station Levels	not specified	3 dB
Signal-to-Noise Ratio	36 dB (minimum)	42 dB (minimum)
Hum	5%	1%
Tap Isolation	18 dB	30 dB
Reflections (Ghosts)		−40 dB

CONSTRUCTIONS & INSTALLATION

shorts and if the feeder is properly terminated, the resistance should be approximately 100 ohms. This equals 75 ohms for the terminator plus the resistance of the cable. A resistance reading of zero ohms indicates an installation error. An open circuit indicates an improperly installed feeder or a missing terminator.

If the taps do not pass dc or show a short to ground, then it is best to proceed directly to the next test, namely testing signal levels at each output to visually confirm the presence of good quality pictures. In general, when inspecting a picture the problems to watch for are (1) snow, (2) beats, (3) cross-mod distortions, (4) ghosting and (5) audio hum and picture crispness. These factors are examined in more detail in the section on troubleshooting in Chapter 10.

In some cases, testing the distribution system is rather simple. For a small scale system that needs only the output power of the headend, all that is necessary is to check that all outlets are receiving adequate levels of signal, about 3 to 10 dBmV. After the last outlet or tap on the system has been examined, the signal levels should be recorded. These values should be included in the "as built" drawings created at the end of the project. Once it has been verified that all taps are active, the task is completed.

Larger-scale private cable systems also require the activation of signal amplifiers. This is generally a simple task because most systems have only a few distribution amplifiers. Check that signal is present at the amplifier input and that power is running into the unit. If necessary use signal pads and equalizers to bring the input to an even level that meets each amplifier's input specifications. Most amplifiers have a means of adjusting their output levels to compensate for differential frequency attenuation on long cable runs. This tilt should be set in order to provide the necessary gain for the high-band channels to reach the end of all distribution lines with levels equal to those of the lower frequency signals. Again, once the presence of signals at all taps and at the end of all lines is verified and levels are recorded, the system activation is completed.

Note that once the headend and distribution systems have been balanced and activated, it is important to document any changes on the system design drawings. These plans then should be formally redrawn to reflect those changes. A copy should be given to the property owner and one should be retained for records. If this system ever needs servicing or if another project having a similar design is being tackled, these documents can prove invaluable.

System Performance Levels and Maintenance

Record keeping is an extremely important component of system maintenance. Many operators fail to keep adequate records of output levels and other factors that later could prove extremely valuable in solving problems. As a result, troubles which are repetitive in nature and which could have been easily solved with the proper historical data are not immediately identified. To the same end, all system levels should be recorded and checked on a regular basis. In contrast, CATV operators must conform to mandated record keeping procedures. Some of the FCC recording forms are listed in Figures 8-10a through 8-10g. Although many FCC regulations that franchised CATV companies must follow do not apply to private cable operators in the United States, voluntary compliance to these rules is wise practice for two reasons. These are well-considered procedures that can aid in managing a service operation. If the private cable industry functions in a professional and efficient manner, regulations that could create an additional record keeping burden may not have to be legislated.

The task of periodic maintenance is the direct responsibility of the chief technician who should design and implement regular inspection procedures that provide an ongoing record of system performance and service. This includes measuring all headend and system levels as well as visually inspecting television outputs on a random basis.

A professional operator should have an in-house capability to repair the basic electronic components. The test bench should consist of a sweep frequency generator, an impedance matched RF detector, a calibrated RF market generator, a variable attenuator and a good oscilloscope. In general, broadband amplifiers are relatively simple to repair while strip amps which are frequency sensitive require the more sophisticated electronic testing setup.

In general, a number of strategies that can lead to technical excellence should be implemented. Developing an ongoing training program for all company technicians and taking advantage of programs offered by various manufacturers is a wise policy. The useful information gleaned from records of all on-the-job observations and troubleshooting calls should also be noted and used to identify any systematic problems. For example, recurring calls for related problems can indicate possible design faults.

Other Requirements for In-House Construction

In addition to requiring the manpower and skills to perform all tasks, having an in-house construction capability brings with it the responsibility of conforming with certain legal requirements. The principle one may be the need for a contractor's license. Without this document, it is illegal to proceed with the construction of a commercial job. In order to obtain most building permits, a contractor's license is also required. In addition, most states and provinces require that a number of years of experience can be documented before successfully applying for a license. It may be possible to hire a licensed contractor as an employee in order to have him license the company. In any case, do not engage in any construction expansion without the proper credentials.

Insurance is also a crucial element. A contractor must have an adequately sized liability policy for protection against possible lawsuits. The potential for damage and injury is especially high when engaging in underground construction. Such risky endeavors create high insurance premiums and can make obtaining insurance virtually impossible for the novice without experience. These same factors also result in increasing the workmen's compensation rate. The combined insurance premiums and resulting increased overhead can end up consuming a large portion of profits. One too many claims can lead to bankruptcy. These are very real factors in the trade-off between having an in-house construction capability and hiring subcontractors.

The payroll for in-house crews can be based on both piece and hourly work. Technicians involved in headend construction should be paid on an hourly basis, as they are not employed solely in construction activities. Those crews that are only engaged in construction should be paid on a piecework schedule. This allows for a tighter control over expenses and eliminates wasted time. Those individuals who are motivated will be rewarded for their hard work and those who are lazy will soon tire of their meager paychecks and quit. Although the use of piecework rates and schedules requires some additional paperwork, it is more than offset by the benefits of having lower wastes and improved cost controls.

Construction Planning

Construction planning for in-house work is identical to that in effect when hiring a subcontractor. However, there is one major difference. The SMATV firm is now totally responsible for completing the job so that there is no subcontractor to blame or yell at when work falls behind schedule. In addition, a detailed and accurate plan for the work force is necessary if the job is to be completed under budget and on time. Lack of planning is the major cause of construction delays.

As discussed earlier, there are a number of structured planning techniques that lend themselves to construction activities. For example, the Program Evaluation and Review Technique (PERT) was developed by the U.S. Navy for managing complex one-time projects and it lends itself very well to cable TV construction planning. In this scheme, the project is subdivided into events and activities. An event is a milestone or specific accomplishment, while an activity is the effort and resources needed to achieve the event. Events and activities are shown in order of occurrence and outlined on a chart. The time frame for completion is indicated at each event and, as a result, the total project duration can be estimated.

PERT charts also discipline managers to create detailed plans and to establish and honor commitments. They also facilitate tracking a project's progress and developing an overall picture of the task. There are a number of computerized versions of PERT charts available that run on PC systems. Again, it cannot be overstated that careful planning is the key to successful project management.

The Final Decision

In the final analysis, the decision whether to use in-house managed or subcontractor labor has to be based on economic considerations. Employing a subcontractor can reduce insurance costs, equipment inventories and payroll expenses and, as well, can recruit expertise and a commitment to get the job done. On the other hand, using in-house labor can offer an operator more control and potentially higher profit margins. The choice rests on a careful examination of needs and future construction projections. If there is not enough work scheduled to justify a full time staff, use a subcontractor. Remember that payroll is an expense that needs to be met every week regardless of the amount of construc-

CONSTRUCTIONS & INSTALLATION

tion activity under way and the resulting cash flow. In contrast, a subcontractor is paid as work objectives are met. Clearly all the evidence must be carefully evaluated in order to choose the best option that suits the particular needs of each company.

FCC Performance Test Log

SYSTEM LOCATION _____   DATE SYSTEM BEGAN OPERATION:
SYSTEM OPERATOR _____   _____
_____   NO. OF SUBSCRIBERS _____
_____   TEST DATE(S) _____

Tests Conducted

FCC Reference	System Parameter	Test Conducted (√)*
76.605(a) (1)	Channel frequency boundaries	
76.605(a) (2)	Visual carrier frequency	
76.605(a) (3)	Aural carrier separation	
76.605(a) (4)	Minimum visual carrier level	
76.605(a) (5)	Visual carrier level variation	
76.605(a) (6)	Aural carrier level	
76.605(a) (7)	Hum and low-frequency disturbance level	
76.605(a) (8)	Channel frequency response	
76.605(a) (9)	Visual carrier-to-noise ratio	
	Visual carrier-to-cochannel ratio	
76.605(a) (10)	Visual carrier-to-coherent disturbance ratio	
76.605(a) (11)	Subscriber terminal isolation	
76.605(a) (12)	Radiation	

*Insert NR to mean not required (and not tested) at this time according to the most recent schedule for testing ordered by the FCC.

Figure 8-10a. FCC Performance Criteria for Cable TV Systems

CONSTRUCTIONS & INSTALLATION

Test Procedures

Where a generally recognized standard test procedure (such as those issued by the NCTA) is used, indicate the issuing agency and the standard number. Where a non-standard test procedure is used, provide a sketch of the test setup and a description of the procedure on the following pages.

System Parameter	Test Procedure	
	Standard	Non-Standard (√)
Visual carrier frequency		
Aural carrier separation		
Minimum visual carrier level		
Visual carrier level variation		
Aural carrier level		
Hum and low-frequency disturbance level		
Channel frequency response		
Visual carrier-to-noise ratio		
Visual carrier-to-cochannel ratio		
Visual carrier-to-coherent disturbance ratio		
Subscriber terminal isolation		
Radiation		
Visual carrier to cross modulation ratio		
Noise in the slot ratio		

Figure 8-10b. FCC Performance Criteria for Cable TV Systems

CONSTRUCTIONS & INSTALLATION

Line Extender Levels

DATE _____

Amplifier # ___ Phase ___ Page ___ of ___   Cascade # _____   Map # _____

BRIDGER Ch 3-

Ch -WW

#										
___ PAD- Eq- ___ IN		3	6	A	E	I	7	10	13
		K	R	W	EE	JJ	RR	WW	
VAC- ___										
VDC- ___		2	6	A	E	I	7	10	13
POWER IN THRU OUT		K	R	W	EE	JJ	RR	WW	
___ PAD- Eq- ___ IN		2	6	A	E	I	7	10	13
		K	R	W	EE	JJ	RR	WW	
VAC- ___										
VDC- ___		2	6	A	E	I	7	10	13
POWER IN THRU OUT		K	R	W	EE	JJ	RR	WW	
___ PAD- Eq- ___ IN		2	6	A	E	I	7	10	13
		K	R	W	EE	JJ	RR	WW	
VAC- ___										
VDC- ___		2	6	A	E	I	7	10	13
POWER IN THRU OUT		K	B	W	EE	JJ	RR	WW	

SIGNATURE _____

Figure 8-10c. FCC Performance Criteria for Cable TV Systems

CONSTRUCTIONS & INSTALLATION

Frequency Accuracy Test
(One Test Required At Headend)

Cable Ch. No.	Broadcast Ch.		Visual Carrier Frequency (MHz)		Aural Carrier Frequency (MHz)	Aural Carrier Separation (MHz)
	Call	Offset	W/O Set Converter	W Set Converter (Where Required)		

Figure 8-10d. FCC Performance Criteria for Cable TV Systems

CONSTRUCTIONS & INSTALLATION

Visual Carrier Level Variation Tests
(Required at Three Locations)

TEST LOCATION NO. _____ : TEST POINT* _____ AFTER _____ TRUNK STATIONS
AND _____ LINE EXTENDERS, LOCATED AT** _____

Cable Ch. No.	Time (Approx. 2 Hr. Intervals) / Visual Carrier Level (dBmV)												Max. Variation Over 24 Hrs. (dB)
2													
3													
4													
5													
6													
A													
B													
C													
D													
E													
F													
G													
H													
I													
7													
8													
9													
10													
11													
12													
13													
J													

*Electrical location: subscriber, terminal, tap, etc.
**Physical location: address, pole no., etc.

Figure 8-10e. FCC Performance Criteria for Cable TV Systems

CONSTRUCTIONS & INSTALLATION

Overall Subscriber Tests
(Required at Three Locations)

TEST LOCATION NO. _____ :TEST POINT* _____ AFTER _____ TRUNK STATIONS AND _____ LINE EXTENDERS.

LOCATED AT** _____

Cable Ch. No.	Visual Carrier Level (dBmV)	Aural Carrier Level (dB Below Visual)	Hum Modulation (%)	Channel Response (± dB)	Carrier-to-Noise Ratio (dB) Measured	Carrier-to-Noise Ratio (dB) Corrected	Carrier-to-Cochannel Ratio (dB)	Carrier-to-Intermod. Ratio (dB)	Subscriber Terminal Isolation (dB)

*Electrical location: subscriber, terminal, tap, etc.
**Physical location: address, pole no., etc.

Figure 8-10f. FCC Performance Criteria for Cable TV Systems

CONSTRUCTIONS & INSTALLATION

Final Proof of Performance

SYSTEM _____ DATE _____ TIME _____ TEMP _____
AMPLIFIER NUMBER_____ LOCATION _____
_____ CHANNEL CASCADE: _____

CH	Levels
2	
3	
4	
5	
6	
A	
B	
C	
D	
E	
F	
G	
H	
I	
7	
8	
9	
10	
11	
12	
13	
J	
K	
L	
M	
N	
P	
R	
S	
T	
U	
V	
W	
AA	
BB	
CC	
DD	
EE	
FF	
GG	
HH	
II	
JJ	
KK	
LL	
MM	
NN	
OO	
PP	
QQ	
RR	
SS	
TT	
UU	
VV	
WW	

Crossmodulation
CH 2 _____
CH A _____
CH 7 _____
CH 13 _____
CH R _____
CH W _____
CH HH _____
CH RR _____
CH WW _____

Second Order
CH 2 _____
CH A _____
CH 7 _____
CH 13 _____
CH R _____
CH W _____
CH HH _____
CH RR _____
CH WW _____

Composite Triple Beat
CH 2 _____
CH A _____
CH 7 _____
CH 13 _____
CH R _____
CH W _____
CH HH _____
CH RR _____
CH WW _____

Frequency vs Response
CH 2 _____
CH A _____
CH 7 _____
CH 13 _____
CH R _____
CH W _____
CH HH _____
CH RR _____
CH WW _____

Carrier to Noise
CH 2 _____
CH A _____
CH 7 _____
CH 13 _____
CH R _____
CH W _____
CH HH _____
CH RR _____
CH WW _____

Figure 8-10g. FCC Performance Criteria for Cable TV Systems

D. LOCAL ORIGINATION and ADDITIONAL SMATV SERVICES

The private cable industry has discovered that once the cable has been installed, numerous services in addition to either satellite or off-air television broadcasts can be offered to subscribers. These extra services can be excellent marketing tools, new information sources or additional pay-TV entertainment. All are the basis for increased profits. SMATV systems have the advantage of being upgradable so that more channels than would ordinarily be required can be transmitted. Frequency diplexing technology takes this flexibility one step further by eliminating the need to rewire systems when an existing system is upgraded to two-way operation.

One-Way Services

Many types of services can be inserted into a private cable system at the headend for distribution to all subscribers. Among the options are character generators for services such as local weather information or community news, FM stereo for either simulcast with video or radio broadcasts, VCR movie playbacks, and video cameras used for surveillance and security. The technology required for adding locally originated signals is generally quite simple (see Figure 8-11).

Character Generators

Character generators which provide both text and video outputs can be low-cost systems which interface at the headend via a modulator (see Figure 8-12). The simplest systems can be operated from a home computer keyboard. The computer's video output is fed into a modulator and then to the distribution system. The accompanying audio is often inserted from a local radio station. Factors such as the total number of pages that can be stored for later insertion and other system features are only limited by the software and the memory of the microprocessor. The more sophisticated character generators can superimpose color text over detailed pictures. Many commercially available teletext services such as weather reports, airline schedules or stock market quotations can also be accessed via phone line modems for SMATV distribution. While the options available to a private cable operator are expanding as the technical capabilities of character generators improve, their costs continue to decline.

Some of the higher resolution character generators are specifically designed to drive high quality video monitors. In these cases, small text may become unreadable and larger letters should be used. Higher resolution units also may have signal bandwidths that exceed the standard 4.5 MHz television channel width and may therefore require low pass filtering before being connected to a headend modulator. If the higher frequencies are not removed before modulation, inserting a character generator signal may beat and interfere with adjacent channels.

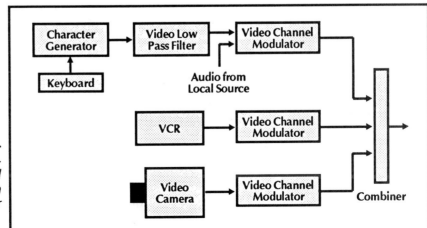

Figure 8-11. Locally Originated Services. *A VCR, video camera or character generator can easily be inserted into an SMATV headend as shown in this diagram. The video lowpass filter may not be necessary in some cases.*

183

CONSTRUCTIONS & INSTALLATION

Figure 8-13. Low-Cost Character Generation System. *This system utilizes an Atari 800 computer with a preprogrammed graphics cartridge. While more expensive units may cost from $2000 to $3000, this system with a $100 computer and a $200 cartridge makes purchase quite inexpensive. This user friendly, menu-driven system features 64 pages of information and an on-going crawl line which appears at the bottom of a television screen for inserting commercials or other subscriber information. Three different type styles are available and flashing characters appearing in multiple colors can be used to attract viewer attention. This system can be modem driven to allow the information field to be changed from a remote location. Cartridges from Cable Graphic Sciences are also available for other computers such as the Radio Shack.*

Video Cameras

A video camera can also be easily connected at the headend via a channel modulator for broadcast into the distribution system. While one-way inserted video camera signals have a rather limited range of capabilities, two-way systems allow the creation of a sophisticated security system having a virtually unlimited number of channels, especially when numerous, higher band and UHF channels are used. If low pass filters are employed at key points in the distribution system or signal inversion or other scrambling techniques are used, these security signals may only be accessible at pre-defined control areas.

Video Cassette Recorders

Insertion of VCRs for playback of feature length movies or local commercials allows an operator to have more control over the quality of programming. Most modulators in consumer brands of VCRs use nearly twice the bandwidth as broadcasted television channels and have rather limited output levels. Therefore, the VCR output should be filtered and inserted into a commercial grade modulator.

Two-Way Services

The capability of a private cable system can be tremendously expanded by the use of two-way services. With a well-conceived design, signals can be added and distributed from any point in the system. Therefore, for example, a cabled hotel might become a teleconferencing center whereby any room could be used to originate as well as tune to live broadcasts from any other room in the complex. A condominium complex could locate security cameras at any drop without the need for additional wiring. Or a two-way, voice communications system could be installed by employing citizen band radios operating in the 26.9 to 27.4 MHz band. The possibilities for creatively using a two-way private cable system are limited only by the imagination of its designer.

The simplest two-way design would use an existing cable network to transmit signals one-way on one frequency and the other way on a different channel (see Figure 8-13). For example, headend signals could be distributed on VHF channels 2 through 5 and return information could be relayed on channel 10. Signals would then be combined at the headend using a 6-way splitter and divided at any drop using a 2-way splitter. However, this simple

CONSTRUCTIONS & INSTALLATION

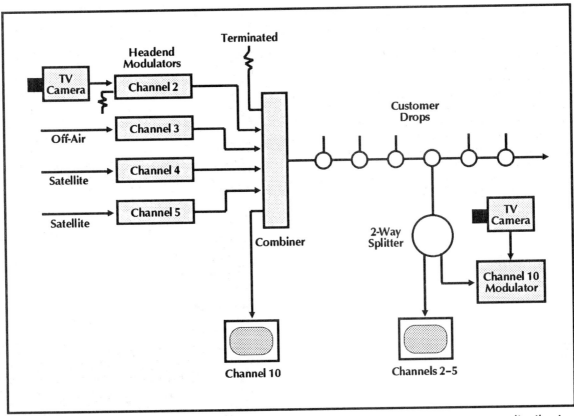

Figure 8-13. A Simple Two-Way Design. *The simplest method to design a two-way distribution system is by using standard splitters and combiners with channels separated in frequency. In this system, channels 2, 3, 4 and 5 are relayed in one direction while signals are returned to the headend or anywhere else in the system on channel 10. Problems with poor isolation between the two directions of transmission can easily arise in such a design.*

design has two inherent disadvantages. The insertion loss of the splitters would be high enough to serious degrade S/N ratios. But even more importantly, the upper and lower channels would be too poorly isolated and cross-talk would most likely occur.

A much better solution is to use a "diplex" filter (see Figures 8-14 and 8-15). This special type of filter consists of one high-pass and one low-pass filter joined together at one end. Signals having frequencies up to the low-pass cut-off are only transmitted from the common port to the low pass port. Similarly, the higher frequency signals are only channeled through the high pass port. The frequency range in between these two cut-off points functions as a guard band which cannot be used to transmit signals. Thus, a common UHF/VHF combiner is, in essence, a diplex filter.

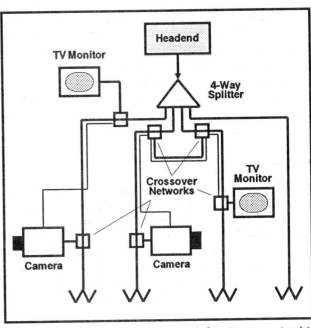

Figure 8-14. A Basic Two-Way Diplex System. *In this system the diplex cross-over networks are connected so that the camera signals are forced to follow only two pre-defined routes to their respective monitors.*

185

CONSTRUCTIONS & INSTALLATION

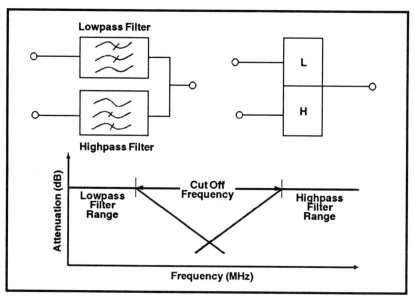

Figure 8-15. A Diplex Filter. *A diplex filter consists of a lowpass and a highpass filter joined together at one common end. The region between the lowpass filter upper cut-off frequency and the highpass filter lower cut-off frequency is a guard band where no transmission should be allowed. (Courtesy of Private Cable Magazine)*

Diplex filters can be designed to have virtually no insertion losses while providing excellent isolation between the two frequency ranges. Two-way systems are typically built to use the standard UHF and higher frequency channels for signal distribution and the subchannel frequencies for transmission in the reverse direction. This choice rarely affects an existing MATV or SMATV system because the VHF and higher channels are typically already used for one-way distribution. The standard "T" channels are listed in Table 8-3.

The use of T-channels for relaying signals upstream and VHF channels for downstream distribution is not a sacred choice. Diplex filters having a variety of split frequencies are available. Therefore, for example, a system could be designed which has the split between the VHF-lo and UHF-hi channels. In this case, the guard band would fall within the FM and midband regions of the spectrum.

Design Considerations

Although two-way distribution systems are very similar to standard SMATV designs, a number of design differences must be understood. First, cable losses at sub-band frequencies are relatively low compared to those occurring on VHF channels. The relative difference in attenuation between the lower and higher sub-band channels is substantially greater than the losses between VHF channels 2 and 13 because the ratio of these frequencies is higher in the lower band. To illustrate, the ratio between the video carrier center frequencies of channels T13 and T7 is 6.14 while the ratio for VHF channels 13 and 2 is just 3.82. The relative differences in tilt also follow this trend. As a result, the difference in tilt between the video and audio subcarrier within a single T-band channel can surprisingly be great enough to cause a "soft" picture and a "noisy" color since the color subcarrier lies 3.58 MHz above the video carrier. This would be particularly the case for channel T7. Its audio carrier at 11.5 MHz is 64% higher in frequency than the 7 MHz video carrier! Using the equation for calculating coaxial cable attenuation in

TABLE 8-3. SUB-VHF, "T" CHANNELS

Channel	Frequency Range (MHz)	Video Carrier (MHz)	Audio Carrier (MHz)
T7	5.75-11.75	7	11.5
T8	11.75-17.75	13	17.5
T9	17.75-23.75	19	23.5
T10	23.75-29.75	25	29.5
T11	29.75-35.75	31	35.5
T12	35.75-41.75	37	41.5
T13	41.75-47.75	43	47.5

CONSTRUCTIONS & INSTALLATION

Appendix B shows that the tilt between these carriers is a whopping 28%.

This large difference in "octaves" or ratios of frequencies make it imperative to also use higher quality "push-pull" amplifiers. This reduces the potential occurrence of second order distortions in those systems where more than one upstream channel is being employed.

Other factors which might normally be overlooked are also deserving of attention. For example, the diplex filter specs should be checked to note the precise location of the guard band. No channels which fall within this frequency region should be used. The frequency response of all passive devices must also be examined to ensure that these can handle frequencies as low as the 5.75 MHz lower edge of channel T7.

Two-way systems must also be well protected from ingress interference. These unwanted signals can be transmitted back to the headend for rebroadcast. In most cases, a good quality low pass filter can minimize any such difficulties.

An extensive two-way sub-channel network is outlined in Figure 8-16 as a design example of this technology.

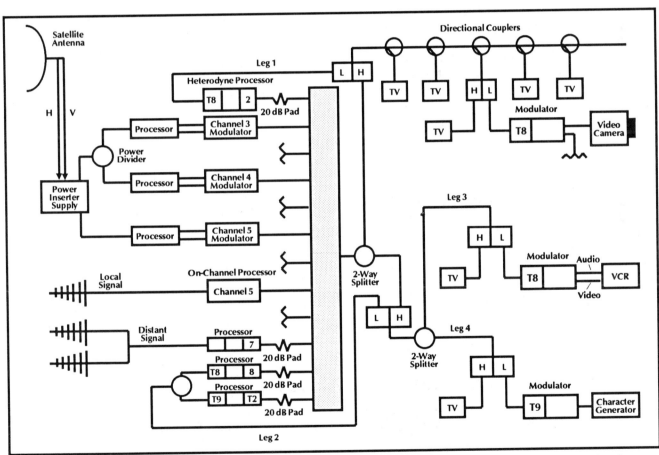

Figure 8-16. An Extensive Two-Way Network. *This distribution system uses three sub-channels to feed return signals to the headend. While each separate run can carry all three T-channels to the headend, there can be no duplication of channels in the two legs at the lower right which both feed through the same diplex filter and 2-way splitter. (Design Courtesy of Nexus Engineering)*

E. STEREO AUDIO SERVICES

Stereos, VCRs and televisions are evolving into "home entertainment centers." Most movies and other video specials that are produced today are accompanied by stereo sound. Stereo televisions and hi-fi VCRs have become popular consumer items.

Private cable systems can be easily designed to carry two forms of high quality stereo. First, numerous audio-only services are transmitted via satellite or off-air from local FM radio stations. These broadcasts include specialty audio programs of all varieties, talk shows and background music. The second type of service is stereo television. Stereo TV has been available to the home satellite TV market for over seven years and is just beginning to make inroads via conventional off-air broadcasts.

In order to understand the options available to an SMATV operator, the method by which an audio signal is transmitted either via satellite or off-air must first be examined.

Satellite Audio Reception

Audio signals can arrive at a satellite receiver in various forms. These messages are usually transmitted as subcarriers modulated onto the main video carrier. This is possible because although the baseband signal which is transmitted over each transponder occupies frequencies from zero to 10 MHz, the video signal only ranges from zero to approximately 4.5 MHz. Therefore, audio subcarriers can be inserted in the remaining space (see Figure 8-17).

The audio signal may also be transmitted by use of a technique known as sound-in-sync whereby a digital audio signal is inserted into either the horizontal or vertical blanking intervals of the standard television signal. This is perfectly feasible because although these blanking intervals occupy nearly 20% of the transmission time, they have little information content. Using sound-in-sync eliminates the need to use separate audio subcarriers and allows the satellite transponder power to be concentrated on the video information, so that its carrier-to-noise power can be increased.

Sound-in-sync digital audio is transmitted as part and parcel of the two principle satellite scrambling systems, the M/A COM VideoCipher II+ and the Oak Orion. Digital audio is also used as an integral component of the television signal in the Scientific Atlanta MAC broadcast format and a number of the encryption systems used in European countries. The decoders used in conjunction with these scrambling systems provide the necessary circuitry to recapture the baseband audio signals.

There are presently four different systems used to transmit stereo sound over satellite circuits: multiplex, Warner Amex, discrete and processed, narrow deviation stereo. Once these audio signals are decoded by either a stereo processor or a satellite receiver with a built-in processor, the baseband composite audio signal, namely the raw audio signal, is available for processing in the SMATV headend.

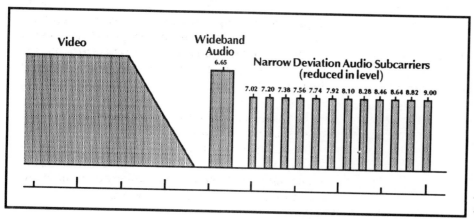

Figure 8-17. Satellite Audio Subcarriers. *While the video carrier occupies the zero to 4.5 MHz region of the baseband signal, audio subcarriers can be transmitted in the remainder of the band. While narrow band, processed audio subcarriers occupy less frequency space than wide band discrete subcarriers, they require more sophisticated signal processing to deliver the same quality output.*

CONSTRUCTIONS & INSTALLATION

Figure 8-18. Multiplex Stereo.
A multiplex stereo circuit processes the left and right audio channels so that both can be relayed on a single audio subcarrier and then reconstructed at the satellite receiver. The input audio is mixed to form the sum and difference signals. A 19 kHz reference signal is used at the receiver to adequately separate the two stereo channels.

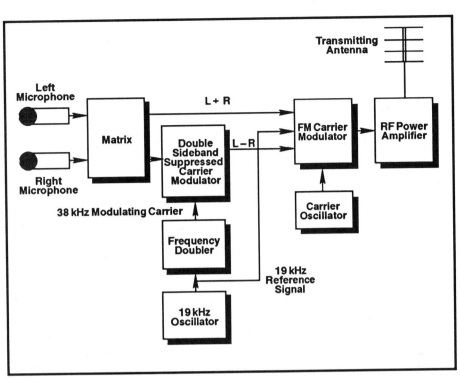

Multiplex Stereo

The multiplex stereo system is similar to that used in regular FM radio broadcasts (see Figure 8-18). First, the left (L) and right (R) audio channels are combined to produce the sum and difference signals, L + R and L − R. Then the difference signal is modulated onto a 38 KHz subcarrier. Finally, the L+ R baseband signal, the modulated L − R difference signal and a 19 KHz reference signal are multiplexed together onto an audio subcarrier. The 19 KHz reference signal is used at the receiver to reconstruct the individual L and R components with the proper relative phase from the sum and difference signals.

Warner Amex Stereo

In the Warner Amex stereo system, the sum and difference signals are transmitted over two separate subcarriers. The L+ R signal is frequency modulated usually onto a 5.8 MHz subcarrier, and the L − R signal onto either a 6.62 or 6.8 MHz subcarrier. Any combination of these or other subcarriers which are located in the audio baseband range can also be used. When signals are demodulated by a satellite receiver, they are matrixed to recover the original left and right audio channel information.

Discrete Stereo

This is the simplest and most widely used satellite stereo system. The left and right channels are not matrixed. Each is simply relayed on a separate audio subcarrier.

Processed Narrow Deviation Stereo

Processed narrow deviation stereo is another form of discrete transmission except that very narrow deviation modulation is used. Deviation is a measure of how much the carrier wave is varied from its center frequency. For example, a frequency modulated 10 kHz carrier which varies between 9 and 11 kHz deviates less than the same carrier varied between 7 and 13 kHz. There is a trade-off involved here. Transmissions having more narrow deviations and which occupy less frequency spectrum unavoidably contain greater amounts of noise and have lower S/N ratios. The audio processing techniques used to improve the S/N ratio, such as companding and spectral companding which are used in conjunction with narrow deviation stereo, are examined in more detail in *Ku-Band Satellite TV*, one of the references.

Processed stereo transmission has one principle advantage over conventional discrete methods. Many more audio subcarriers can be relayed since the deviation is reduced and less bandwidth is used. A satellite receiver must have a narrow band audio filter to receive such audio broadcasts.

Off-Air Stereo

The multichannel television sound (MTS) system was approved in late 1984 by the Broadcast Television System Committee (BTSC) for use with off-air television. The BTSC system permits the simultaneous transmission of a 15 Hertz to 15 kHz stereo audio signal along with a completely separate audio program (SAP) all within the conventional 6 MHz television channel bandwidth. The SAP provides audio diversity allowing, for example, transmission of a second language or a tutorial channel.

MTS is designed to be compatible with the audio circuits in existing television receivers, just like color transmissions are compatible with black/white sets. A multiplex method is used (see Figure 8-19). The left and right audio channels are converted into sum and difference signals. The L+R replaces the conventional monophonic signal and provides compatibility with non-stereo equipped televisions. When stereo transmission is used, the L−R signal is multiplexed onto a subcarrier having a frequency of twice the reference frequency, which equals the color TV line scan frequency of 15.734 kHz. The separate audio program (SAP) is multiplexed onto a subcarrier having a frequency five times that of the reference signal. Neither the difference signal nor the SAP are detected by non-stereo equipped television receivers. Finally, the baseband sum signal and the modulated difference and SAP subcarriers are combined. This signal is used to modulate the standard audio subcarrier within a standard television channel bandwidth. Advanced audio signal processing is used to ensure that the stereo signal is transmitted with as little noise as possible.

Receiving MTS Stereo

Prior to the introduction of MTS, the only method to deliver off-air stereo broadcasts was via a "simulcast." The stereo soundtrack was transmitted via an FM radio station to stereo receivers at the same time that the television was viewed. Today, although most new TV sets are now capable of detecting and reproducing MTS stereo sound, the vast majority still in place are not so equipped.

This situation is similar to that encountered in the earlier days of the home satellite market when few satellite receivers had a built-in stereo capability. The solution then was the addition of a separate stereo processor for retrofit with older model TVRO receivers. Such decoders as well as stereo-ready satellite receivers have outputs which can drive the left and right speakers of a standard stereo amplifier.

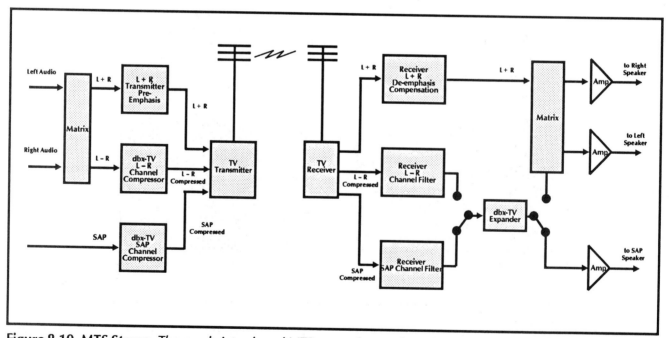

Figure 8-19. MTS Stereo. *The newly introduced MTS stereo format for off-air broadcasts uses a type of multiplex stereo with a reference carrier frequency equal to the television line scanning frequency. The left and right audio signals are processed into sum and difference channels. The sum signal receives standard 75 microsecond preemphasis. The difference and SAP signals are compressed and spectrally companded before transmission. This judicious use of noise reduction techniques results in stereo quality comparable to FM radio broadcasts even when television signals are of marginal quality.*

CONSTRUCTIONS & INSTALLATION

Among the types of MTS decoders available for retrofit on non-stereo ready televisions, the universal decoder is the most useful for SMATV systems. This device uses a probe that attaches to the exterior wall of a television to detect the audio subcarrier. Two units which can be used by SMATV subscribers to decode MTS stereo detected at the headend are illustrated in Figures 8-20 and 8-21. Many higher quality consumer VCRs now feature built-in tuners that are also designed to receive MTS stereo for playback on a conventional sound system.

Stereo Processing and Distribution

In order to receive stereo, an SMATV headend must either have satellite receivers which are capable of detecting the various forms of stereo or must have add-on stereo processors. These latter components process the unfiltered baseband composite audio output from a satellite receiver and produce the left and right unmodulated audio signals. The left and right channels are then fed into an stereo FM modulator.

Figure 8-20. Starsound Stereo Adapter. *This low-cost stereo decoder receives and demodulates an MTS encoded signal relayed off-air or over a cable distribution system. It can be used as an add-on to the output of converters and decoders or with unscrambled signals to create stereo sound for non-stereo televisions. (Courtesy of*

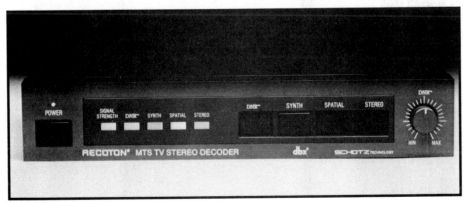

Figure 8-21. Recoton MTS Television Stereo Decoder. *This device uses a small, puck-like probe that attaches to the exterior bottom or rear of any non-stereo equipped TV set and eavesdrops on the tuner RF signal. In those few cases where the internal circuitry is too well shielded, the probe must be installed in an internal location. A standard MPX jack output is also provided for those newer model TVs which have such output connectors. This decoder is designed to provide full dynamic, full frequency stereo sound. (Courtesy of Recoton Corporation)*

CONSTRUCTIONS & INSTALLATION

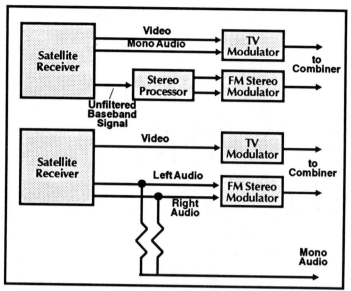

Figure 8-22. Satellite Stereo Reception. *A satellite receiver not having stereo capability can be fed into a separate stereo processor and then into an FM stereo modulator. In those cases where a processor is built into a receiver, the monophonic audio signal can easily be recreated by combining the left and right audio channels.*

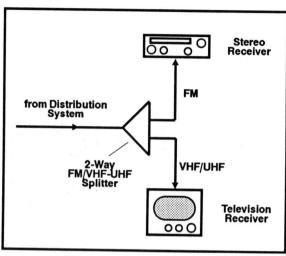

Figure 8-23. Stereo Simulcasts. *A stereo simulcast can be created at a subscriber drop by use of a 2-way splitter and 75-to-300 matching transformer for the stereo receiver.*

A separate stereo modulator is required for each FM channel which is distributed (see Figure 8-22). The audio FM modulator output should be set at approximately 16 dB below the video carrier levels just like in-channel audio subcarrier levels. Stereo can be relayed within the 88 to 108 MHz FM band. One hundred 0.2 MHz-wide channels, starting with a center frequency of 88.1 MHz which is just above VHF channel 6, are assigned in this band. At the subscriber drop, a 2-way splitter is used to route the signal to a standard audio receiver and the television set (see Figure 8-23). The stereo that is heard on a hi-fi system is, in effect, simulcast with the pictures viewed on a television set.

If MTS stereo is transmitted along with the off-air video signal, it is distributed throughout a private cable network. However, most subscribers are generally not equipped with stereo-ready televisions, MTS decoders or MTS-compatible VCRs. Therefore, when an operator wishes to distribute local MTS stereo audio, the best policy is to decode the off-air signal at the headend and then to use a stereo FM modulator.

CHAPTER 9. SYSTEM OPERATIONS

A private cable system can be successful only if customers are signed up and retained. The system operator must assume that each subscriber's opinion is a matter of economic life and death. In most cases it is. The subscriber is the boss. Unlike a franchise cable operation, it is rare that an exclusive or special right to operate in a territory or even in a building complex is given by a legislative body. The SMATV system is installed simply because the property owner was convinced that he or she would get a better deal than by working with the local franchise operator. An operator's reputation as well as his or her continuing existence as a successful business entity depends on the ability to offer and produce a superior level of service.

Superior technical service is the cornerstone of a good reputation. Not only must the video and audio quality be excellent, but customer relations and service must be conducted in a professional manner. When a customer requests a change in service level or has a complaint, the organization must be prepared to respond quickly and professionally. Satisfied customers are the key to success.

Many operators make the common mistake of underestimating the need to offer quality service and technical support. Since the technical department is usually viewed as a liability with little potential for generating revenue, there is always pressure to try to minimize its size. This can easily be a serious error in judgment that ultimately directly affects the operator's overall reputation. After all, the major obstacle that cable companies face is obviously customer dissatisfaction and as a result customer churn, namely the loss of subscribers. The lack of service will certainly reduce the subscriber base and the poor company reputation developed over time will unquestionably hinder marketing efforts. A budget that reflects realistic customer support needs should be created and funded.

Effectively operating a private cable system requires a combination of excellent technical support staff plus an office structured to efficiently coordinate and dispatch technicians. This smooth integration of office and field personnel is necessary to achieve high levels of customer satisfaction. If a customer has a complaint and it is handled promptly and professionally, it will be perceived as a minor irritation. However, if repeated calls were necessary to simply arrange an appointment for some time in the not-too-immediate future, the client's resulting anger could seriously affect his judgment when the end-of-month bill is due.

SYSTEM OPERATIONS

A. THE CUSTOMER RELATIONS OFFICE

The customer relations office is the heart of a cable operation. Requests for service, complaints and new customer sign-ups are all funneled through this location. All customer records and billing information as well as service and technical work orders originate here. In most cases, this is the hub and the busiest area in a private cable organization. Given the importance of the customer relations office and its direct effect on the health of an organization, every effort and consideration must be expended to insure that it runs smoothly and efficiently.

The first consideration should be the number and quality of the staff. The staffing requirements are determined by details of the type of operating system. A two-way, addressable system of superior design and quality will require less manpower to implement service than an older one-way, negative trap system. In any case, after a system level of approximately 500 subscribers is attained, at least one person should be dedicated full-time to customer service. As the subscriber count grows, the rule of thumb is to add another representative for every 3000 subs. Obviously, if the phone lines are ringing off the hook with customers anxious to sign up or report a problem, add more people. Every consumer knows that customer dissatisfaction is aggravated out of proportion by long waits on the phone.

The customer service representative must handle many different situations and occasionally must solve outright dilemmas. One moment he or she may have to present a sales pitch to a prospective customer and the next minute attempt to placate an irate subscriber who's TV set has lost its signal in the middle of her favorite soap opera. The customer rep is responsible for updating the billing system when changes in service levels occur. In addition, he or she may have to create work orders for the service department and track their progress. In a large franchise operation, these different duties might be delegated to highly specialized representatives that deal only with narrow areas such as new subscribers or complaints, but in private cable operations, the constraint of limited manpower resources require that the customer service representative wear many hats. Choose representatives carefully. Look for a person with good public relations skills coupled with ample patience. Customer reps carry the responsibility for maintaining the company reputation. The image they project to subscribers directly mirrors the corporate countenance.

After the operation has been staffed, these personnel must be provided with the proper tools of the trade. The telephone and the computer are the two most important instruments in the customer relations office. They are the backbone of communications and information management systems necessary to smoothly run a cable operation. Without adequate phones lines and a well designed computer system, a management crisis of pyramidal proportions could easily arise.

The phone system is the lifeline to customers. Proper usage planning is essential to maximize its benefits. As a bare minimum, each customer rep must have his own line. These lines should be installed in a rotary configuration so that any operator can answer any line at any time. In addition, these telephones must be dedicated to customer relations. If not, the time may come when other departments tie up lines with their calls. Choose the phone system carefully. Remember that the goal of the organization is growth. Keep this in mind when planning telephone needs and designing the systems. Do not get locked into a system that has to be replaced every time an additional thousand subscribers is added. A system that is based on modular components and therefore allows for expansion may be more expensive at first than four single line sets, but it will prove cost effective and more productive in the long run. Purchase the system from a reputable contractor and be sure to get an extended service contract. A little planning and careful consideration is important during the initial acquisition of a phone system which can go a long way to maximizing customer satisfaction and growth in subscriber numbers.

B. COMPUTER SYSTEMS

The computer has revolutionized record keeping in the cable industry. Large, time-sharing systems such as Cabledata serve thousands of cable companies across the country. These systems can track customer service level, billing status, user history as well as company marketing efforts and results. Other brands of software and computer systems control the technical operation and activate or inactivate serve to customers directly under control from the main office. Until recently, these feature were limited to those companies that owned expensive mainframe computers and that were supported by a large customer base. It is common knowledge that the technology has evolved at an astonishing pace. Now systems having as few as 500 subscribers can realize the same benefits on PC sized computer systems as ten year old mainframes (see Figure 9-1).

Many manufacturers are now marketing extensive software packages that support from 500 to more than 10,000 subscribers. These control systems can run on microcomputers, minicomputers and small mainframes. They track all customer data, prepare and print billing forms and envelopes, monitor payments and output service orders for the technical department. In addition, they perform bookkeeping functions and prepare financial statements. Computer controllers offer the private cable operator a relatively inexpensive means of accurately controlling billing and paperwork with a minimum amount of effort.

When choosing a software package, compare its features, price and performance. Prepare an ideal standard work order and customer bill, and list the other features required for managing the operation. Then compare this mock-up to the actual product. Does it perform the required tasks? Are the on-screen operations easy to effect with a minimum of key strokes? Are the visual displays user friendly and easy to operate? Functions that are hard to understand or employ will soon fall into disuse and will prove to be a waste of money. Look for systems that allow tracking of groups of multiple installations or perhaps larger franchise areas. In this manner, separate files can be created for each SMATV system and the chance for confusion in the customer relations office is minimized. Today, most software manufacturers offer extended support and training. Be sure to take advantage of these services to maximize the return on the software investment.

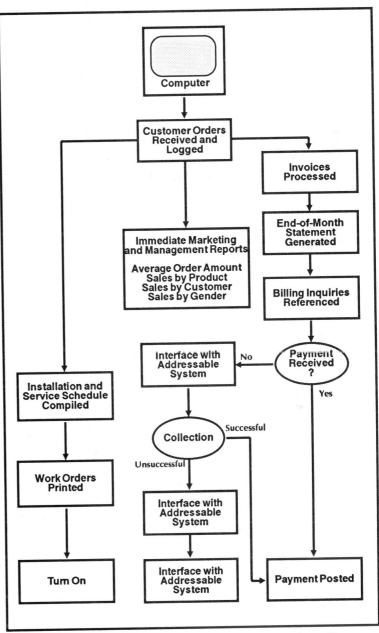

Figure 9-1. A Computer Billing System. *This chart illustrates some of the functions performed by an in-house SMATV computer billing system.*

SYSTEM OPERATIONS

Fortunately, most control software can operate on microcomputers. In most cases, a system having a modern 386 or 486 computer with 100 or 200 megabytes of hard disk and extended random access memory suffices. Memory requirements are related to the number of subscribers. In addition, 3 remote terminals with emulation software allow 3 representatives to share the computer system. As the subscribers base increases past a point, microcomputers begin to be inadequate to handle the load. At these levels, the choice of a mainframe system such as the IBM System 1 is a better investment for computing capacity. The use of a mainframe system at this level will allow for virtually unlimited future expansion as the customer base grows. Of course, the point past which a microcomputer is not sufficient keeps moving to higher subscriber numbers as computer technology advances.

C. THE TECHNICAL SUPPORT GROUP

The quality of service provided by the technical staff directly reflects upon the ability of an SMATV company to satisfy customers. While both the sales and technical staff are the only employees who directly contact customers, the technician must actually correct system problems and failures. He or she is a direct representative and the company is clearly judged by his or her actions. If the technical department can satisfy the needs of customers, the bulk of operational battles can be won. If they cannot, the competition will probably be won by another private cable firm or the local franchise operator.

Staffing

In a franchise CATV operation, the technical department is composed of and divided into personnel having a variety of specialties. There are installers, general technicians, trunk technicians as well as chief technicians. This division of labor is a result of the need to service a vast distribution network required to feed a major cable system. In contrast, in a private cable operation the distribution system is relatively much smaller. As a result, operations can be supported by a reduced number of technicians and duties can be consolidated. By eliminating the trunk technician and having the chief tech function in a dual supervisory role, the installer and technician can be combined to manage one job function, namely the installer/tech.

The installer/tech is the backbone of the service organization. He manages all the customer hook-ups and also services level changes on non-remote systems. Any customer trouble calls and certain routine maintenance chores are handled by the installer/tech. He interfaces most directly with the subscriber in the field.

The best way to fill an installer/tech's position is to hire an experienced cable TV installer and then train him in the details of the SMATV systems under control. This employee should participate in the construction of headends and distribution plants. It is wise to concentrate tasks on systems that he has directly helped install. The chief technician or technical director should develop training programs that familiarize the installer/tech with all phases of the technical operations. If this person is trained to perform the job correctly, it will probably be managed as expected.

The chief technician is directly responsible for all system technical operations. Headend construction and adjustment as well as system maintenance fall under his control. All technical problems that cannot be solved by the installer/tech by virtue of his position automatically become the chief tech's responsibility. These duties also include supervising, training and educating the installer/techs as well as conducting employee evaluations. The chief tech answers directly to upper management, usually the operations director.

The amount of technical support required is directly proportional to the number of subscribers. As is the case with office staff, the technical characteristics of the systems managed determine the ratio of technicians to customers. By the stage when the first 200 subscribers are connected, a full-time installer/tech is usually required. For every additional 1000 to 2000 subscribers that go on line, another installer/tech should be hired. The chief technician's responsibility can be handled at first by those directing the overall technical operations. However, when the number of installer/techs reaches 4 or 5, a separate position for chief technician should be created to directly manage this work force. At this stage, the chief tech can take over day-to-day system operations and can lend support to the technical director on design and construction matters.

D. THE OVERALL MANAGEMENT STRUCTURE

As the operations team continues to grow in size, the need for structured management begins to manifest itself. In the beginnings of a private cable operation, many different jobs can be performed by a few people wearing many hats. However, as the business grows, it becomes essential to increase the work force to accomplish company goals of customer growth and profitability. This expansion of staff necessitates developing an effective and efficient management team to oversee company growth.

The first steps in establishing an effective management structure are simple. Regular hours of operations must be set and followed. Having a successful operation mandates that the phone lines are answered 24 hours a day by an in-house employee or an answering service. A service can effectively answer phones after hours and on weekends. When the company grows large enough in size, a night person can then be hired to answer phones and dispatch service personnel in times of emergency. As new employees are hired they should learn standard rules and codes of conduct. It is not necessary to be harsh, just consistent and rational. Ambiguity in employee relationships and conduct is an invitation to trouble.

Establish procedures for taking orders and dispatching service personnel. The underlying logical structure of the computer software can provide the format necessary to model a management system. Create scheduling procedures that staff can follow with ease. For example, do not schedule service calls at an exact time. This strategy can result in missed appointments and customer dissatisfaction. It is much a better policy to schedule service calls during a time period from, for example, either 9 AM to 12 PM or from 1 PM to 5 PM. This gives technicians flexible schedules that they can meet even if problems arise. The end result will be fewer missed appointments. Create a priority system for calls received after hours and on weekends. Do not roll trucks and crews because one subscriber is having minor troubles with his set. If 4 or 5 customers from the same system call with similar complaints, then an outage is likely and it is time to respond. Establish a rotation system with the technical staff that delegates the night dispatch responsibility to one person each night. The after-hours answering service should be trained to recognize system outages and to contact management when a problem becomes apparent. Then management can have the opportunity to exercise their responsibility to dispatch service personnel.

Both office and technical personnel should come under the control of an upper management director. This person should be responsible for both technical management and systems operation. As the system grows, this position can probably be split into two jobs with both managers reporting to a general manager.

In a typical, small-scale private cable operation, an efficient delegation of responsibility is very important. The need to perform many different tasks with limited manpower resources creates a situation where others must be relied on to accomplish tasks effectively. Staff the organization with good people who are both self starters and highly motivated. A high level of "esprit de corps" will set the tone for the efforts of the entire staff.

E. MARKETING THE PROGRAMMING

Certainly not the least important task is marketing the off-air and satellite programming, the bottom line for which all the hard work has been performed. An operational system obviously needs a good subscriber base to maintain its profitability. It is a fantasy to assume that apartment or condominium owners will pay more for a service than customers in an adjacent franchised apartment building. A private cable operator must be truly competitive with local CATV systems. This is especially important given the "forced access" controversy that is now raging. Lessons from the franchise cable industry are valuable. It is a wise strategy to learn from and to apply some of their well-tested marketing techniques. Ironically, many cable executives will agree that marketing is usually the weakest, yet the most important component of their operations.

Marketing must be made an integral part of any private cable business. Its importance should be recognized so that sufficient funds are budgeted. It is clearly crucial to achieve projected sales goals. These objectives are supported by projecting a professional image. All forms of communication should be clear and presented with quality whether on stationery, via the telephone or in person.

SMATV systems are judged to be profitable by experienced private cable brokers and buyers on the basis of the success that an operator has in signing and retaining customers, not solely on the potential income of the system or the excellence of its design. However, in reflecting back to the basic sales contract, if the potential income is poor because in the original agreement the SMATV owner/operator had given too large a portion of the income to the building owner(s), the perceived value would certainly diminish.

The first decision is to decide whether the marketing will be managed in-house or by an outside marketing contractor. There are many organizations that conduct marketing for franchised cable operators via door-to-door, telephone or mail campaigns. They work on a straight commission basis and are usually, but not always, very experienced in cable marketing strategies. Many of these firms have often conducted extensive marketing research in a particular area. They can quickly answer questions about the stability of the occupancy rate, the average age and income of the tenants or the average number of children per family. The lack of marketing information could lead to ludicrous situations where, for example, an operator pours money into selling The Disney Channel to residents of a singles-only condominium complex.

Marketing research conducted after a system is operational can help in evaluating a company's service policy and pricing structure, in determining the advisability of rate increases and in developing a better picture of the interest in various types of programming and specials. Well-conducted marketing research may also effectively uncover unusual situations. For example, it may be discovered that a substantial amount of piracy exists in one condominium complex by comparing the sales penetration with those in other similar operations. This situation should of course be dealt with quickly and effectively to protect revenues. Incidentally, marketing research can also provide extremely useful information about the tastes and habits of a particular customer base prior to bidding on or constructing a private cable system.

Hiring an outside marketing contractor may completely eliminate the need for an in-house sales department and its overhead. As is the case when hiring a construction contractor, the best method to locate and select a contract sales firm is through the local franchise operation. A marketing company should be carefully chosen and each sale must be verified before any commissions are paid. Paying on a commission-only basis will aid in determining the final cost of sales. By carefully monitoring, supervising and directing the marketing activities in the field, a contract sales force could be a very effective way to reach and sell a large number of prospective subscribers.

One key to a successful marketing effort is the degree of support that is received from the resident building manager. This person or group can be the most effective sales force when actively involved in the sales efforts. This person or group should be provided with all the materials needed to sign up walk-in customers as well as with movie posters and other point-of-purchase items to display in the on-the-site office. The support received from the resident management staff can make an enormous difference in the number and quality of sign-ups.

One of the more effective methods to market SMATV services or glean information is by door-to-door sales and interviews, a process that should be conducted at least once every 90 days. This is often a relatively easy but rather time consuming job because all units are close together in a large multi-unit complex and a mass effort can be easily coordinated from a central location such as the rental office. Questions can generally be asked freely and positive face-to-face sales techniques can be employed. These efforts should be supported by a comprehensive information package complete with rates and programming lists. If possible, this sales tool should include a sample cable guide and any other aids that the various programmers supply. Installations can be scheduled at the time of sales if one or two technicians are available to perform instant connects. Selling the customer and then immediately performing the installation while the interest level is at its highest point is a very successful strategy.

While the system is being installed, building occupants are generally quite curious and interested. This can be used as an effective marketing tool. But even before this point, potential subscribers should be informed about when and how the SMATV system will be constructed by a preliminary information package that can be handed out by the building manager or the construction foreman. The residents should know who the company is and should be graciously informed of potential inconveniences. Excitement can be created before system turn-on by giving some preliminary details about services to be offered. Early sign-ups can also be encouraged by offering free installations or other financial inducements. Remember, this is the time when residents are excited about the new system. Taking advantage of this excitement can increase the initial subscriber base. However, it is also important to realize that this excitement can be easily diffused if the system is not running perfectly when it is first demonstrated to subscribers.

In general, every opportunity available to make the private cable company visible on the project should be seized. Residents can be invited to a "launch party" to officially turn on the system where they are given items such as free promotional posters. When and if additional services are added, fliers offering introductory specials on the new channel can be an effective sales tool. Programmer suppliers will also have special items to entice new subscribers. These promotions are a useful way to take advantage of each programmer's marketing budget.

**SYSTEM OPERATIONS**

CHAPTER 10. TROUBLESHOOTING and TEST INSTRUMENTS

A. TROUBLESHOOTING

Every SMATV system will eventually have operational problems of one sort or another that need immediate attention. Difficulties will occur in spite of how well a system has been installed or how professionally it is operated. Operating an SMATV system has the additional handicap of being in part a public relations job. Sometimes what has been perceived as a technical problem is simply a matter of customer education. Other times, a difficulty may be caused by the subscriber but blamed on the operator. Occasionally, line equipment will be damaged by careless gardeners and construction personnel. Of course, there will always be equipment failures even when the highest quality gear has been installed. No man-made device will last forever. The manner in which system problems are diagnosed and repaired is a true indication of the quality of a private cable organization.

The majority of complaints result from subscriber errors. Their lack of familiarity with set-top equipment and how the equipment functions are usually the main cause of first-time trouble calls. If an installer does not show each new subscriber how to properly operate the converter, chances are good that shortly after being hooked up he or she will be on the phone complaining. It is important to instruct each installer/tech to explain pertinent aspects of the system to each new customer. This should include a clear demonstration of all the functions of the converter as well as verifying that the customer truly understands its operation. The customer must comprehend that the converter only works if the TV is tuned to one specific channel, usually VHF channel 3. A clear explanation will generally eliminate the chance of a service truck being called out late at night to merely change the channel on a customer's set! In addition, if all office personnel are thoroughly trained in recognizing minor problems, they can usually talk a customer through a basic television adjustment and, in the process, eliminate a service call.

The Basic Approach

An operator must have a consistent and methodical approach to diagnosing and solving system difficulties in a minimal amount of time. The diagnostic procedure can itself be an interesting and rewarding exercise if a well-planned strategy has been studied and practiced. In order for a technician to be an effective troubleshooter, he should thoroughly understand all aspects of the system under study. Optimally the person who had originally been involved in its installation should also be responsible for its repair. Having a familiarity with how a system operated when it was first built certainly makes it easier to understand and repair any problems which might later arise.

TROUBLESHOOTING & TEST INSTRUMENTS

The most logical approach to troubleshooting is to follow the simplest route to the most rapid solution. The first and most obvious step in troubleshooting is an in-depth customer interview, preferably over the telephone, as a time and cost saving measure. It is important to establish a standard questionnaire that each subscriber should answer when complaints are voiced. Then even if the customer rep cannot solve a problem over the phone, technicians will be able to guess at a solution before arriving at the site.

The interview can include such obvious questions as:

- When did the problem begin?
- Describe the problem.
- Are all channels equally affected?
- Has the television set recently been moved?
- Is the sound or picture most severely affected?
- Is the set tuned to channel 3?
- Does the TV set still receive local stations?
- Do your neighbors have the same problem?
- Has the television set recently been purchased?

If the initial interview had determined that a site visit were necessary, the logical point to begin testing is usually at the customer drop. Of course, if numerous customers on every leg of the distribution system had been complaining, it would be more logical to begin the detective work at the headend or at another central point upstream from the customer drops. This strategy is explored in more detail later in this chapter in Section D.

Test Instruments

The proper tools and test instruments must be available to do site surveys, installations and repairs. While coaxial cable crimpers, screwdrivers, wrenches and other standard tools are clearly necessary, a signal level meter, a spectrum analyzer, a portable TV or monitor and a multimeter are indispensable. Instruments that combine a number of these functions are also available.

A multimeter, also known as a volt-ohm meter (VOM), is an inexpensive component that can help in identifying numerous problems (see Figure 10-1). It can be used at each subscriber drop to verify electrical continuity in jumpers or to check ac or dc voltage levels. It can also be employed on line gear to confirm the presence or absence of proper voltages and connector continuity.

A field strength meter converts radio frequency (RF) power into a voltage reading and can be effectively used to test numerous points in the headend and distribution system. Field strength meters (FSMs) come in all shapes and sizes and range in price from $300 to $2,000 or more. At least one high quality field strength meter or spectrum analyzer should be on hand for the chief technician's use during headend construction and system balancing and maintenance. However, an FSM that indicates the levels on all system channels within ±2 dB is perfectly adequate for installer/techs. Even if the operations budget is tight, it is essential to have at least one good meter to avoid "working in the dark."

Figure 10-1. A Multimeter. *A multimeter is frequently either in the field or on a test bench. It has the capability of reading ac and dc currents and voltages as well as taking resistance measurements.*

TROUBLESHOOTING & TEST INSTRUMENTS

A 5" color portable television or monitor serves three purposes. First, it provides a visual confirmation of system performance to confirm the results that are indicated on a field strength meter. Second, it can be employed to detect ghosts and other more subtle picture distortions that are not indicated by a FSM. Third, it is crucial to effectively demonstrate the picture quality that is available to a subscriber. There will be many occasions when a customer will have an old or defective television that simply does not produce a decent picture. The customer will usually swear to high heaven that it has played fine for the last 20 years and that there is something wrong with the SMATV system. This conflict can easily be resolved by merely connecting the portable TV to their outlet to demonstrate that the picture is fine. If the test television plays well and the signal levels are adequate, the only conclusion that can be reached is that the customer's set needs repair. Without a test set, all the meter readings in the world would not convince a subscriber of this fact.

If the budget permits, the test kit should include a spectrum analyzer. This instrument displays the frequency structure of a signal versus time and has numerous diagnostic and installation uses. For example, a portable satellite antenna/reception system in combination with an analyzer can be invaluable in finding an optimal site for a satellite antenna and in testing for terrestrial interference. It allows a tech to see power levels from just a single channel or the relationship between channels spanning the entire range of broadcast frequencies. Although spectrum analyzers can cost up to $50,000 or more and have numerous sophisticated features, there are a number of much more reasonably priced units marketed specifically for the home satellite and SMATV industry. These low-cost analyzers have built-in tuners that span the frequency range from zero to 4.2 GHz and above. Such a wide testing range provides the versatility needed for private cable work. Most vendors of spectrum analyzers are glad to offer useful operational information and can demonstrate numerous other ways to apply these instruments to the SMATV industry.

B. A CATALOG of SYSTEM PROBLEMS

The ultimate method to judge the operational quality of an SMATV system is to observe the television pictures and listen to the sound. The types of distortion which appear can then often be easily traced to various system problems. These include inadequate S/N ratio, a high audio-to-video carrier ratio, hum modulation, signal ingress, co-channel or adjacent channel interference, ghosts, sync compression and other types of interfering signals.

Inadequate S/N Ratio

When noise levels are too high, television receivers will detect the noise, not the desired signal. Noise generated within a distribution system from cascaded amplifiers and other broadband sources such as poor connections result in "snowy" pictures with so-called "sparklies" (see Figure 10-2). Snowy pictures also result when signal levels leaving the headend are too low. The measure of these effects is the signal-to-noise ratio. If the S/N ratio exceeds the 36 dB FCC specified level or the "good engineering practice 42 dB level (see Table 8-2) pictures should be excellent. It cannot be over-stressed that once the S/N ratio has dropped too low no amount of amplification will restore it to acceptable levels.

Externally generated noise has a different set of effects such as random noise bars and other distortions. These topics are discussed below.

Figure 10-2. The Effect of Inadequate S/N Ratio. *Snowy pictures as illustrated here result from inadequate levels of S/N ratio.(Courtesy of Sencore)*

High Audio-to-Video Carrier Ratio

When the level of the audio carrier exceeds about −13 dB relative to the adjacent channel video carrier, it may interfere with the video signal and cause beats to appear on the television screen (see Figure 10-3). In general the audio carrier should be from 13 to 17 dB below the video level.

Figure 10-3. Co-Channel and Adjacent Channel Interference. *This illustrates the typical interference pattern caused by either co-channel or adjacent channel interference. (Courtesy of Sencore)*

Hum Modulation

Hum, also known as hum modulation, results when the ac power waveform from a power supply leaks into a broadcast signal and introduces unwanted 60 or 120 Hz modulation. Its source is usually a faulty capacitor in an amplifier's dc power supply or an inadequate ac voltage at the power supply input. 60 Hz modulation generally indicates a problem in a half wave supply; 120 Hz difficulty in a full wave power source. The result is an excessive ac ripple in the dc power output. Hum can be also introduced by a corroded connector or a broken or loose ground that results in a ground loop somewhere in the headend or distribution system, or ingress from a strong ac field of a nearby power transmission line. The former condition is also an invitation to other types of ingress interference.

Hum is visible as a white, gray or black horizontal bar(s) that slowly rolls through a television picture (see Figure 10-4). Their intensity is directly related to the amplitude or percentage of hum modulation of the video carrier. 60 and 120 Hertz hum causes one and two bars to appear, respectively.

Regular measurement of hum is a wise policy because a slow increase in hum from one test to the next can be a good indication of developing problems. Extended time tests may uncover system components whose performance varies with time or other parameters. For example, a hum level that changes noticeably with temperature could indicate a deteriorating filter capacitor in a power supply. Hum modulation levels should be tested on each channel since a faulty processor or preamp may cause increased levels on just one channel.

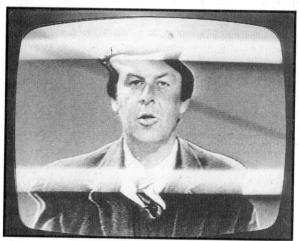

Figure 10-4. Hum Modulation. *A moderate level of hum is usually visible as bars that roll through a television picture. (Courtesy of Sencore)*

Signal Egress and Ingress

Signals leaking from a private cable system, signal egress, can interfere with other communicators such as air traffic controllers or local ham radio operators who share portions of the same frequency band. Egress can also cause others difficulties if left unchecked. The escaping RF signal can be detected by anyone possessing a receiver tuned to the correct frequency, including non-subscribers. Leakage can also lower system S/N ratio and result in picture degradation. In the worst case, an SMATV system having severe leaks runs the risk of being shut down by regulators such as the FCC.

When signals inadvertently are broadcasted from a distribution system, chances are excellent that interference from amateur or aircraft navigational radios or even co-channel television broadcasts can also ingress. Therefore, if there is leakage from a loose, broken or defective cable or connector, from corroded grounds or from broken or loose amplifier housings, subscribers may also receive some unwanted "off-the-air" information. Patrolling for leaks has the added advantage in detecting "pirates." Often when they rig into a distribution system they often use poorly installed connectors which can have substantial and detectable losses.

Signal egress is measured by the field intensity around a cable or device. The greater the field intensity, measured in microvolts per meter, the greater the amount of RF energy picked up by a dipole antenna that is tuned to the frequency of this RF signal. A properly shielded distribution system typically radiates less than 2 microvolts/meter of signal while a defective system may radiate level as high as 100 microvolts/meter or higher. The FCC limits are outlined in Table 10-1. The methods used to measure egress field intensity are described in the section on field strength meters that follows.

TABLE 10-1. FCC MAXIMUM PERMITTED RADIATION LIMITS from a CATV SYSTEM

Frequency (MHz)	Maximum Radiation (mV/meter)	Distance (feet)
<54	15	100
54 – 216	20	10
>216	15	100

Interference from Video and FM Broadcasts

The symptoms of interference originating from video or FM broadcasts can range from mild cases of herringbones on the TV screen to serious picture distortions. The cause of beat interference is often difficult to identify from its symptoms and some careful detective work is usually required to find a solution.

The effects of co-channel and adjacent channel interference are similar. Co-channel interference or cross modulation results when an off-air television signal enters the system and beats against a channel sharing the same frequency. The visual effects are two superimposed pictures and a beat pattern as illustrated in Figure 10-3. It can also be characterized by vertical bars which move back and forth on the television screen. One simple method to eliminate co-channel interference is to offset the distributed channel from the interfering signal. Alternatively, and probably more appropriately in smaller scale SMATV systems, the source of ingress must be located and cured.

Adjacent channel interference has effects that depend upon the location of the interfering channel. If the upper adjacent channel is the culprit, the video carrier that is closer in frequency than the audio carrier produces beats and a superimposed picture upon the desired channel. When the lower adjacent channel interferes, sound beats appear in the picture. This type of interference can be cured by balancing signal levels if the unwanted signal is being distributed or by eliminating ingress if an off-air adjacent channel is interfering.

Beat interference can often be traced to improperly set levels in UHF-to-VHF converters or modulators. An easy method to isolate such a problem is to disconnect each suspected component in turn, beginning with converters and followed by the modulators. When the culprit is identified, check (1) that the picture carrier level is balanced with other channels, (2) that the sound carrier of the lower adjacent channel is at least 13 dB below the picture level, (3) that the device is operating within its specified input and output ranges and (4) that spurious outputs are at a sufficiently low level. Excessive spurious levels could mean that the component is defective. Or it may mean that a converter is receiving signals in addition to the desired frequency so that an input trap or filter might be necessary.

Overloading and the resulting beating can often be caused by strong FM signals (see Figure 10-5). In systems using broadband amplification, this interference can affect a single VHF-low channel, usually channel 6, any single VHF-high channel, or a combination of all VHF channels. In contrast, when using cut-to-channel strip amplifiers for off-air processing, FM overloading is usually limited to channels 5 or 6 that lie just below the FM band. The amount of interference and the number of channels affected is directly related to the strength and number of FM signals. Of course, when poor quality coax is installed the problem will be worsened.

The first step in identifying FM overload is to confirm that these signals are substantially more powerful than the nearby VHF television broadcast levels. If FM stations are not being distributed on the pri-

TROUBLESHOOTING & TEST INSTRUMENTS

Figure 10-5. Beats Caused by Ingress from an FM Radio Station. *Signals that ingress into a SMATV system can cause beats as illustrated here. (Courtesy of Sencore)*

Figure 10-6. Ghosts. *A ghost appears as a second image either to the left or right of the primary television picture. (Courtesy of Sencore)*

vate cable system, a single bandpass filter can be used to attenuate the whole 88 to 108 MHz FM band by about 20 dB. If a preamp had been used, this trap should be installed before its input. Note that the insertion loss of the bandpass filter will decrease the system S/N ratio. If FM signals are being distributed, single channel adjustable traps can be installed. Some traps have two adjustments to minimize the effects of temperature drift.

If interference is intermittent, it is usually not from an FM or video source but more likely is caused by some low frequency transmitter such as a Citizen Band or Amateur Band radio overloading a preamp or another headend processing component. A FSM and earphone can be used to tune to and identify the interfering carrier. If the source is lower in frequency than channel 2, a cross-over network can be used to separate this subchannel band from the TV band. The network can be installed at the antenna downlead and the subchannel output should simply be terminated. Such a filter rejects any signals below 54 MHz in frequency. In those cases where the interference is identified as in-band, a trap can easily be installed.

Ghosting and Ringing

Ghosting and ringing are two types of interference that occur on an RF distribution system when an RF signal interferes with itself. Although the causes of these two phenomena differ, the visual results are similar, namely a second picture image that is displaced either to the left or right of the desired picture (see Figure 10-6).

Ghosting

A ghost is observed as a single secondary weaker image to the right of the main picture. A location at the right implies that the second signal has arrived a little later and therefore is traced a little later. Such ghosting is caused when a reflected signal is detected by the off-air antenna after the main signal arrives from the television transmitter. By measuring the on-screen distance between the ghost and the main image, one can determine the extra distance it had to travel. This then gives some indication of the position of the reflecting object. For example, a ghost separated by 1/4 inch from the main image on a 21 inch screen required a time delay equal to 800 feet. This calculation is graphically presented in Figure 10-7.

When ghosting occurs, the approximate location of the reflecting object can be determined by knowing the distance to the television transmitter. For example, if the transmitter were 10 miles away and the difference in distance were 800 feet, the reflecting object could be 400 feet behind the transmitting or receiving antenna, just under 800 feet directly to either side of the transmitting and receiving antennas, or, when in between the antennas, 4500 feet to either side of the centerline between the two (see Figures 10-8). The requirement that the difference in distance between the main and reflected signals is a fixed 400 feet determines that the reflecting object falls somewhere on an elliptical outline. This elliptical figure can easily be drawn on a local map first by inserting a pin at the location of each antenna and then by looping string having an extra length equivalent to 400 feet around the pins. A sharp pencil will trace an ellipse when the string is pulled taut.

The ghost-causing reflected signal can arrive from any angle. Although this graphical method does not localize the reflector, a little investigative work can often be successful. If the off-air antenna has a very narrow main lobe, the reflecting object is probably in one of three locations, (1) almost directly in front of the antenna, (2) relatively far away in the direction of the transmitter or (3) behind but relatively close to the receiving antenna. If the reflector is in front of the antenna, it probably has to be something massive like a bridge or another large structure. In this case, lowering the height of the antenna may well improve reception. If slightly rotating the receiving antenna causes a change in the ghosting, then the object is probably relatively close to the site. Slightly adjusting the position of the antenna or finding a nearby ghost-free installation site might eliminate the ghosting. If not, a horizontal stacked antenna array might be used to cancel the interference (please see Chapter 5 Section D for more details). In rare cases, if the ghost is substantially stronger than the main signal, it may be worthwhile to maximize reception only from the reflected signal and attempt to eliminate the one arriving directly from the transmitter.

Occasionally, the ghost image leads the main signal. This means that there is direct or "local" signal pickup. A strong local channel is ingressing directly into the television receiver. This condition can usually be eliminated by properly shielding cables, tightening up the connectors and using good quality matching transformers. When both leading and lagging ghosts simultaneously occur, each problem must be solved separately.

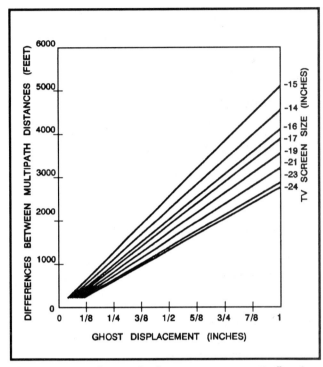

Figure 10-7. Ghost Displacement versus Reflection Distance. *This graph shows the differences in multipath distances between a reflected and boresighted off-air signal as a function of the amount of displacement between the ghost and main picture for various sized television sets.*

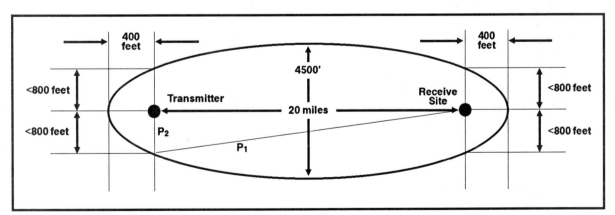

Figure 10-8. **Locating a Reflector of Off-Air Signals.** *The location of an object which reflects an off-air signal can be narrowed down to an elliptical path once the distance between the boresighted and reflected signal has been determined from Figure 10-3. In this case, it is assumed that this difference in paths is 800 feet. The objects can lie just under 800 feet to either side of the receiving or transmitting antennas because a small part of this extra 800 feet, P_2, is added in as the signal travels a distance just over 20 miles along the path marked P_1. See the text for additional details.*

TROUBLESHOOTING & TEST INSTRUMENTS

Ghosting that has other characteristics also occurs. When reflected signals arrive with time delays equal to 230 feet or less, a separate image is not distinguishable. The result is a fuzzy, out-of-focus picture. This is also usually caused by time delays or discontinuities in the cable system but may also be the result of signals reflected from nearby objects. Occasionally, picture "flutter" is observed. This is caused by reflections from passing aircraft and is characterized by an erratically changing distance between the main image and the ghost. A partial solution to this problem is to stack antennas vertically. In addition, "tuning" ghosting sometimes is observed when the fine tuning knob on a television is adjusted. This is caused by a mismatch between the tuner and the IF section in the receiver and is a good sign that the set should be serviced.

Ringing

Multiple ghosts that appear to the right of the main picture, diminish in brightness and are evenly spaced are produced by a phenomena known as ringing. Ringing results when reflections from a poorly matched device, a pinched cable, an improper termination or water within a cable produce secondary images in a distribution system. Their source can usually be diagnosed by inserting a fixed attenuator, for example a 6 dB pad, into the line at various locations until the problem disappears. This patchwork approach can be an effective remedy because it improves the match of the faulty device. In those cases where there is extra signal level to spare, the attenuator may be left in place. But clearly the optimal solution would be to replace the poorly matched device.

Improperly terminated lines can also be checked by using a VOM to measure resistance between the center conductor and ground in the feeder lines at the headend. If an open or a short circuit is detected, this indicates that a line has been improperly terminated or that a defective device or cable is installed. The problematic component can be traced by step-by-step testing with the VOM in the direction of the impedance mismatch.

Sync Compression

Sync compression results when any active component in a headend or distribution system is driven to a point beyond its dynamic range so that it does not pass the complete RF signal without distorting it and clipping off the high amplitude portions. Since both the horizontal and vertical sync pulses in the television signal are the highest amplitude portions, these are clipped and subsequently the television receiver cannot lock. As a result the picture rolls both horizontally and vertically.

Sync compression can occur in numerous locations from the source of the video signal to distribution amplifiers. The signal may be clipped in an overdriven amplifier, in an under-powered amplifier or in an amp having an out-of-adjustment AGC circuit.

Fortunately, sync compression can be relatively easily identified and cured. The first step is to check input levels to the channel processor and the launch amplifier. If necessary these may have to be reduced while viewing a monitor to visually confirm if pictures return to normal. To check if sync compression is resulting from a faulty or poorly adjusted AGC, set the amplifier gain control to manual and reduce the gain while viewing the monitor. If a normal picture returns, the AGC must be readjusted.

Of course, it should not be overlooked that a loss of horizontal sync and the resulting picture rolling may also stem from a malfunction in a subscriber's television set.

Other Sources of Interference

Interfering signals may arise from a number of unexpected sources including electrical discharges from lightning or ignition systems, mobile radios or electrical motors (see Figure 10-9).

Electrical Interference

Electrical interference is characterized by two horizontal bands of snow which may remain steady or move up and down on the screen. When such interference is intermittent but fairly regular, it is probably caused by the electric motor in a refrigerator, furnace or other appliance. If it is random, the source could be a man-operated machine. Electrical interference is caused by the arcing of worn motor brushes and is more intense at lower frequencies. The offending machine can be isolated by using a FSM/test antenna as a probe. In those cases where the signal is particularly weak, an earphone can be used to listen to the interference.

Continuous electric interference is generally caused by power line defects such as cracked insulators, arcing transformers or poor electrical joints.

TROUBLESHOOTING & TEST INSTRUMENTS

This problem is usually more pronounced during damp weather. The local power company should be alerted in these cases.

Ignition Noise

Ignition noise is rarely a problem except when detecting low power signals. It causes small, randomly spaced streaks on the picture when the input signal-to-noise ratio is too low. The obvious solution is to increase antenna gain.

Figure 10-9. Interference from a Brush-Type Electrical Motor. *This interference is caused by ingress of signals from a brush type electrical motor. (Courtesy of Sencore)*

C. THE FIELD STRENGTH METER

A field strength meter (FSM), also known as the signal level meter (SLM), is an indispensable tool used to monitor and maintain signal voltages in broadband communication systems (see Figures 10-10, 10-11 and 10-12). Regular system performance tests insure that adequate signals reach all points in a system. Levels that vary from one test to another indicate problems that may be slowing developing.

Field strength meters detect RF energy and display the measurement in decibels relative to one millivolt (dBmV). They incorporate built-in tuners that can focus on any individual channel or frequency as well as output meters or LED read-outs to indicate signal levels. Some feature built-in television monitors and also serve as multimeters.

Most field strength meters have several scales printed onto their face-plates below the front panel dial. Each scale spans a range from zero to a specified maximum voltage level. This range is set by use of internal attenuation pads. Although the range can be manually adjusted, some brands feature automatic range setting. Some brands also have built-in step attenuators to pad down and therefore permit measurement of unusually high signal levels. It is important to adjust all measurements to take into account any additional outboard attenuators that may be in the signal path. On most digital meters, this calculation is performed automatically.

The SLM is an installer's first and most important piece of test equipment. Its use and care should become second nature. It is wise practice to own a high quality brand and to have it re-calibrated once every three months by a certified technician or by its manufacturer (see Figure 10-13). Note that if a FSM is dropped or shocked, it should be re-calibrated.

A signal level meter is not only capable of indicating signal levels, but can also be used to detect and measure hum, signal-to-noise ratio or the level of signals leaking into and out from a distribution system. When used in conjunction with other pieces of test equipment, it can be used to measure signal frequency, indicate return loss or test the frequency response of the system. A signal level meter can even be combined with an oscilloscope and sweep generator to mimic the functions of a spectrum analyzer. These capabilities are all explored in the following section.

TROUBLESHOOTING & TEST INSTRUMENTS

Figure 10-10. The Sam 3 Signal Level Meter. *This high-end meter is suitable for use by lead technicians in system set-up and evaluation. It has microprocessor control and a direct entry key pad to accurately enter frequency settings. (Courtesy of Wavetek)*

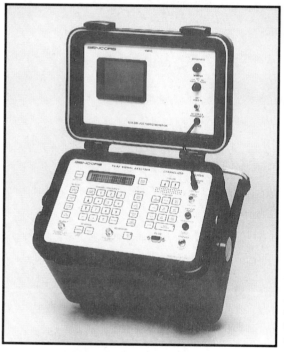

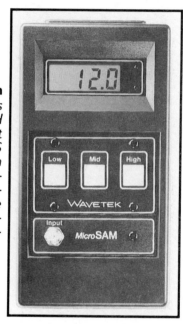

Figure 10-11. MicroSam Signal Level Meter. *This is a low cost meter used to evaluate signal levels at a cable drop or MMDS downlead. It is preset with three different frequencies to provide an adequate test range while eliminating tuning errors. (Courtesy of Wavetek)*

Figure 10-12. Signal Level Meter. *The Sencore SL750-A signal level meter features a 750 MHz bandwidth and a color monitor and a pilot test signal for system balancing. It has a 0.5% accuracy on RF level tests. (Courtesy of Sencore)*

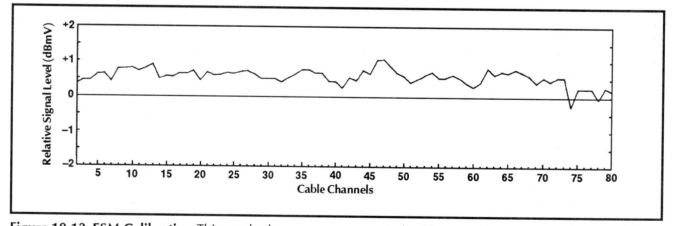

Figure 10-13. FSM Calibration. *This graph, the computer-generated calibration of a Sencore FS74 field strength meter, shows the difference between a standard 0 dBmV (1000 mV) signal and the readings taken by the FSM. This difference is less than 1 dB at all points. This calibration chart can be used to correct readings taken in the field. For example, if the graph shows the reading to be 0.35 dB, this amount is simply subtracted from the reading.*

TROUBLESHOOTING & TEST INSTRUMENTS

System Performance Tests

A FSM can be used to measure a variety of system parameters including video and audio carrier levels, the ratio of audio to video carrier levels, signal-to-noise ratio, hum modulation, signal egress and return loss (see Figure 10-14). This instrument can also be used to test passive components such as splitters, taps and coaxial cables. These functions are outlined below.

Signal Levels

Audio and video carrier levels on each channel leaving the headend, at any point in the distribution system and at the subscriber drop can be directly measured with a FSM, either in dBmV or microvolts.

Video carrier levels at any point in a distribution system should fall within 1 to 2 dB of adjacent channels. The level should vary from the most powerful to the weakest carrier by less than approximately 7 dB. Adhering to these standards will serve to minimize problems with cross modulation. A FSM can also be used at a subscriber drop to ensure that levels are within the 3 to 10 dBmV rating, well below the approximately 5000 microvolts that can overload a television receiver. These recommended levels have been previously outlined in Table 8-2.

The audio to video signal level ratio can also be measured with a FSM. This can be determined by measuring the video carrier level and then the audio carrier level and subtracting the two readings. However some FSM meters are capable of directly measuring the A/V ratio for each channel. This ratio should fall between −13 and −17 dB to prevent the sound carrier from interfering with the video carrier resulting in sound beats in the video.

Signal-to-Noise Ratio

Measuring S/N ratio is quite straightforward but this process can have some unexpected pitfalls. First the signal level of the video carrier is read. Then the FSM is tuned to some "free space" in a portion of the spectrum where no signal is present and the S/N ratio is found by either manually or automatically subtracting the two readings. The pitfalls in this process come in determining to which noise level the signal should be compared.

System noise may vary widely throughout the broadcast band as shown in Figure 10-15. Amplifiers and headend signal processors are the two principle sources of noise in a distribution system. While amplifiers generate a fairly constant amount of noise across the band, processors produce noise in a fairly narrow range. If S/N ratio is being measured on a channel that has a noisy processor and if a nearby channel with a low noise level is used as a reference,

Figure 10-14. Using a Wavetek SLM. *This technician is using the SAM-III to test levels at a line extender in an apartment complex. This meter's capabilities include measuring signal and hum levels, S/N ratio, frequency and return loss. (Courtesy of Private Cable Magazine)*

TROUBLESHOOTING & TEST INSTRUMENTS

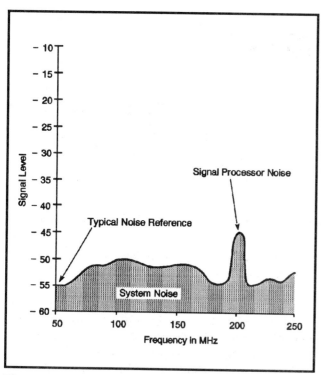

Figure 10-15. System Noise Variation. *System noise can vary quite widely through the broadcast band, often due to noisy processors. The difference between the typical noise reference and the noise at 200 MHz above is a large 15 dB. (Courtesy of Sencore)*

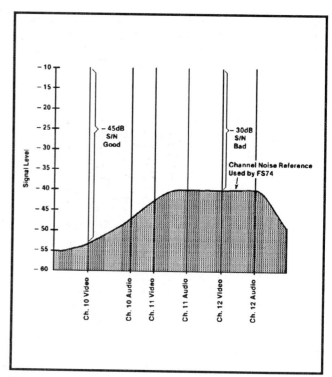

Figure 10-16. Variation of S/N Ratio. *The S/N ratio can vary across the broadcast band if the reference noise level also varies even when all channels have identical output levels. (Courtesy of Sencore)*

the S/N ratio would be underestimated (see Figure 10-16). A better method to determine S/N ratio is to compare the signal to the noise present during the vertical blanking interval of the active channel in question.

Some FSMs have the capability of automatically measuring S/N ratio using either an adjacent "quiet" channel or the same channel vertical blanking interval noise as the reference. In the latter case, the meter must add approximately 13 dB to the noise value to yield the equivalent noise in a signal of 4 MHz bandwidth, as required by the FCC.

Hum

While some FSMs can directly measure levels of hum on active channels, others may require that the video and audio carriers be removed and replaced with a continuous wave carrier. The FSM is simply connected to the output of a tap, a subscriber drop, the test point on an amplifier or any other port in the system. A hum level in excess of 2% is usually considered objectionable. If a particular brand of meter is not equipped for measuring hum, the same can be accomplished in combination with a oscilloscope connected to the signal level meter's video output.

Signal Leakage and Return Loss

There are numerous methods that can be used to detect and measure leaks. The simplest device to simply find leaks is a tunable FM receiver that emits a tone when egress radiation is detected. More advanced and more convenient hand-held detectors are also available. For example, one such unit automatically scans the FM band until a leak is found. It then stops scanning and emits a warning tone. This small-scale tester can be worn on the belt of a technician doing routine preventive maintenance.

However, one of the most sensitive and therefore effective ways to both detect and measure leaks is to connect a "site-survey" dipole antenna, preamp and bandpass filter to the input RF connection of an FSM (see Figure 10-17 and 10-18). First adjust each arm of the dipole to 1/4 of the wavelength of the signal to be detected (see Appendix E for a list of wavelengths and Appendix B for a formula used to calculate the length of the dipole antenna. Note that if this dipole is adjusted to receive a local off-air channel, the measurement will clearly not be one of egress). Then rotate the antenna in a horizontal plane parallel to the cable and about 10 feet above the ground and away from a suspected leak. The di-

TROUBLESHOOTING & TEST INSTRUMENTS

Figure 10-17. Configuration for Leak Testing. *A calibrate dipole, bandpass filter, possibly a preamp and either a SLM or spectrum analyzer are used to detect and measure leaks on a distribution system.*

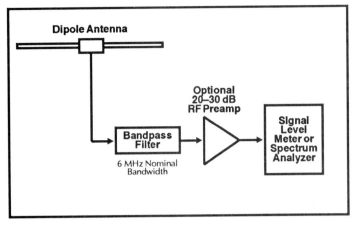

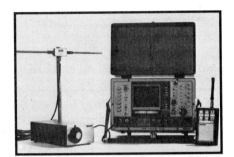

Figure 10-18. Apparatus to Test for Leaks. *This combination of test dipole antenna and SLM can be used as a high sensitivity instrument to test for leaks (Courtesy of Private Cable Magazine)*

pole should be supported and positioned on a non-conductive rod. A metal support could act as a second antenna and either concentrate or reflect the signal. The highest reading should be used as the measure of radiation.

When extremely accurate measurements are required a calibrated dipole can be used. However, even a standard TV rabbit ear antenna can be used. In this case, the 300 ohm cable should be removed and replaced with 75 ohm coax connected directly to the telescoping arms.

Such a sensitive testing apparatus is often required because borderline levels of egress can range from 30 dBmV to as low as 50 dBmV. In this case, the bandpass filter provides selectivity so that the equipment can discriminate between the leak and other higher power signals which may also be present in the vicinity.

The FSM readings, in microvolts, can be converted to field strength reading, in microvolts per meter by the method outlined at the conclusion of Appendix B.

Return Loss

A field strength meter can also be used to measure return loss. Reflective losses can occur at any circuit junction as a result of impedance mismatches. Return loss is defined as the ratio of the original to the reflected signal. So a perfect junction would have zero reflective losses and therefore an infinite return loss. The return loss would be zero if all the signal power were reflected. For example, if 1 millivolt (0 dBmV) enters a device having a 15 dB return loss, the reflected signal would be 15 dB lower in value. Thus 15 dBmV or 0.18 millivolts would be reflected and the remainder of the signal, 0.82 millivolts, would enter the device.

Return loss can be measured using a signal level meter, a signal generator and a return loss bridge which is basically a three-way divider. Both the meter and generator are first set to the same frequency. The signal generator is then connected to the input of the bridge and the SLM to the intermediate port. Then two measurements are made: one with the output port terminated; and one with the device under test connected. The return loss is the difference between the two measurements.

Other Measurements

A FSM can be used for other types of measurements. These are limited by the imagination of the technician. For example, signal frequency can be precisely measured via the configuration outlined in Figure 10-19.

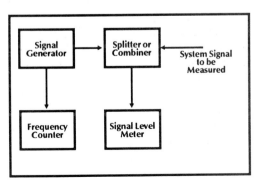

Figure 10-19. Precisely Measuring Frequency. *This simple arrangement of signal generator and SLM can be used to detect the point of "zero beat" and therefore be used to precisely measure frequency on a distribution system.*

TROUBLESHOOTING & TEST INSTRUMENTS

Testing Passive System Components

A field strength meter can be used to test passive devices such as splitters, taps and coaxial cables. This is a useful function considering that these system components can develop problems such as excessive signal attenuation or poor isolation between ports. Those techniques for splitters and taps are described here.

Splitters

The insertion loss of a splitter can be easily measured with a FSM as follows. First record the input signal level. Then terminate all but one of the output ports and measure the signal at the free output port (see Figure 10-20).

To measure isolation between two ports, first terminate the input port and all but the output ports under test. Then apply a signal of known voltage to one of the output ports and measure the value that leaks through to the second output port (see Figure 10-21).

Taps and Directional Couplers

The three parameters that define a directional coupler, tap loss, isolation and insertion loss can be easily measured with a field strength meter. These test are quite similar to those described above. To measure tap loss, simply terminate the output port, apply a signal of known value to the input port and measure the level at the tap output. The difference in level is the tap loss (see Figure 10-22).

To measure insertion loss, terminate the tap output, apply a signal of known value to the input and measure the output voltage. The difference in these readings is the insertion loss.

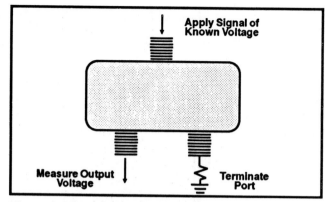

Figure 10-20. Measuring Splitter Insertion Loss. Splitter insertion loss can be measured with a FSM as indicated in this illustration.

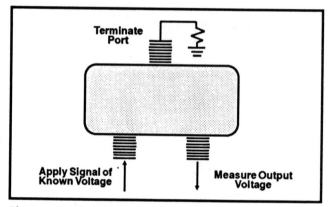

Figure 10-21. Measuring Splitter Isolation Loss. Splitter isolation loss can be measured with a FSM as indicated in this illustration.

To measure isolation loss, terminate the tap input, apply a signal of known value to the tap output port and measure the output voltage. The difference in these readings is the isolation loss.

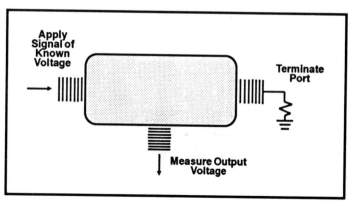

Figure 10-22. Measuring Tap Loss. Tap loss can be measured with a FSM as indicated in this illustra-

D. THE SPECTRUM ANALYZER

Spectrum analyzers are powerful diagnostic tools designed so that incoming signals in a range of frequency bands can be observed and measured (see Figure 10-23). These instruments plot a time-varying graph of signal power across 50 ohms or voltage across 75 ohms versus frequency. Frequency increases from the left to right on a CRT screen while signal amplitude is presented as a series of peaks and valleys. The vertical display is usually calibrated in dBm or dBmV. A spectrum analyzer, a "frequency domain" instrument, can be considered the companion to an oscilloscope that plots signal amplitude in the "time domain" as a function of time, not frequency.

An analyzer can be very useful in visualizing inputs throughout a communication system ranging from the satellite C-band signals to off-air carriers to lower frequency RF signals traveling through a distribution system. For example, incoming microwaves located anywhere within the entire band of satellite communication frequencies can be observed at once or fine details of the frequency and amplitude structure of any single carrier can be studied in isolation.

An analyzer has numerous uses in studying MATV, SMATV and wireless systems. It can be used to:

- View individual or groups of channels on a distribution system, a headend or at the output of a preamp or downconverter
- Separate an individual channel into audio and video carriers
- Measure the frequency and amplitude of any audio or video carrier
- Measure peak FM deviation of an audio carrier and AM modulation depth of a video carrier
- Locate and observe both ingress and egress interference
- Display co-channel interference and other spurious signals within a channel
- Test for composite triple beat and second-order distortions
- Determine tilt and gain flatness at any point in a system
- Measure S/N ratio (see Figure 10-24)

A spectrum analyzer can also be quite useful in observing signals at any point in a satellite reception system. For example, when the signal from an LNB is fed into this instrument, microwaves of any chosen polarity can be detected and measured. By scanning a feed or an antenna across the sky and observing any variations in detected power, an interfering

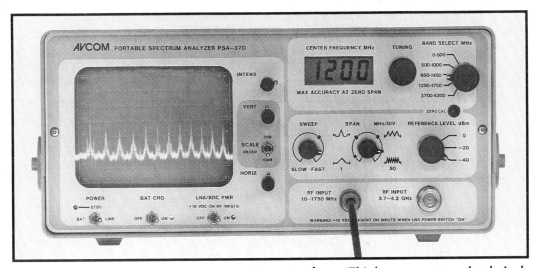

Figure 10-23. The Avcom PSA-37D Spectrum Analyzer. *This battery-operated, relatively low cost analyzer accepts input signal from 10 to 1750 MHz and from 3.7 to 4.2 GHz in five bands. It has both an LCD digital and CRT display. A built-in supply powers LNAs and LNBs. (Courtesy of Avcom of Virginia)*

TROUBLESHOOTING & TEST INSTRUMENTS

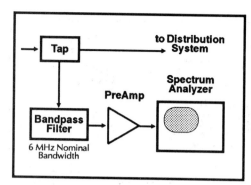
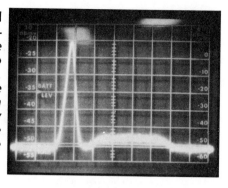

Figure 10-24. Measuring S/N Ratio. *The test setup for measuring S/ ratio is shown here. The preamp can be incorporated to increase the analyzer's sensitivity. A correction must be made for the effect the bandpass filter has on narrowing bandwidth. The display shows noise in the vicinity of the signal. (Photo courtesy of Private Cable Magazine)*

source can also be pinpointed. This capability makes an analyzer invaluable in aiming an antenna and in peaking a feed. An LNB typically generates a signal having powers ranging from 60 to 80 dBmV while the highest level TVRO signal at a receiver's modulator output is on the order of 40 dBmV. Most analyzers can detect RF signals produced at any point in a typical satellite reception system.

Uplink facilities also use on-line analyzers to monitor the power, frequency, stability and composition of signals relayed to communication spacecraft (see Figure 10-25). Powers are generally equalized across all transponders on-board a broadcast satellite. This is seen as a series of equal peaks spread across the entire C or Ku-band frequency range. A spectrum analyzer monitoring the output of an LNB should see the same, equal distribution of transponder power.

The ability of an analyzer to create a visual display of signals over a wide range of frequencies makes it a valuable troubleshooting instrument that can pinpoint headend, distribution system or component problems. This tool can also be directly connected into the output of an off-air antenna or LNB or into the input or output of any cable or electronic component in a headend or distribution system. Signals can therefore be traced en route from an antenna through the distribution system and finally to each television set in order to identify any faulty components or cable breaks.

Two common difficulties characteristics of satellite reception systems observed on an analyzer are tilt and wipe-out. For example, if transponder power appears to decrease across the satellite band on the instrument's display, the feedhorn or LNB may be faulty. If the problem is a bad component, it might be identified by switching in a healthy one. Tilt across the transponder bandwidth can also be traced to cable lines which have increased attenuation at higher frequencies. Another common problem, transponder wipe-out, that can also be traced with an analyzer, appears as a sharp reduction in power of one or a group of transponders. One source of this problem may be an impedance mismatch which causes signals in a rather narrow frequency band to be reflected back to the antenna. Something as simple as a missing terminator on an unused splitter port or a faulty connector could be the culprit. These concepts are explored in more detail in Chapters 4, 5, and 6.

Measuring TI Levels

An analyzer is an invaluable tool during a site survey when mapping and measuring any source of terrestrial interference. It can be used to detect the presence of even extremely weak interference down to 60 dB relative to the power of the both satellite and local broadcasts. Many difficulties can be avoided by knowing what signals are being detected by all the satellite and off-air antennas before the headend design is completed.

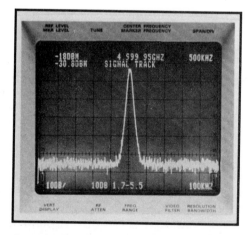

Figure 10-25. Spectrum Analyzer Frequency Display. *The analyzer screen on this spectrum analyzer shows the amplitude and frequency of a 4.99 GHz signal. (Courtesy of Tektronix)*

TROUBLESHOOTING & TEST INSTRUMENTS

Figure 10-26. Test Setup for Measuring TI. *This combination of equipment can precisely determine interfering carrier frequencies and polarizations and localize sources of TI.*

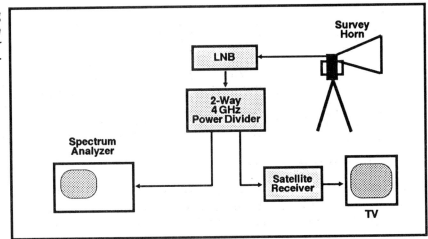

Non-television, co-channel carriers can usually be distinguished from other interfering signals because they have center frequencies which are offset 10 or 20 kHz from the off-air center frequencies. TI carriers generated from line-of-sight microwave links are also offset from the center frequencies of satellite transponders. Measurements of such interfering signals can either be taken at the headend output or directly off-air.

Precise measurements describing any terrestrial microwave interference should also be gathered at the location where a satellite antenna will be installed. A survey horn mounted on a stable, modified camera tripod can be used to obtain the elevation and azimuth to interfering sources (see Figure 10-26). This horn should be positioned as closely as possible to the focal center of the proposed permanent antenna. With the horn targeted towards the satellite(s) of interest, the spectrum analyzer can identify the presence of interfering carriers. This instrument can be installed in series with a device called a frequency tracer to help in identifying TI carrier type, center frequencies, relative amplitudes, polarities and bandwidths. These parameters should be recorded in the site survey log.

All this information should be factored into a report which includes detailed site maps, drawings showing surrounding buildings and other structures which may serve as shields, antenna mounting recommendations as well as TI sources and levels.

E. POINT-BY-POINT DIAGNOSTICS

Troubleshooting and repair may be initiated by customer complaints of by information gathered from periodic measurements of parameters such as hum modulation and S/N ratios. Once a problem is evident, the first step is to determine how much of the system has been affected. In the problem isolated to one subscriber, does it affect a group of subscribers or in one or more locations or is it system wide. This information and a brief analysis of the as-built maps should lead to a rapid identification of the general location of the fault. For example, if just one subscriber is affected, the problem almost certainly lies at or just outside of the customer premises. If all customers are affected, most likely the fault can be traced to the headend or a central distribution point. If a group of subscribers are affected, a local distribution amplifier, connector or power supply may be defective.

A system-wide problem that is traced to the headend can be further diagnosed by determining which channels are faulty. For example, if a group of satellite channels are not functioning properly, the problem may lie with the associated satellite antenna, LNB, group combiner or any component common to all these channels. Faults with a single channel may be traced to one modulator or a defective connector.

A brief catalog of and solution to problems that may be encountered at both the customer drop and the headend are listed below. The diagnostic procedure that has been briefly outlined above should eliminate the need for many of these tests.

TROUBLESHOOTING & TEST INSTRUMENTS

The Customer Drop

As a rule, the further downstream from the headend one progresses, the more likely are the chances of encountering operational problems. When the initial analysis suggests a problem at or near the customer drop, troubleshooting should begin at a customer's television set and then should progress step-by-step upstream towards the headend. This involves systematically checking the operation of each device and the continuity of all cables. The following is a suggested, basic troubleshooting procedure:

1. Check the customer's television set to make sure that the power is on and that it is tuned to the correct converter output channel.

2. Check all connections to the converter and the television set input. Read signal levels at this point.

3. If the appropriate 3 to 10 dBmV signal levels are present at this point, hook up a test TV set to verify the quality of pictures. If a good picture is seen, show it to the customer and suggest that his television set be serviced. Interference problems are often caused by overloading a TV set, and snow by signals that are too weak. In a well designed and properly operating system, these pictures should look as good on the customer's set as at the headend output.

4. If no signal is detected at the customer's set, check the drop box. Verify that the customer is hooked up. Use the signal strength meter and test set to verify that a signal is available at the box. If a signal is present at the output of the tap, disconnect the customer's drop at both ends and look for a short or open circuit with a volt/ohm meter. Replace the drop if it is faulty.

5. If a signal is not detected at the box, check all of its distribution devices and fittings. Check the center seize screws on all devices that use a hardline input. Examine the point where the line enters the first device in the box. If a signal is present, check each device previously examined for continuity and secure fittings. However, if a signal is not detected, the technique of halving the system should be employed to determine the faulty component. Look at the system plan and locate a device or junction box that is halfway between the headend and the problem area. Check this point for proper signal level and component operation. If it is functioning well, then the problem lies between that point and the original location. If not, the problem lies somewhere between that point and the headend. Continue with this technique and check points that are halfway between the previous area examined and the location of the original problem. For larger systems the time saved with this method can be substantial.

6. At each device or junction point, check its output for the presence of a signal. If a signal is available, disconnect the distribution line from the back of the box and check for continuity. Replace it as needed. If no signal is present at its output, check this device. Check splitters for input and output levels at all ports and check all connections. If there is an input but no output levels and if the connections are in good shape, replace the device.

7. If the device suspected to be faulty is an amplifier, verify that the input level falls within the specified range. If a level is measured but there is no output signal, check the input line for voltage. Most distribution devices are line powered and an interruption of power shuts off the amplifier. Examine the fuse inside the amplifier and verify with a VOM that it is not open. If power is not present, check the output of the power supply with a VOM and replace or repair it as needed. If a signal and voltage are present at the amplifier input, replace the amplifier module.

8. If a problem still exists, work back towards the headend one station at a time. Check the operation of all devices and verify the continuity of all lines.

9. If a problem is present in the headend, check its performance against construction as-builts. Compare the existing output and balance levels against those recorded in the construction documents. Finally, re-balance the system to correct any deficiencies.

In general, poor fittings are the most likely cause of system failures. A poor fitting will often manifest itself as a vertical hum bar in the middle of a picture. Sometimes ringing or ghosting will also be visible. Fittings that have been disconnected or that have their central conductors "sucked out" commonly occur when TV sets are often moved. Center conductor suck out is slang for the condition that occurs when the inner coax conductor has withdrawn far enough into the dielectric to break contact at an F-connector. Learning how to correctly install fittings during initial construction as well as during a trouble call goes a long way towards eliminating many operational problems.

Malfunctions in video recorders are also the source of numerous service calls. Many people simply do not know how to hook a VCR up to a cable converter. Usually they attempt to install it according to the manufacturer's directions and end up with a poor signal and bad video quality. It is wise to prepare a hand-out sheet that details the correct method of connecting to the SMATV line and converter. This document should be part of every new subscriber's information package. Since VCRs are generally used by customers to record programs that would otherwise be missed, they can only get the most for their entertainment dollar if these devices are correctly installed.

Troubleshooting the Headend

In those cases where more than one complaint relates to the same problem, the fault often lies in a distribution line or the headend. The first step is to localize the trouble area by mapping and correlating the addresses of the trouble calls. If these customers are all served by a specific line amplifier or coax route, chances are excellent that the fault has been isolated. But if the problem is rather general in scope, it is a wise policy to begin the troubleshooting process at the headend. Remember that this set of difficulties is in stark contrast to the typical case when only one customer complains about having bad pictures. This most often suggests that the problem is localized at their television set or drop box.

Satellite Signals

Troubleshooting a satellite headend can conveniently begin at the receiver output. When a television monitor is connected, the pictures should be clear. If not, the problem could be caused by any number of factors including TI, antenna pointing errors or blockage, a loose connector, a faulty LNB, a deformed dish or even a severed cable. Troubleshooting a satellite system is a relatively simple process when using a device such as a test receiver or spectrum analyzer.

If an adequate signal is present at the satellite receiver input and output, the next step is to check each modulator (see Table 10-1). If the modulators are operating up to par, a good quality picture should be observed on the portable television set. Note that the input signal may have to be attenuated in order not to overload the television receiver. The video or audio modulation controls can be adjusted if the picture is too light or dark or if the sound is too quiet or loud. If interference is observed on any channel, only snow should be observed when the modulator is disconnected. If not, the source of the difficulty is an ingressing signal. If the modulator is turned off, the ingress should not disappear. Spurious modulator outputs should be tested by an FSM or spectrum analyzer. In this case, if the modulator is switched off, the spurious outputs should disappear.

Off-Air Signals

Signals should be first tested at the downlead of each off-air antenna with a portable TV. If pictures are substantially worse than had originally been observed during installation, check for antenna blockage, pointing errors as well as disconnected or damaged cables, matching transformers, preamps or filters. If all appears to be fine on the test set, the FSM meter can be used to verify that off-air signals are still at their original levels. These tests should also show the presence of interference or other spurious signals.

TABLE 10-1. SATELLITE RECEIVER SIGNAL LEVELS and PICTURE QUALITY

Input C/N Ratio (dB)	Video Description
4	Extremely noisy, audio noise, tearing
5	A little better, sparklies
6	Watchable but sparklies
7	Near threshold, fewer sparklies
Threshold	A few sparklies
9	Sparklies only on saturated colors
10	Good picture quality
11	Good picture quality with fade margin
12	CATV headend quality
16	Local broadcast quality
18	Network quality

TROUBLESHOOTING & TEST INSTRUMENTS

Next, connect the portable television to the output of each processor. If there is any noticeable difference between the pictures before and after each processor, the FSM should be used to obtain more detailed measurements. At this point, it is useful to re-balance the audio and video signal levels as well as to check for spurious outputs in excess of 60 dB relative to the video carrier. If the processor outputs are not clean, inadequate filtering may be to blame.

Similar procedures can be followed for troubleshooting both locally originated and FM signals. The quality of the FM stereo can be easily traced from the output of a satellite modulator or from an omnidirectional FM antenna with a set of headphones and a stereo amplifier. Narrow deviation stereo carriers can be particularly susceptible to drifting in the stereo processor. These units, like all other headend components, should be burned in for at least a day to stabilize their outputs. Faulty or poor quality receivers, modulators, processors or other headend equipment that does drift can cause unnecessary headaches.

The Combiner

Tracing the headend signals through the combiner can proceed in much the same fashion. Once the satellite and off-air signals are combined, there should be no differences between the individual and combined signals on each active channel. If any discrepancies are noticed, each channel can be disconnected in turn in an attempt to identify the culprit. Once a clean output has been verified on each channel and the launch amplifier is working well, the headend should deliver an adequate signal to the distribution system.

Equipment Repair and Periodic Maintenance

By far the best way to avoid problems is by adhering to a regular schedule of system maintenance. The money spent in using the proper test equipment to periodically monitor SMATV system conditions is well spent. For example, two common difficulties, a decrease in signal level and increase in hum or cross modulation over time, can be tracked and failures can actually be predicted. Assuming amplifiers had been properly tested and the result had been documented before installation (please see Section C of Chapter 8 for more details), routine tests of S/N ratio, cross modulation and hum modulation can be made on a periodic basis. If any troublesome trends are noted, action can be taken before, not after, a failure occurs.

SECTION II

WIRELESS CABLE/MMDS SYSTEMS

Your One Stop Source

Whether it's Headend Equipment, Distribution Products, Connectors, Splitters, Couplers, Diplexers or any kind of Amplifiers...

PICO MACOM
"We Have It All"

For your technical questions call the leaders in filling your complete SMATV equipment requirements.

The PR2400IRD has been tested and licensed by General Instrument Corporation to meet the performance standards for the

VIDEO*CIPHER*® II PLUS
Commercial Descrambler Module.

Features
- Front Panel Module Access
- Threshold Extension Detection
- 70 MHz IF Loop
- 2 Year Warranty

PICO MACOM, INC.
Lakeview Terrace, CA 91342 • (818) 897-0028 • (800) 421-6511

VideoCipher® is a registered trademark of General Instrument Corporation.
VIDEOCIPHER II PLUS™ Decoder Module warranted seperately by General Instrument Corporation.

CHAPTER 11. WIRELESS CABLE SYSTEM OVERVIEW

Wireless cable, also known as a multichannel multipoint distribution system (MMDS), is a ground-based microwave broadcast system. It is, in effect, analogous to a satellite broadcasting system in which the satellite is replaced by a microwave transmitter located at ground level. In many situations, this technology can be a viable alternative to both satellite and cable television audio/video distribution methods. Much smaller receive antennas can be used because power levels are substantially greater than those received from satellite.

Wireless cable TV was developed in the United States in response to the legal, political and bureaucratic complexity and other constraints to developing cAble television franchises. The construction of cable television systems, especially in urban and large suburban markets has occurred very slowly and has been quite time consuming. Wireless cable systems, while not a universal panacea, are generally financially, technically and administratively more efficient and can be more quickly and easily installed, launched and operated than wired cable networks..

Wireless cable technology potentially allows entrepreneurs an opportunity to enter the subscription television business quickly, with a capital investment that is typically 50 to 65% of that required for a traditional hard-line cable television system. As a result, wireless cable has become the buzz-word of today's communications investment community. Viewed by some as the ultimate television service, and feared by others as the death knell for franchised cable, it is neither. Wireless cable is an effective transmission medium that works well in many areas, but is not technically appropriate for others.

In an effort to provide readers with a clear understanding of the complex factors that influence this business, the material in this chapter is organized into three sections: a technical overview, history and frequency allocations, channel availability and regulations. In this chapters that follow many of these issues are examined in more detail.

Some of the arguments in favor of wireless cable are summarized below. The limitations of this technology become apparent later in this and subsequent chapters.

- Wireless technology can effectively compete with conventional cable television (CATV) distribution of entertainment. Any programming that is available for purchase and resale can be distributed over a wireless cable system.

- Capital costs are much lower, less than half of those required to install a wired CATV system. So programming can be offered at competitive rates and excess capital is available for more effective customer service.

- A wireless system can be on-line and operational within a relatively short six to nine month period following regulatory approval.

- Subscribers can generally be served by a central transmitting antenna, two or three fill in or repeating antennas and, if necessary, direct cabling of small areas and buildings. Microwave powers involved are very low, under 1 watt at any receiving site. Both outlying areas and customers in the immediate vicinity of the transmitting antenna can be served at once upon system turn-on. Cable networks builds could take years to reach these same outlying areas.

WIRELESS – SYSTEM OVERVIEW

- This is a "wireless" system with no need for time consuming and disruptive excavation of public thoroughfares. Labor is required only at the transmitting site, the repeating antennas and at the end-user locations. The receiving equipment consists of a small antenna that is cabled to a downconverter. Signals are fed indoors to a converter, similar to a cable TV converter box that sits on television sets.
- A wireless cable company can be effectively and efficiently operated according to a solid body of operational experience and tested management principles.
- Every subscriber has a unique address so that service can be activated or canceled from the main control office, thus eliminating the need for service calls.
- The number of channels is limited only by the amount of allocated frequency spectrum. For example, an eight channel NTSC or PAL wireless cable broadcast system requires as little as 64 or 56 MHz of bandwidth, respectively.
- Once a wireless transmission system has been installed, signals can be delivered to multi-unit dwellings via wireless at substantially lower costs than those received directly via satellite at SMATV headends.

A. TECHNICAL OVERVIEW

A wireless cable TV station is a local delivery service that transmits multiple channels of video or data via a central microwave antenna to small, inexpensive roof-top antennas installed at each subscriber location within a designated service area. Since the frequencies used are many times higher than those employed in conventional off-air UHF or VHF television, standard TV sets cannot detect the microwave signals. A wireless cable system is, in effect, an over-the-air cable TV network that does not require the installation of expensive and often intrusive coaxial cable lines to reach subscribers. As a result no public thoroughfares or routes of access need be disturbed.

In contrast, a conventional cable network requires utility pole, make-ready engineering, possible rearranging of existing pole line utilities to accommodate a cable TV line, underground and street trenching when no pole line is available and a comprehensive design, often with hundreds or even thousands of miles of cable to access each potential subscriber.

A wireless cable system consists of the program origination and transmission facilities and subscriber reception equipment. While satellite and off-air broadcasts, programming from stand-alone local facilities such as video tapes or any other conceivable sources may be received at and transmitted from the wireless headend, most programming for wireless cable systems is generally received from broadcast satellites (see Figure 11-1). Services can include full color television, high speed data, high definition television, facsimile and multiple stereo FM broadcast signals. In the United States, the RF spectrum in the 2150 to 2686 MHz frequency band has been allocated for use for wireless signal transmission, permitting delivery of potentially up to 33 channels of combined satellite, local origination, educational and local VHF/UHF programming to subscribers within a given service area.

Since a wireless system is a line-of-sight service, the transmitting antennas must be located on a tall building, tower or mountain, if at all possible, to maximize the area of signal coverage. The transmit site building is configured to include adequate space, climate control, fire and vandalism protection and stand-by power facilities.

While the range of a transmitting antenna can vary up to 35 miles (56 kilometers) depending upon the broadcast power, each subscriber antenna must have a "view" of the transmitting antenna because trees, buildings, high terrain or other obstructions absorb microwave signals. Those areas which are "screened" and do not have a clear line-of-sight view of the transmitter can be served by secondary transmission sites, via repeaters, or can be cabled directly. Wireless cable transmission power, which lies in the perfectly safe 1 to 100 watt range, is substantially below the hundreds of kilowatts typically required by a conventional off-air VHF or UHF broadcast television station. A properly designed,

designed, low power, wireless cable system can serve most if not all subscribers in the market area.

Signals are received by end users via a small antenna, typically on the order of 25 centimeters in diameter. A downconverter mounted onto the high gain microwave receiving antenna lowers the signal frequencies to those that are recognizable by standard or cable-ready TV sets, typically to those in the VHF band. A small diameter coaxial cable then relays the signals to a set-top converter. This decoder, like the standard cable TV box, enables a subscriber to select any number of channels for viewing on a standard television set. Downconversion is a necessary step to minimize cabling losses that would be substantial at higher microwave frequencies.

A conventional off-air antenna is often incorporated into a wireless cable reception system to deliver local VHF/UHF signals. This allows an operator to offer additional channels without using valuable wireless cable bandwidth. However, if a wireless system is located in an area with poor off-air reception, the wireless operator may have no choice but to transmit at least the major local channels on the microwave system.

The set-top decoder is the customer control center. At a minimum, it should combine off-air and wireless channels into a single, multi-channel format and must be able to provide channel mapping functions for use of ITFS frequencies (more on this in Section C). It also descrambles secured signal transmissions and is addressable from the operations center. More advanced boxes offer advanced features such as pay-per-view and on-board character generation. Depending on its complexity, the wireless converter can perform a number of functions that rival and surpass the services offered by traditional franchised cable. A well-designed system with modern converters can provide the following features:

- A unique address for each receiving unit or subscriber.
- Tamper proof and secure program delivery.
- Various combinations or tiers of television programs. For example, a basic set of channels could be available for a fixed cost while other combinations of pay-TV channels could be purchased on an individual basis.
- Pay-per-view or pay-per-day services remotely controlled by the operator. For example, one or more first-run movies or special events might be purchased for a flat fee

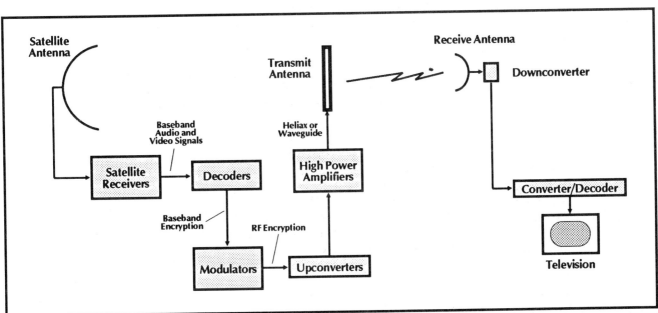

Figure 11-1. Schematic of a Wireless Cable System. *Audio and video signals are received from a variety of sources including satellite, video tapes and direct feeds. In the case illustrated here, a satellite antenna receives signals from a program supplier and transmitted over a geostationary satellite. Upon reception they are processed by satellite receivers. The raw video and audio signals are each passed to modulators and decoders. Then each television channel is amplified to a power ranging from one to 100 watts. Transmitter outputs are combined and passed into either heliax or elliptical waveguide for relay up the tower to an antenna. The transmission antenna radiates these signals, often in a omnidirectional pattern, to an area typically within 15 to 30 miles from the tower. At the customer site a small microwave antenna detects the signals then passes them via a downconverter to an indoors converter/decoder. The output of this set-top is fed into a television set.*

by placing a telephone call to the operations center. Modern converters now use store-and-forward methods whereby impulse purchases can be made at the touch of a button on the converter. This reduces customer service reps (CSR) manpower requirements at the head office and increases buy rates.

- Remote disconnect any or all channels and/or subscribers to expedite collection of fees.

- Feeding of programming to multiple-dwelling units, as a stand-alone system or as an adjunct to an existing MATV or SMATV system. In multi-family dwellings, just one receive antenna can be used to detect wireless signals. The signal can then be amplified and split for distribution via a network to serve each subscriber. The MATV/SMATV network can carry just a single wireless channel or can offer a full menu of programming, one that can compete with most franchised cable offerings.

- Additional potential future communication services. Since the frequencies allocated to the wireless cable station can ultimately be employed for two-way communications, the unique subscriber digital code can also double as a telephone number for cellular telephone or data communications, paging, fascimile or other services.

While a wireless cable converter can do everything that a cable TV converter can do, and often more, both systems are often carefully regulated. For example, in the United States the total number of channels that may be transmitted is controlled by the FCC and MMDS systems are subject to a myriad of rules and regulations.

B. HISTORY

Wireless cable technology has evolved from two similar television distribution services that were created during the 1960s in North America. In that decade, both Canada and the United States established an Instructional Television Fixed Service (ITFS) using the 2500 to 2686 MHz band. In the United States the original ITFS rules were established by the Federal Communication Commission (FCC) in 1963, allocating a 31 channel (NTSC) spectrum band in the frequency range from 2500 to 2690 MHz for local distribution service. Twenty eight of the thirty one channels were assigned to ITFS service.

Today there are several hundred ITFS systems in use in Canada and the United States, all which are licensed to non-profit educational entities. These systems are generally used for the distribution of instructional programming to classrooms on a "video library" or centralized lecture basis.

In the late 1960s, the FCC realized the commercial potential for a local broadband distribution service. This regulatory board subsequently allocated two channels just above 2150 MHz in each of the major metropolitan areas for a service which it termed multipoint distribution service (MDS). A single channel, MDS 1, was allocated to all markets; a second channel, MDS 2, was allocated in the top 50 markets based on population.

From the first user in 1971, MDS service in the United States quickly evolved into a means of distributing pay television programming in large cities to apartments, hotels and later to individual residences. One of the most notable of these services was HBO. This company which began as an MDS operator on the east coast also operated a system in the San Francisco bay area. The Z Channel was another large operator that provided services in the Los Angeles basin area in the 1970s and early 1980s.

During the same time period, another form of television service, subscription television service (STV), was introduced. STV utilized the UHF band, not the microwave band, for transmission. Two STV operators, ON-TV and Select TV experienced tremendous growth in major markets such as Los Angeles and Chicago.

Single channel subscription services were extremely popular in their heyday and subscriber counts were high in the larger markets. Although faced with numerous technical problems including inferior picture quality and easily defeated, primitive scrambling systems that led to massive theft of services, these systems experienced tremendous financial growth fueled by the public's insatiable appetite for viewing uncut premium films at home.

By the early 1980s, MDS and ITFS technology was serving over one million North American subscribers. Technical characteristics such as geographical coverage, propagation reliability, signal quality and installation costs have been well documented during its 20 years of service history.

The rapid growth of cable television signaled an end to the single channel MDS success story. As ever more cities were connected to franchised cable in the late 1970s and early 1980s, subscribers who were paying $20 to $30 U.S. for just a single channel were suddenly able to receive 12 to 24 channels for the same price. Most MDS and STV markets disappeared almost overnight due to the mass exodus of their subscribers to multi-channel franchised cable.

Some of the more persistent MDS operators, who were clearly aware of the "handwriting on the wall," began to search for ways to expand their channel line-ups. They accomplished this by leasing unused air time on the 28 ITFS (instructional television fixed services) frequencies allocated for educational broadcasting. Given MDS industry encouragement and the static situation existing with applications for new ITFS licenses, in 1983 the FCC established a "multichannel" MDS (MMDS) service, using the 2500 to 2686 MHz band on a shared basis with ITFS users. They reassigned the E and F ITFS groups to MMDS and a lottery systems for obtaining licenses was established. The FCC also allocated three channels in the OFS (operational fixed services) band. Applications poured in for all major markets throughout the United States. Within several months over 16,500 applications for MMDS licenses were received by the FCC. The lottery procedure for license selection has taken a number of years to complete. Many of those who received awards were ill-equipped to follow through with creation of MMDS systems.

Currently 33 channels are available in the United States for wireless cable broadcasting. Of these, 31 channels are generally used in most markets. With the recent restructuring of licensing rules by the FCC, a wireless cable operator today can amass 20 or more channels to compete favorably with franchised cable. The recent successes of a number of multi-channel wireless cable systems in the United States, has shown that the public will respond favorably to a franchised cable alternative that provides good service at a competitive price.

Wireless systems are not always successful. For example, while MMDS operations are now being established in some larger American cities, many are already heavily wired for cable television. As a result, wireless operations have encountered stiff competition with entrenched CATV service providers. In both Canada and the United States most cable television networks use the existing telephone and utility pole aerial cable infrastructure. The coaxial cable is "overlashed" to the existing pole and wire messenger structure. This form of cable television construction, which is not possible in most European countries, has proved quite cost effective. Construction costs for cable distribution systems in North America are therefore much lower than in Europe and subscriber rates reflect these economies. Attractive connection and monthly rates encourage subscriber penetrations which make cable television services economically viable. However, in Europe and many other areas of the world with an older base of buildings and utility infrastructure, wireless has significant cost advantages compared to cable distribution. Coaxial and fiber optic cable systems have generally proven too costly to achieve financial viability in these locations unless heavily subsidized by governments or the utility providers.

Until recently, another major impediment to MMDS development in the United States has been the difficulty in obtaining programming. Some satellite services that are available to cable operators for resale to subscribers were withheld from wireless operators. Not co-incidentally, some of the programming services are owned by cable multiple-system operators. A number of legal initiatives as well as substantial political pressure applied by the wireless industry served to break this program barrier.

International Wireless Development

Numerous wireless cable systems have been developed and are operating in many countries around the globe. Others are in planning stages or under construction. (In general, wireless systems are favored over CATV networks in locations where the necessary cadre of well-trained technicians is not available, where tearing up century-old structures would be difficult or where the lower up-front capital costs of wireless are advantageous).

WIRELESS – SYSTEM OVERVIEW

For example, during the past four years MMDS technology has been implemented in a number of Caribbean, Central and South American locations as well as in Australia. Major system are being planned for several locations in southeast Asia. The technology is now well developed and has been tested in numerous operating environments.

In Ireland, in the summer of 1986 an established cable television operator in Cork filed a proposal with the Department of Communications of the Republic of Ireland to install a three site MMDS system in Cork County that would permit extension of cable television services to some additional 60,000 households throughout heavily populated valleys of that country. Consideration of this application, and the potential for use of this technology throughout Ireland as a means of extending cable services economically to smaller towns and even rural homes, led to MMDS approval by the Irish government for use on a national basis. A national plan, using two eleven channel groups was established, with applications now being processed for the first licenses in that country.

In 1984 the Canadian Department of Communications undertook technical standards definitions relating to establishment of MMDS systems. In March of 1985 the first wireless license was awarded to the Manitoba Telephone System (MTS). Having extended cable television service to virtually every community over 1,000 in population throughout the province of Manitoba using coaxial cable distribution on existing aerial utility poles, the MTS sought means of extending multichannel television service to smaller communities or clusters of communities in rural and isolated areas, where household densities and headend costs made cable television economics unfavorable.

In Canada, the Canadian Radio Television and Telecommunications Commission's (CRTC) grants broadcasting licenses. The CRTC has been very slow to award MMDS operating licenses, probably because of concerns related to potential competition between MMDS and existing CATV operations, that already serve over 86% of all television households in Canada. As a result, MDS has never been licensed as a separate service and the instructional television systems have been, until recently, the sole providers of video distribution services in the 2.0 to 2.7 GHz microwave spectrum.

Technical History

Low power wireless technology has followed in the wake of developments that made low-cost reception of satellite microwave broadcasts possible, itself an outgrowth of the recent development of powerful, linear semiconductor devices using GaAsFET (gallium arsenide field effect transistor) technology. Such transistors have also been used in cellular radio and single sideband microwave systems and have resulted in the conversion of MDS, ITFS and MMDS components from vacuum tubes to solid state devices.

Improvements in GaAsFET power amplifiers during the past few years now allow the broadband amplification of a number of television channel at SHF (super high frequencies), with per carrier output powers in the neighborhood of 1 watt. Intermodulation products with the newer devices are maintained at a low enough level relative to carrier power to keep signal distortion below an acceptable level. As a result the use of broadband amplifiers in conjunction with low cost input modulators and a broadband upconverter has become a feasible reality. The principle advantage gained with this technology is a dramatic reduction in transmit site costs. While conventional wireless systems require a separate transmitter, including individual aural and visual amplifier chains for each channel, the broadband, low power approach utilizes a single upconverter and single broadband amplifier for all visual and aural carriers in the signal package, normally eight channels or less. The only individual channelized equipment is the input modulators or heterodyne processors, which are relatively inexpensive VHF devices that have been produced for many years for the cable television, MATV and SMATV industries.

The capacity of wireless systems is now poised for a dramatic expansion as video compression systems are introduced during the next two to five years. Since more than one compressed video signal can occupy a standard channel bandwidth, the allocated wireless bandwidth can now be used to relay more than one television channel. This development certainly improves the economic viability of existing and proposed wireless systems.

C. FREQUENCY ALLOCATIONS, CHANNEL AVAILABILITY and REGULATIONS

Many high quality television channels can be transmitted over a wireless cable system with the only limitation being the availability of frequency space. For example, each NTSC-formatted television channel requires a 6 MHz frequency band. Thus 200 MHz of bandwidth could be used to relay as many as 33 NTSC television channels.

For technical and marketing reasons, co-location or channel consolidation of multiple groups of channels on a contiguous frequency band is favored. This strategy allows economy of scale in transmission construction costs, permits an operator to offer service packages closely competitive with other multi-channel delivery systems and helps to minimize interference between adjacent channels. However, given that the frequencies fall with the same general contiguous bands, the costs to process and transmit non-adjacent channels are generally lower than those for adjacent channels.

The Optimal Frequency Range?

The optimal range for wireless transmission frequencies is considered to lie in 1.9 to 2.7 GHz band for technical and economic reasons. This range has become the de-facto standard because off-the-shelf components used at these frequencies are widely available at very reasonable costs. While the standard wireless cable frequency range used in North America and many other parts of the world lies primarily in the 2.5 to 2.7 GHz range, the full 1.9 to 2.7 GHz band has been employed. For example, an Australian system in now transmits one channel in the 1.9 to 2.2 GHz band and the original MDS channels in the United States fall in the 2.15 to 2.162 GHz range.

Is the 1.9 to 2.7 GHz band the optimal frequency range? Can higher microwave frequencies be successfully employed? The roadblocks to implementation of higher frequency wireless systems are explored below.

Transmission

The transmission of wireless signals above normal microwave frequencies has a number of limitations. First, as frequency increases microwave signals tend to travel in more defined straight-line paths. This characteristic is used to a definite advantage in line-of-sight point-to-point microwave relays. However, as frequency rises, increasingly more power must be pumped into an omni-directional antenna to maintain a given transmission range, namely, for a given power output antenna range decreases with increasing frequency. As transmission range decreases the market area or number of potential subscribers and the potential return on invested capital decreases for a given transmitter cost.

This generally decreased coverage at higher frequencies would eventually necessitate a move to frequency modulation and a cellular coverage approach, with consequent increased receive complexity and cost. A wireless system operating at 12 GHz, for example, would have to employ multiple transmit antennas each serving a cell of a rather limited area. The antennas in each cell would have to be carefully arranged so that signals relayed to a customer within one cell would not be received by another in an adjacent cell. If this occurred ghosts would appear on the subscriber's television screen.

Omni-directional wireless cable antennas have been developed and are commercially available for use in the 1.9 to 2.7 GHz range. Developing an omni-directional antenna that transmits at higher frequencies would be more expensive. Most higher frequency signals are generally transmitted on line-of-sight paths via a parabolic or horn antennas, both of which are quite directional. While a number of firms manufacture line-of-sight components that operate at frequencies up to 28 GHz, only a handful have the capabilities to manufacture omni-directional antennas designed for frequencies much above 2.7 GHz.

Second, amplifiers and transmitters that generate sufficient linear, AM power at higher frequencies to provide requisite coverage of an extended market area are not as common, nor readily available and therefore more costly.

Third, as the microwave signal frequency increases, signal loss due to "rain fade" or absorption by water in the form of rain, fog or vapor, dust and other atmospheric materials increases. At frequencies much above 7 GHz this becomes an important factor in limiting transmission range. Rain fade has the net result of decreasing transmitter antenna range, especially during torrential storms or in heavily polluted areas.

Reception

A wireless system designed to operate at frequencies above the standard band would face a number of interesting technical and economic problems. First, as the frequency of a microwave signal increases, the beamwidth of a receive antenna decreases. Consequently, at higher frequencies it must be held in a more stable position and must not be swayed by winds or other forces to ensure maximum power reception. This requirement for improved pointing accuracy becomes more important with increasing signal frequencies and would translate to a need for a more accurate, perhaps more costly antenna. In addition, since wireless antennas are often mounted on relatively unstable supports such as long poles, achieving acceptable pointing accuracy may be difficult and installations could be more expensive at higher frequencies. Maintaining a low end-user cost, the major cost factor in a wireless system, is critical to an acceptable return on investment.

Second, downconverters are not readily available at frequencies outside the conventional range. The lower the difference between the standard wireless frequencies and the proposed range, the easier would be the task of developing and manufacturing a non-standard downconverter. Wireless set-tops require a very stable frequency input, generally provided by a device known as a prescalar, one component within the phase lock loop amplifier. These function well only up to approximately 3 MHz. At higher frequencies, a more expensive dielectric resonant oscillator would be required. However, set-tops converters would also have to incorporate stable automatic frequency control to manage the less stable frequency input, of course at a cost premium. This oscillator stability issue limits the use of low cost AM downconverters to frequencies of at very most 12 GHz. Downconverters designed for operation at frequencies above 12 GHz would require a very different and more expensive "line and wire" construction.

Frequency Availability and the Licensing Procedure in the U. S.

Licensing a wireless cable system in some regions of the world may ultimately prove to be the most difficult part of the business. In the United States, for example, extensive FCC regulations, existing license holders, and unscrupulous "applications mills" have impeded the development of many systems. Potential operators wishing to participate in the industry must have patience and perseverance to succeed in obtaining permits for the 20 or more channels required for a competitive system.

Frequencies

The wireless cable spectrum is comprised of a set of four frequencies divided into nine separate groups for a total of 33 channels. These are outlined in Figure 11-2 and Table 11-1. For comparison purposes Tables 11-2 and 11-3 present the wireless channel allocations in use in both Australia and Ireland, respectively.

The first American group of channels, the original multipoint distribution service (MDS), was assigned to commercial video ventures. It consists of two channels, MDS 1 and MDS 2, although MDS 2 is only available in the top 50 markets. Very few MDS 1 channels are available at this time as almost all of the markets for this group were applied for and assigned in the 1970s.

The ITFS group, contains the largest number of available channels, five groups of four channels (ITFS A, B, C, D and G) for a total of 20 channels. Although this group was reserved specifically for educational usage, the FCC does allow a commercial operator to lease excess broadcast time on these bands. Many schools and institutions still have their own distant learning systems in locations across the country.

Past requirements mandated that each channel be used for a minimum of 20 hours per week for educational broadcasting, but recent FCC rule changes allow for a minimum of 12 hours per week for the first two years of operation. Subsequently, the minimum then increases to 20 hours per week per channel.

With the use of a technique called channel-mapping, an operator can create three full-time channels from a group of four ITFS frequencies. Channel mapping switches the video program to whichever fre-

WIRELESS – SYSTEM OVERVIEW

quency is not in use. Over the course of a broadcast numerous channels may be used. To avoid having a subscriber tune the set-top converter to match the frequency in use, the box automatically changes channels in response to a data command transmitted with the video. The converter continues to show the same number in the channel display to avoid subscriber confusion. The use of ITFS channels in conjunction with channel mapping is one of the best way to substantially increase the number of channels in a MMDS line-up. Video compression is certainly the other up-and-coming option.

MMDS allocated frequencies contains two groups, E and F, of four channels each, for a total of eight channels. As mentioned above, these channels were originally included with the ITFS group until the FCC reassigned them and made them available for commercial use. Since this took place in 1983 numerous application have been filed for most of the major markets nationwide. In some of the smaller markets the opportunity to acquire one or both of these license still exists. However, these are now also being quickly assigned due to the recent changes in FCC application procedures that allow a single applicant to file for both the MMDS E and F group. Previous to this change, no single entity could hold both licenses.

The OFS frequencies complete the menu of channels available to a wireless cable operator. These provide an operator with three additional frequency slots at the top of the wireless cable spectrum.

MDS group signals are transmitted in an inverted NTSC video 6 MHz format in which the aural carrier is located at a lower frequency than the video (see Figure 11-3). As a result, the MDS channels downconvert to a frequency located two MHz off the standard center frequency format. This may cause problems if a set-top box is not used or with some head-end demodulators that may not be able to tune away from this offset.

In addition, an MDS 2A channel that is 4.5MHz wide and is used for data and other non-video transmissions has been allocated. All of the other ITFS, MMDS and OFS channels are broadcasted in the standard NTSC format. These channels readily

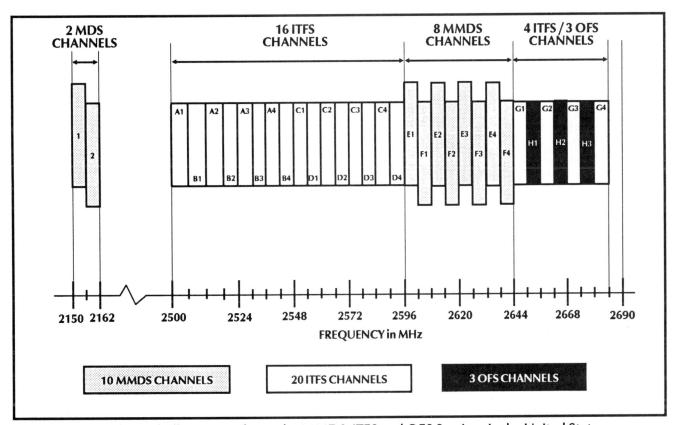

Figure 11-2. Channel Allocation Scheme for MMDS, ITFS and OFS Services in the United States.

231

WIRELESS – SYSTEM OVERVIEW

downconvert to standard NTSC channel assignments with a converter using a local oscillator frequency of 2278 MHz. Table 1 illustrates this point and provides channel conversion assignments.

The aural carrier on all channels is typically adjusted to 10 to 20 dB below the level of the visual carrier. The exact value of the relative amplitude of the aural to visual carriers depends upon the type of signal encryption technology employed.

TABLE 11-1. MDS/MMDS/ITFS/OFS WIRELESS CHANNEL ASSIGNMENTS for the U. S. (Downconverter L.O. of 2278 MHz)

Channel Designation	Bandwidth (MHz)	Carrier Center Frequency (MHz)			Downconverted Channels (MHz)		
		Luminance	Chrominance	Audio	Channel	Video	Audio
MDS CHANNELS							
MDS1	2150-2156	2154.75	2151.17	2150.25	14	123.25	127.75
MDS2	2156-2162	2160.75	2157.17	2156.25	99	117.25	121.75
MMDS/ITFS CHANNELS							
ITFS A1	2500-2506	2501.25	2504.83	2505.75	24	223.25	227.75
ITFS A2	2512-2518	2513.25	2516.83	2517.25	26	235.25	239.75
ITFS A3	2524-2530	2525.25	2528.83	2529.75	28	247.25	251.75
ITFS A4	2536-2542	2537.25	2540.83	2541.75	30	259.25	263.75
ITFS B1	2506-2512	2507.25	2510.83	2511.75	25	229.25	233.75
ITFS B2	2518-2524	2519.25	2522.83	2523.75	27	241.25	245.75
ITFS B3	2530-2536	2531.25	2534.83	2535.75	29	253.25	257.75
ITFS B4	2542-2548	2543.25	2547.83	2547.75	31	265.25	269.75
ITFS C1	2548-2554	2549.25	2552.83	2553.75	32	271.25	275.75
ITFS C2	2560-2566	2561.25	2564.83	2565.75	34	283.25	287.75
ITFS C3	2572-2578	2573.25	2576.83	2577.75	36	295.25	299.75
ITFS C4	2584-2590	2585.25	2588.83	2589.75	38	307.25	311.75
ITFS D1	2554-2560	2555.25	2558.83	2559.75	33	277.25	281.75
ITFS D2	2566-2572	2567.25	2570.83	2571.75	35	289.25	293.75
ITFS D3	2578-2584	2579.25	2582.83	2583.75	37	301.25	305.75
ITFS D4	2590-2596	2591.25	2694.83	2595.75	39	313.25	317.75
MMDS E1	2596-2602	2597.25	2600.83	2601.75	40	319.25	323.75
MMDS E2	2608-2614	2609.25	2612.83	2613.75	42	331.25	335.75
MMDS E3	2620-2626	2621.25	2624.83	2625.75	44	343.25	347.75
MMDS E4	2632-2638	2633.25	2636.83	2637.75	46	355.25	359.75
MMDS F1	2602-2608	2603.25	2606.83	2607.75	41	325.25	329.75
MMDS F2	2614-2620	2615.25	2618.83	2619.75	43	337.25	341.75
MMDS F3	2626-2632	2627.25	2630.83	2631.75	45	349.25	353.75
MMDS F4	2638-2644	2639.25	2642.83	2643.75	47	361.25	365.75
ITFS G1	2644-2650	2645.25	2648.83	2649.75	48	367.25	371.75
ITFS G2	2656-2662	2657.25	2660.83	2661.75	50	379.25	383.75
ITFS G3	2668-2674	2669.25	2672.83	2673.75	52	391.25	395.75
ITFS G4	2680-2686	2681.25	2685.83	2685.75	54	403.25	407.75
OFS CHANNELS							
OFS H1	2650-2656	2650.25	2654.83	2655.75	49	373.25	377.75
OFS H2	2662-2668	2663.25	2666.83	2667.75	51	385.25	389.75
OFS H3	2674-2680	2674.25	2678.83	2679.75	53	397.25	401.75

WIRELESS – SYSTEM OVERVIEW

TABLE 11-2. WIRELESS CHANNEL ASSIGNMENTS for AUSTRALIA

Group	Channel Number	Bandwidth (MHz)	Carrier Center Frequency (MHz) Visual	Aural
1A	A1	2276-2083	2177.25	2082.75
	A2	2090-2097	2091.25	2096.75
	A3	2104-2111	2105.25	2110.75
2A	A2	2083-2090	2084.25	2089.75
	A4	2097-2104	2098.25	2103.75
1B	B1	2302-2309	2303.25	2308.75
	B3	2316-2323	2317.25	2322.75
	B5	2330-2337	2331.25	2336.75
	B7	2344-2351	2345.25	2350.75
2B	B2	2309-2316	2310.25	2315.75
	B4	2323-2330	2324.25	2329.75
	B6	2337-2344	233825	2343.75
	B8	2351-2358	2342.25	2357.75
3B	B9	2358-2365	2359.25	2364.75
	B11	2372-2379	2373.25	2378.75
	B13	2386-2393	2387.25	2392.75
4B	B10	2365-2372	2366.25	2371.75
	B12	2379-2386	2380.25	2385.75
	B14	2393-2400	2394.25	2399.75

TABLE 11-3. WIRELESS CHANNEL ASSIGNMENTS for IRELAND

Group	Channel Number	Bandwidth (MHz)	Carrier Center Frequency (MHz) Visual	Aural
AP	AP-1	2500-2508	2501.25	2507.25
	AP-2	2516-2524	2517.25	2523.25
	AP-3	2532-2540	2533.25	2539.25
BP	BP-1	2508-2516	2509.25	2515.25
	BP-2	2524-2532	2525.25	2531.25
	BP-3	2540-2548	2541.25	2547.25
	BP-4	2556-2564	2557.25	2563.25
CP	CP-1	2564-2572	2565.25	2571.25
	CP-2	2580-2588	2582.25	2587.25
	CP-3	2596-2604	2597.25	2603.25
	CP-4	2612-2620	2613.25	2619.25
DP	DP-1	2572-2580	2573.25	2579.25
	DP-2	2588-2596	2589.25	2595.25
	DP-3	2604-2612	2605.25	2611.25
	DP-4	2620-2628	2621.25	2627.25
EP	EP-1	2628-2636	2629.25	2635.25
	EP-2	2644-2652	2645.25	2651.25
	EP-3	2660-2668	2661.25	2667.25
	EP-4	2676-2684	2677.25	2683.25
FP	FP-1	2636-2644	2637.25	2643.25
	FP-2	2652-2660	2653.25	2659.25
	FP-3	2668-2674	2669.25	2675.25
	FP-4	2684-2692	2685.25	2691.25
FP	GP-1	2692-2700	2693.25	2699.25

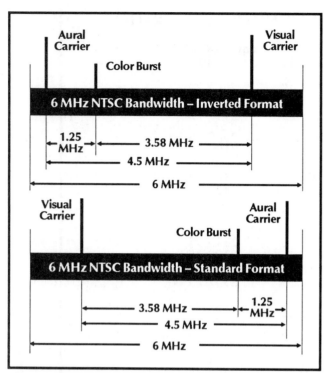

Figure 11-3. Standard NTSC and Inverted MDS Channel Formats. *Conventional NTSC UHF and VHF broadcasts use the arrangement of audio, luminance and chrominance carriers as indicated in the top diagram. This broadcast format is also used for MMDS, ITFS and OFS channels. In contrast, the carriers are "inverted" in MDS transmissions as indicated in the bottom diagram.*

The License Application Process

The process for applying for any of the available channels in a specified market area is arduous at best and, at worst, can be extremely frustrating. Although an individual is permitted to prepare his own application, the use of a reputable firm specializing in the preparation of these applications is highly recommended. Legal and broadcast engineering firms that are experienced in the various procedures and requirements of a license application can save a novice applicant substantial time in dealing with the FCC.

The FCC has recently enacted several rule changes that have had a favorable impact on the wireless cable industry. The most important of these is the one-day filing rule that has eliminated the public notice waiting period and prevents another applicant from filling for the same market once its economic value has become apparent. A conflict only arises if two separate parties file for the same area on the same date. In that instance, the FCC conducts a lottery drawing to decide the winning application.

Another rule that has been enacted prevents franchised cable operators from holding an MMDS license in their franchised area. The FCC has also increased the permissible MMDS power levels to 100 watts, with the maximum allowable output now being defined in terms of EIRP (effective isotropically radiated power), the power radiated from the antenna. Note that this power level must still meet minimum interference criteria before it can be used.

A careful strategy should be developed for entrepreneurs wishing to enter the wireless cable business. The best options are markets that make financial sense, namely that have a wide base of potential subscribers. Since most of the major markets are either presently under development or at least have all available frequencies allocated, the most likely candidates are smaller rural markets of 25,000 to 50,000 television households. Factors including whether or not a local cable franchise has been granted, which areas are uncabled, population demographics, the local terrain, transmit tower site availability and other technical considerations must all be evaluated before filing an application.

A reputable consulting or engineering firm can provide the type of technical analysis required and can help in developing a business plan. Table 15-1 lists several reputable firms that specialize in license acquisition and project evaluation.

Among the first steps in the process are the filing of a microwave license application, FCC form 494, along with a qualification report, FCC form 430. In addition, an interference analysis must be completed to prove that the project will not cause microwave interference with adjoining service areas. The use of qualified professionals in this endeavor can insure that all forms are correctly prepared and are not returned by the FCC.

A prospective applicant is cautioned to steer clear of "license mills" that promise exaggerated financial rewards for "minimum risk," usually $5,000 to $10,000 or more. These operations point to the tremendous profits made during the cellular telephone license applications and draw a false corollary between that industry and the wireless cable business. In reality, a single group of channels is only worth a fraction of the profits a cellular license could bring. These unscrupulous individuals further compound the problem by filing multiple applications on behalf of all their "clients" in a single market. They do not care who wins the license as they have collected as much as $10,000 or more from 50 or more applicants with only one winner possible.

The unethical practices of these firms has created massive congestion at the FCC and is slowing the development of the wireless cable industry. Furthermore, the greed they stimulate in those fortunate to actually win a license has priced those channels outside the realm of reasonable business potential.

Nevertheless, professional services are available for all phases of license evaluation and acquisition. A qualified firm can provide a prospective wireless cable operator with such services as license searches, shadow maps that predict terrain interference, terrain profile maps and tower studies to identify available transmitting tower sites. A reputable firm will also represent only one client per site and will notify a prospective client of conflicts of interest in existing applications. If licenses already exist for the desired area, the consultant may be able to help in negotiating lease arrangements with a holder who does not intend to develop the system.

In summary, given that the overall licensing process involves detailed engineering and legal knowledge, a novice wireless cable operator can save time and frustration by retaining the services of experienced professionals.

Comparison with Cable and Conventional TV

Wireless cable television has a number of distinct advantages compared to both cable and conventional broadcast television. Cable television distribution plants can transmit typically up to 62 channels of 6 MHz programming over a state-of-the-art coaxial cable system. Conventional off-air television broadcast stations have even more limited channel capacities. Wireless cable systems are limited only by the allocated frequency space.

The limitations of CATV and broadcast equipment and spectrum does not easily allow upgrade to new services such as high definition television (HDTV) broadcasts (although video compression techniques will change this situation). When HDTV and reception equipment becomes available, wireless cable systems will be relatively easily retrofitted to deliver these innovative signals.

Each wireless subscriber is served independently with high quality television programming. A conventional cable signal passes through many miles of cable plant and many amplifiers so the potential for disruption of service is greater than for a wireless cable system. When difficulties do occur, thousands of subscribers are potentially affected. In those cases when a wireless cable subscriber's reception has operational difficulties, only that particular subscriber is affected. The transmitting equipment is highly reliable and, if necessary, redundant backup can be available if problems do arise.

Equipment and Cost Comparison

The average per subscriber capital cost to build cable television plants ranges from a low of $600 up to $1000 depending upon labor costs and other factors such as excavation difficulties. The largest component of this cost is the capital to build cable television plants, the distribution system, which ranges from a low of $15,000 U.S. to a high of $38,000 U.S. per mile of cable.

Wireless cable systems must incur a fixed cost for installation of transmission equipment and for the operations center plus a variable cost for installing equipment at each subscriber location. An average 50,000 subscriber cable TV system would cost approximately $50 million U.S. while a comparable wireless cable system serving the same number of subscribers would cost $15 million U.S. A wireless cable system can be quickly installed and is capable of serving all potential subscribers in the service area while a cable system can, at times, take years to complete.

WIRELESS – SYSTEM OVERVIEW

CHAPTER 12. WIRELESS CABLE COMPONENTS

A technical description of each component that comprises a wireless system is presented in this Chapter. While the equipment discussed is in general use within the wireless industry certain aspects of these technologies, such as addressability, have applications in both wired and wireless cable TV.

A wireless system transmits multiple channels of television and other broadband services between a central transmitter and receive sites throughout a local service area. Signals are encoded prior to broadcast and subsequently received and decoded at each end user location.

A. THE TRANSMITTER FACILITY

The transmitter facility generally consists of one or more microwave broadcast antennas and the system headend components, namely the signal reception and processing equipment. The business office and warehouse sites are almost never co-located at the transmitter (see Figure 12-1 and 12-2).

Transmission Antennas

Transmission antennas can be located on towers, buildings or hills (see Figures 12-3 and 12-4). They are sited to maximize the desired line-of-sight coverage in the intended market area. As antenna height increases, the range and line- of-sight range increases. However, increasing height also increases the losses incurred as signals travel along waveguides or coaxial cable from the channel transmitters to the antenna. An optimal situation would be to have the signal processing equipment located on top of a tall building, just at the base of the transmission antenna.

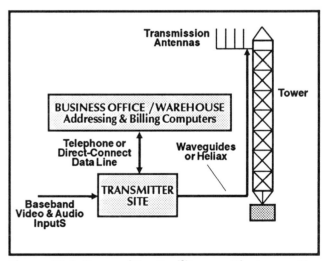

Figure 12-1. Transmitter Facility. *The transmitter facility consists of one or more microwave broadcast antennas, the system headend composed of processing equipment, satellite antennas and other program sources, and often the business office and warehouse sites.*

WIRELESS CABLE COMPONENTS

Transmitter power outputs of 1, 10, 20, 50 and 100 watts are common and permit an adequate signal coverage up to a radius of 50 km and beyond under clear weather conditions. However, wireless transmissions are more vulnerable to interference than conventional lower frequency over-the-air television broadcasts. Any physical obstructions, such as foliage or buildings, can reduce received signal levels. A well-designed market survey, including a careful selection of the transmitter site as well as a study of the local terrain is crucial in identifying potentially low signal areas within the service area.

Transmitting antennas that generate either omnidirectional or cardioid radiation patterns typically with 13 to 16 dBi gain relay signals throughout the desired service area (see Figure 12-5 and 12-6).

In general coverage is maximized by co-locating all transmitting antennas at the same location and by orienting them so they relay signals of just a single polarity. If necessary, however, signals of both polarities can be relayed at the expense of reduced receive antenna gain. To detect both polarities, a receive antenna would have to be oriented at 45° to both to vertical and horizontal signals resulting in a 3 dB loss of gain. This effectively reduces the transmission range for a given receive antenna gain.

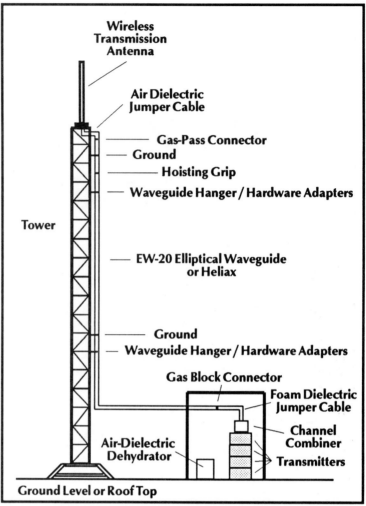

Figure 12-2. Schematic of Transmitter Facility. *This illustrates the components in a typical wireless transmitter facility. Foam dielectric jumpers, typically 1/2" in diameter, connect the transmitters with the combiner and the combiner with the transmission line. A 7/8" air dielectric jumper connects the antenna to the transmission line. Air dielectric cable is used for this purpose to protect the antenna from stresses generated by any movements in this relatively long transmission line.*

TABLE 12-1. TYPICAL MMDS TRANSMIT ANTENNA SPECIFICATIONS

Operating Frequency Range: 2500 – 2686 MHz

Bandwidth: 186 MHz

Radiation Pattern: As required for specific application. Omnidirectional, cardioid and peanut patterns

Antenna Gain: Related to the required pattern

Downtilt Angle: Antenna designed to provide a 0.3° downward tilt

Polarization: Either horizontal or vertical polarization of the E-field in propagated wave, as required

Return Loss (VSWR): A return loss of 18 dB or better over any 8 MHz RF channel in the specified frequency range is required, with 30 m of 2.3 cm, 50 ohm, air dielectric coaxial cable transmission line equipped with female N-connectors at the transmitter end

Input Power: The antenna and transmission line capable of handling an input RF power rating 600 w

WIRELESS CABLE COMPONENTS

Figure 12-3. Transmit Antennas. *These two wireless transmit antennas are sited on a building top. Also pictured here are two satellite reception antennas and a small microwave link that is used to receive incoming STL line-of-sight signals carrying ITFS broadcasts. (Photo by Steve Berkoff)*

Figure 12-4. Transmit Antennas. *MMDS Transmit Site. This five channel MMDS system is installed in a remote broadcast site and serves the small town of Demming, New Mexico. (Photo by Steve Berkoff)*

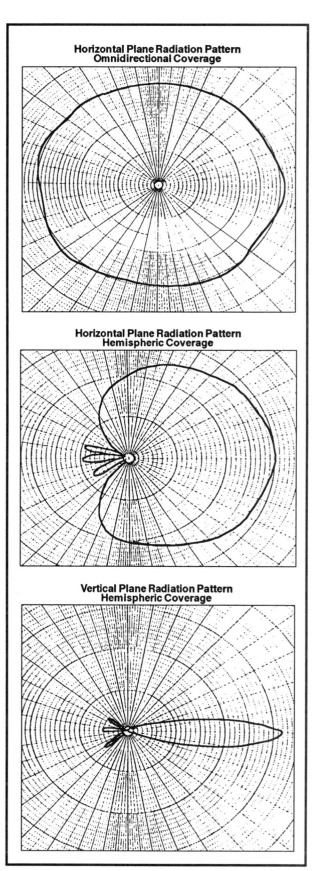

Figure 12-5. Antenna Radiation Patterns. *Three typical radiation patterns for a wireless transmit antenna are illustrated here (see also Figure 12-6).*

WIRELESS CABLE COMPONENTS

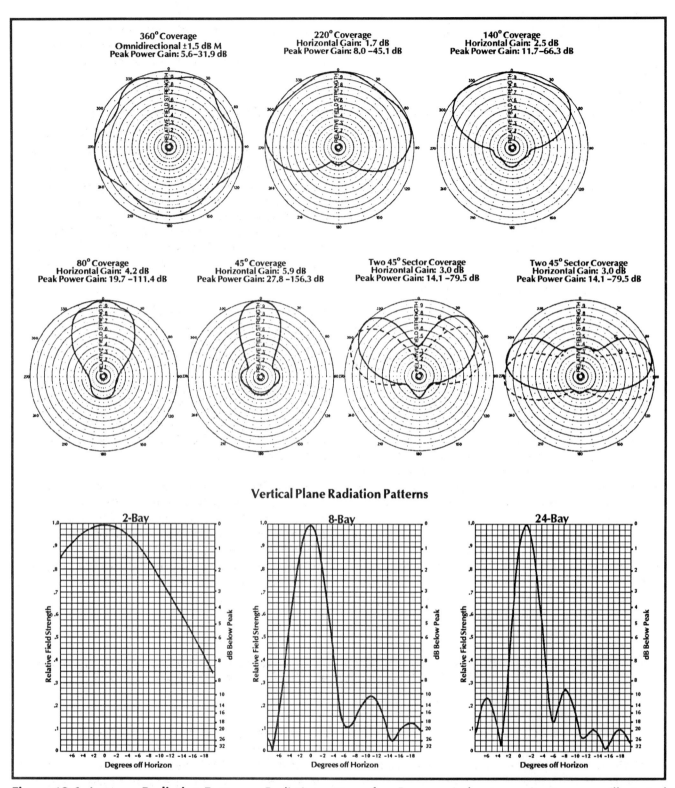

Figure 12-6. Antenna Radiation Patterns. *Radiation patterns for a Bogner wireless transmit antenna are illustrated here. The top seven illustrations illustrate the horizontal coverage patterns for a 360 degree view with the antenna at the center. Only eight of the twelve antenna patterns that Bognar has available are illustrated here. The bottom three graphs show the vertical patterns expressed in degrees above and below the horizon. Notice that as the number of bays increases the signal becomes more directional. (Courtesy of Bogner Broadcast Equipment Company)*

WIRELESS CABLE COMPONENTS

The Transmitter Site

Geographic Considerations and Location.

The transmitter site should be selected once the primary market targeted for delivery of the wireless cable service has been defined. Among the general criteria in the selection process are:

- sufficient elevation to maximize range
- adequate space for all components that are required at the site
- availability of an existing usable tower
- accessibility to the transmitter
- provisions for required utility services
- proximity to business office facilities.

A transmitter designed for maximum, cost efficient signal propagation is normally located on the highest terrain within or contiguous to the proposed service area, generally on top of a mountain or a tall building, where sufficient space for all associated facilities is available.

If the transmitting antenna(s) is to be installed on a high tower, efforts should be made to select a site where existing RF towers that have available mounting space are located. Otherwise adequate space at the selected site must be available for constructing a new tower.

The site must be accessible to engineering and maintenance personnel. It must be near the business office to facilitate proper system monitoring and routine maintenance as well as to reduce operational expenses relating to expensive telephone, data or point-to-point microwave relay stations. In addition, power and telephone facilities should be available at the selected site to avoid construction time delays.

Space and Facilities Requirements.

The space required for the transmit site varies according to the number of RF channels used, program origination requirements and tower space needs. In general, eight RF channel transmitters, climate control equipment, a stand-by generator, power conditioning units, fire protection systems and component storage/repair facilities require approximately 200 m^2 of building space. The transmit building must be maintained within the following ranges:

Voltage	105 – 125 volts RMS / 50 Hz
	210 – 260 volts RMS / 50 Hz
Temperature	0 to 40°C
Relative Humidity	5 to 90% (noncondensing)

The approximate total operating power is 13 and 45 kilowatts for 10 and 100 watt transmitters, respectively. These figures correspond to a power consumption of 45,000 and 154,000 BTUs per hour, respectively.

If satellite reception antennas are to be used, adequate space must be provided based on the size, number of antennas to be constructed and the geographic latitude of the site. This latter factor determines look angles and rotation arc, all of which have an impact upon space requirements. For example, a single 7 meter parabolic antenna requires approximately 60 m^2 of land.

B. SIGNAL PROCESSING and TRANSMISSION

Overview

The program material for a wireless system may be created in a local studio control room or imported via such media as coaxial cable, land-based microwave links or satellite receiving stations. The types and sources of programming to be transmitted over the wireless system as well as the physical location of the transmitter facility in relationship to the business office and studio facilities define the programming source or sources

Once the programming has been received at the transmitter site, the associated signals are descrambled, demodulated and routed to an audio/video (A/V) switcher (see Figure 12-8). The outputs of this A/V switcher are then fed into the appropriate signal processing equipment that generally incorporates baseband or RF encoding, Comband processing or a combination of these systems. These types of encryption or scrambling systems currently in use are discussed in the following section.

WIRELESS CABLE COMPONENTS

The encoding chain of equipment at the transmit site translates the input baseband signals to "clean" microwave signals in the 1.9 to 2.7 GHz range. The components include a video and optional audio encoder and an IF modulator for each channel (see Figures 12-8, 12-9, 12-10 and 12-11). Modulation equipment is available to broadcast in NTSC, PAL or SECAM formats.

Each encoder accepts the baseband or RF video signals from the A/V switcher and produces an encoded signal. An optional audio encoder, generally within the same enclosure as the video encoder, scrambles the baseband audio signal. All encoders are under the control of an addressing computer. An IF modulator then accepts the encoded baseband video and audio inputs and modulates and produces either a separate audio and video or a combined signal for output to the transmitter(s). Any channel not intended to be encoded is fed directly into the modulator and bypasses an encoder. Each transmitter includes an upconverter that raises the signal frequency to the microwave range and an amplifier that generates power, typically 1, 10, 20, 50 or 100 watts. The transmitters then feed the signals into a channel and group combiner and subsequently into waveguides or heliax for relay up the tower to the transmit antenna.

Recent improvements in GaAsFET power amplifiers have resulted in low power equipment that allows the simultaneous amplification of typically eight wireless channels (see Figure 12-12). This simplifies the design and lowers the cost for low power

Figure 12-7. VideoCipher II Processing Equipment. *These VideoCipher II decoders are used at the Corpus Christi MMDS system headend to decode signals received via satellite. (Photo Steve Berkoff)*

wireless systems, also known as LMMDS systems. Such amplifiers usually operate at a power of 1 watt.

Ancillary hardware at the processing site includes hanging kits, connectors, equipment racks, dehydrators as well as climate controllers (see Figures 12-2, 12-13, 12-14 and 12-15).

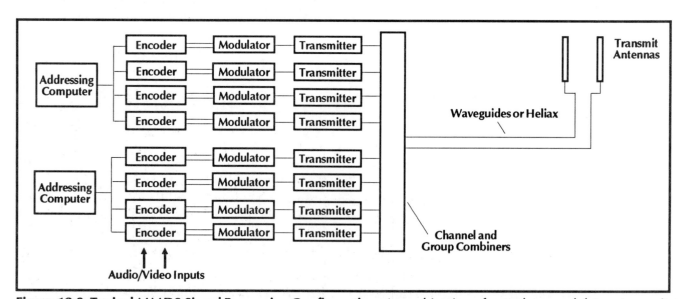

Figure 12-8. Typical MMDS Signal Processing Configuration. *A combination of encoders, modulators, transmitters and amplifiers as well as channel and group combiners, waveguides and transmit antennas make up this wireless signal processing configuration.*

WIRELESS CABLE COMPONENTS

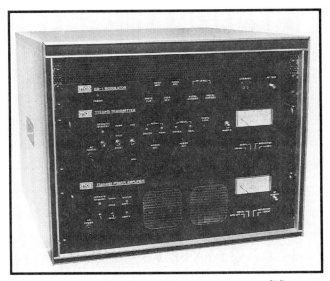

Figure 12-9. EMCEE 50 Watt Power Amplifier, 10 Watt Transmitter and Modulator. *This illustrates a wireless modulator, a 10 watt transmitter and a 50 watt power transmitter. The transmitter is a fully synthesized GaAsFET frequency agile device that separately amplifies the visual and aural carriers. It can be configured for NTSC, PAL or SECAM formats and has an output frequency of 2.5 to 2.7 GHz. Up to 10 of these units can fit on a single headend rack. (Courtesy of EMCEE Broadcast Products)*

Figure 12-10. 10 Watt MMDS Transmitter. *This internally diplexed frequency agile unit employs GaAsFET technology to separately amplify visual and aural carriers. It reduces harmonic radiation, in-band intermodulation products and spurious out-of-band products to −70, −60 and −60 dB, respectively, relative to the desired carrier. (Courtesy of CED)*

Figure 12-11. Two Views of a Headend Rack. *These two photos show front and back views of two wireless transmitter headends containing Comwave transmitter equipment. (Courtesy of Comwave)*

WIRELESS CABLE COMPONENTS

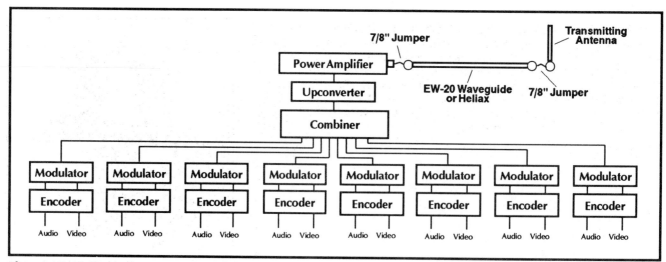

Figure 12-12. Typical Low Power MMDS (LMMDS) Signal Processing Configuration.

Figure 12-13. EMCEE Wireless Headend.
(Courtesy EMCEE Broadcast Products)

Figure 12-14. Wireless Broadcast Facility. *These photos show the the modulators, power amplifiers, transmitters, combiners and other headend equipment at the Corpus Christi wireless headend. (Courtesy Steve Berkoff)*

WIRELESS CABLE COMPONENTS

Figure 12-15. Directional Channel Combiner. *This microwave device combines four non-adjacent channels for input to a common antenna line while maintaining about 40 dB isolation between channels. By patching in additional units up to sixteen non-adjacent channels can be combined. Any non-adjacent channel with bandwidth up to 8 MHz in the 2.5 to 2.7 GHz band is fed into the combiner via an N-connector. The output is via a waveguide. (Courtesy of Microwave Filter Company)*

Microwave Transmitters Technical Details

The transmitter site processing equipment is a key element to the overall performance of a wireless system since broadcast quality is never any better than the transmitted signal. The objective in to transmit a signal with a minimum of distortion and noise (see Figure 12-16).

The performance of microwave transmitters can be evaluated by a number of parameters including the method of diplexing audio and video carriers, control of output power levels, linearity, frequency stability, crosstalk and feedback. Typical transmitter and combiner specifications are outlined in Tables 12-2 and 12-3.

Diplexing Method

Audio and video carriers can either be combined and upconverter together, known as internal diplexing, or can be routed separately and combined just before the transmission antenna, a process known as external diplexing (see Figure 12-17).

While internal diplexing is the least costly method at powers below about 10 watts it can result in out-of-band interference products, a problem when adjacent channels are broadcasted. Externally diplexed carriers share the same oscillator to maintain the exact frequency relationship. This type of processing equipment does not suffer from creation of interference products and can more easily be upgraded to higher power operation.

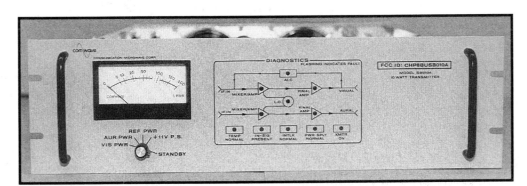

Figure 12-16. Comwave 10-Watt Transmitter and Television Modulator. *This 10-watt synthesized transmitter (top two photos) outputs any 6, 7 or 8 MHz channel in the 2.0 to 2.7 MHz band and is stereo and scrambler ready. The bottom photo is the Comwave TVM-100 television modulator. (Courtesy of Comwave)*

WIRELESS CABLE COMPONENTS

Control of Output Power Levels

Transmitter power requires an automatic leveling control. If power levels fluctuate too widely, video flashing and streaking may occur at the receive site. In order to control power output it must be quantified. The traditional measure of output power is the peak sync power since maximum output occurs regularly during the transmission of the sync pulse. However, scrambling systems that suppress or remove the sync pulses, or worse that randomly suppress or remove these pulses, makes it more difficult to track output power since peak power occurs at other points in the video waveform. While this shortcoming can be managed in well-designed and maintained systems that incorporate modern GaAsFET amplifiers, transmitters should be specified and tested for operation with any particular scrambling system that may be chosen.

Linearity

A perfect transmitter should increase the power level of its input signal without modifying its form. However, no real-world device is perfectly linear. Any non-linear performance generates mixing or intermodulation products that result from interactions between the audio and video carriers. Second order products that stem from both the sum and difference of the two carriers or their harmonics generate signals with frequencies that fall well beyond those of interest and can usually be easily filtered. However, third order and other higher level odd numbered products that result from differences in frequencies can cause distortions unless they fall about 55 dB below the level of the desired signal. (In general, fifth order and higher products are too weak to have any negative effects.) Externally diplexed transmitters which process the audio and video separately do not generate intermodulation products.

Frequency Stability

Most transmitters are designed to have frequency stability of ±25 kHz. A crystal oscillator in the upconverter mixer provides the required stability and clean phase noise sidebands.

Feedback and Crosstalk

Microwave transmitters generate high power signals. Any energy that leaks out and couples into input levels of the system can create unacceptable variations in output power levels. This feedback would be noticeable as power changes when objects move in the vicinity of the transmitter. Adequate system layout and shielding should eliminate such unsafe situations.

Crosstalk or intermodulation occurs when small amounts of power from two or more transmitters couple together. A circulator, a ferrite device that passes power in just one direction, installed on the output of each transmitter should eliminate any crosstalk.

Addressing and Encoding

Modern wireless headend systems can be purchased with computer controlled video processors that provide scrambling, encoding and addressing information to manages all end-user converters. This data, including the system identification number, is typically inserted within the video bandwidth portion of each wireless channel. The addressing facility controls the channel menu for each subscriber as well as all pay-per-view purchases.

The addressing computer, typically a well-equipped 386 or 486 PC system, directly controls all headend video processors. It usually supports direct interfaces to billing computers, although some addressing computers are designed to support billing in a stand-alone configuration. Some of the general features that are available include:

- Control of many thousands of converters
- Multi-user/multi-task operation

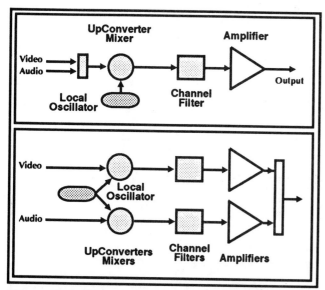

Figure 12-17. Internal and External Diplexing. In external diplexing (bottom) the local oscillator is shared between the separate audio and video routes.

WIRELESS CABLE COMPONENTS

TABLE 12-2. TYPICAL WIRELESS CABLE SIGNAL COMBINER SPECS

Signal combiner(s) is(are) used to amalgamate the diplexed aural and visual carrier outputs of each MMDS transmitter for input to a common transmission line feeding a common transmitting antenna. Channels are normally supplied to a common antenna on a "first adjacent" basis with 16 MHz separation between carriers.

Input/output Impedance: 50 ohms

VSWR (maximum): 1.25 to 1

Power Handling Capability/RF Channel: 100 watts

First Adjacent Channel Rejection (16 MHz Carrier Separation): 25 dB

Insertion Loss (maximum): 1.5 dB

Intermodulation Distortion: For a visual to aural carrier ratio of 10 dB the "2a-b" products appearing at −6.0 MHz and +12.0 MHz with respect to the visual carrier should be >65 dB below peak visual carrier level.

Connectors: 50 ohm type female N-connector for both input and output

TABLE 12-3. TYPICAL WIRELESS CABLE TRANSMITTER SPECIFICATIONS RF SPECS

Output Power: 40 watts peak visual, 4 watts average aural

Type of Emission: 625 line PAL

Carrier Frequency Stability of Upconverter/Amplifier over 0 to $40^{\circ}C$ Temperature Range: ± 500 Hz

Harmonic Emissions: −60 dB

Spurious Products: −60 dB

Intermodulation Products: −70 dB

Incidental Carrier Phase Modulation: 2 degrees

AFC: ± 0.2 dB output power stability with baseband input format. With RF input format transmitters, the allowable frequency variation at the transmitter VHF input must be at least 0.005% of VHF carrier nominal frequency, for retention of overall transmitter output stability within ± 500 Hz.

RF Input Levels: In an RF input format, the allowable range of visual carrier input levels, assuming a 15 dB maximum visual to aural carrier ratio, must be at least minus 10 dBmV to +20 dBmV.

RF Input Impedance: Visual/Aural −75 ohms unbalanced

Input Radio Frequency: Standard PAL RF carrier frequencies

BASEBAND SPECIFICATIONS

Differential Gain (10 to 90% APL): 3%

Differential Phase (10 to 90% APL): ± 2°

S/N (pp visual to RMS noise CCIR weighted: >55 dB

Hum and Noise: <50 dB

K Factor (2T pulse): 2%

Envelope Delay: ± 30 nanoseconds

Low Frequency Linearity: 2%

Audio Frequency Response: ± 0.5 dB ref. frequency 400 Hz at ± 50 kHz peak deviation 50Hz−15 kHz

Audio SNR: 45 dB unweighted

Audio Pre-emphasis Characteristic: 50 microseconds

Audio Distortion: <1% reference frequency 400 Hz @ +8 dBm

PROTECTION

Transmitters employ requisite protection circuitry to prevent damage due to high S/N at antenna terminals. Diagnostic warning of high VSWR is provided with each transmitter. Surge protection is incorporated to facilitate suppression of transients due to lightning or other sources. The SHF output, VHF input (if applicable) and power mains ports all protected

WIRELESS CABLE COMPONENTS

- Multi-hub/remote hub operation
- Programmable event tagging and automatic channel menu control
- User-defined, multi-level security password
- Database access via converter serial number and subscriber account number, name and phone number
- Generation of management reports
- On-line back-up utilities
- Channel mapping control

Most controllers/headend processors available today offer combinations of most of these features. For example, the Tocom Micro Addressable Control System (Micro-ACS) includes a number of these features (see Figure 12-18). It controls as many as 250,000 subscribers from up to 15 remote sites. The subscriber database contains optional specific information on each subscriber including account number, name, address and phone number. Reports may be generated on any of these fields. Subscriber information may be entered or changed from one or more operator terminals. The database for each converter includes authorized channel packages, pay-per-view events and other control information. Thirty-two independent channel packages that include any combination of programs and are defined by the wireless operator, are available. More than 65,000 IPPV program tags are also available.

A channel map may be downloaded to the processor from the controller. This feature effectively turns four ITFS channels into three entertainment and one ITFS channel while still adhering to FCC timing requirements for ITFS programming. An unlimited number of barker channels can be controlled. As a consequence, when a subscriber tunes to an un-authorized channel, the converter automatically re-tunes to the barker channel. Barkers may be used to advertise for a premium service, PPV or IPPV events, or give purchase instructions.

The Tokom Micro-ACS software permits an operators to control on-line and off-line functions. The on-line functions include:

- Add or delete converters from inventory
- Modify subscriber demographics
- Modify converter services
- Create or modify channel program schedules

- Report converter services, customer demographics and impulse pay-per-view (IPPV) information
- Initialize converters
- Back-up the database
- Maintain security information
- Manage hub controllers
- Load scrambler configuration

Another illustration is the Comband CT-1200 encoder/modulator which controls the CT-1000 converters and scrambling system. Like the Tocom system, it is designed specifically for wireless cable environments and provides channel mapping, barker channels and subscriber control. A Hewlett-Packard HP-100, A-400 series computer provides system control. The addressing computer sends all addressing commands and downloads channel mapping, channel level, and pay-per-view files to the headend processor/controller. In the event of a failure in the computer or the communication link between the computer and controller, this provides redundancy since the controller can use all files.

The address code of each converter is known only to the addressing computer. These cannot be read or duplicated by operating personnel. Individual address codes are randomly generated by the controller and downloaded to set-top converters. During installation, a small hand-held IR remote, the technician initialization unit (ITU) can set all the converter parameters. Different configurations are possible for different conditions. For example, if the converters are used as head-end decoders in an SMATV system, they can be programmed to operate in an "auto-on" mode to ensure that they revert to the proper frequency in the event of a power failure.

Figure 12-18. Tocom Encoders. *These encoders are used to encrypt the wireless signal before transmission. Up to eight channels are controlled per enclosure in this baseband encoding system. The headend computer interfaces via an RS-232 port and channel control data is included with the signal of each controlled channel.(Photo by Steve Berkoff)*

The Comband Encoding System – Details

The Comband encoding system consists of a unique video processor, IF modulator and data modulator (see Figure 12-19 as well as Figures 12-23 and 12-24). In this NTSC example, the video processor takes the standard video and produces the Comband formatted signal. A video processor accepts up to eight inputs and produces four Comband video A and B baseband outputs. Time base correction and frame synchronization are performed by the video processor and a special video enhancement technique is used to improve the apparent resolution of the signal. Although, the unit also provides an external generator lock input, when no input is connected the processor automatically switches to an internal reference oscillator.

The Comband IF modulator accepts the video A and B outputs from the processor and the two baseband audio inputs. It then produces three IF outputs, one containing the video A information that is AM modulated onto a 45.75 MHz carrier, the second containing the video B information AM modulated on a 41.3 MHz carrier and the third output containing the audio A information FM modulated on a 46.75 MHz carrier and the audio B information FM modulated on a 46.4 MHz carrier. The modulator also accepts an IF carrier at one or both of the audio carrier frequencies with FSK addressing data as modulation. Control lines which connect between the IF modulator and data modulator switch the FSK carrier with the audio carrier at times when addressing data is scheduled to be broadcast.

The data modulator accepts the addressing data from a computer interface unit, modulates an IF carrier in an FSK format and provides this output to the Comband IF modulator. The data modulator also controls the switching of the Comband IF modulator audio outputs. When addressing data is ready to be transmitted into the system, the data modulator causes the IF modulator to switch from the audio IF output to the IF input from the data modulator.

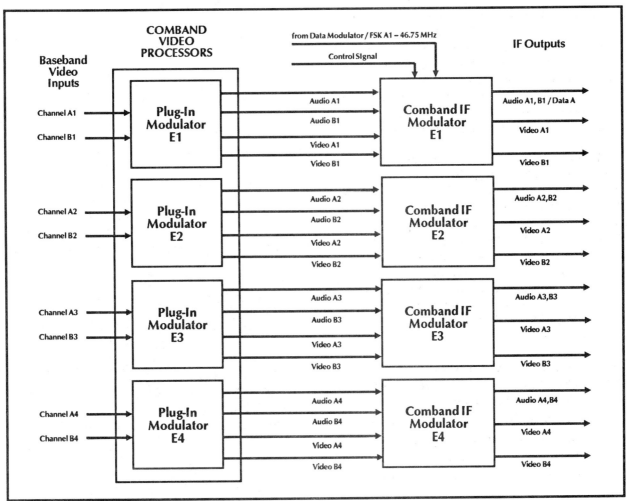

Figure 12-19. E-Channel Group Processors and Modulators for Comband Compressed Format.

Evaluating and Maintaining Headend System Performance

The adequate operation of a wireless cable system is primarily dependent upon the transmitted carrier-to-noise power ratio, the S/N ratio. In addition to maintaining a sufficiently high S/N ratio, spurious signals that generally result from amplifier distortions must be minimized and power must be monitored (see Figure 12-20). With channel spacing of 6 MHz between visual carriers, as is the case with NTSC broadcasts, the transmission equipment must be of good quality and have linear outputs. Furthermore, the signal must be adequately filtered so that all unwanted sideband energy and spurious signals fall more than 60 dB below the visual carrier of adjacent channels. Most modern transmitters process aural and visual carriers separately to minimize spurious outputs.

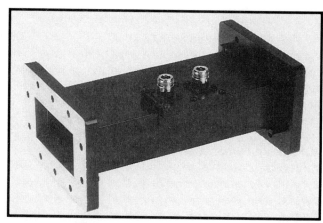

Figure 12-20. Wireless Cable Loop Coupler. *This device is installed between the combiner and transmit antenna to monitor power output and VSWR. It insertion loss is 0.2 dB, VSWR is 1.15 and bandwidth is 2.5 to 2.7 GHz. (Courtesy of Microwave Filter Company)*

Transmission Formats and Signal Security

Broadcast Formats

Until recently, the most common transmission format incorporated into wireless cable equipment has been NTSC. Today converter manufacturers, including Zenith, Comband and Jerrold, also offer SECAM and PAL (phase alternating lines) compatible units. These broadcast formats are prominent in Europe and other areas of the world.

A hybrid NTSC format, the Comband bandwidth compressed system, is presently in limited use in the North America. It differs from the NTSC standard by virtue of the composite make-up of the transmitted signal. Comband components use the standard 6 MHz NTSC bandwidth and are capable of combining two discrete programming sources within a single NTSC channel bandwidth.

Digital video compression systems that transmit two or more channels within a conventional single channel bandwidth, as does the analog Comband two-channel system, are presently under development. These will have the capability to provide excellent quality, bandwidth efficient audio and video signals.

Signal Security

Signal security technologies presently available to wireless operators are RF and baseband encoding systems as well as the Comband compressed system format. Early single channel and limited multi-channel wireless systems did not utilize security. Reception was restricted simply because the microwave signals were at frequencies far above the range of any television set. Piracy has, however, become an increasingly significant problem without implementation of more active security methods. This coupled with the requirement to include features such as channel tiering and pay-per-view entertainment make security a necessary component of any modern multi-channel wireless system.

Single channel baseband and RF scrambling systems such as those offered by Zenith, Comband or Jerrold both yield an economical approach to delivering wireless signals on a one-program-per-channel basis. Either the audio, video or both signals can be encoded (see Figure 12-21).

The Zenith baseband ZTAC system scrambles by dynamically varying the timing of sync suppression and video inversion and by encrypting the control information, relayed in a digital format. If necessary, the security of this system can be further enhanced by encoding the audio. This is accomplished at baseband by translating the audio spectrum in such a way that it remains unintelligible when received by a normal television receiver. The Zenith PM system uses scrambling at RF frequencies – the sync is dynamically suppressed and both the audio and color subcarriers are inverted. The audio is masked, a process that renders it difficult to decipher.

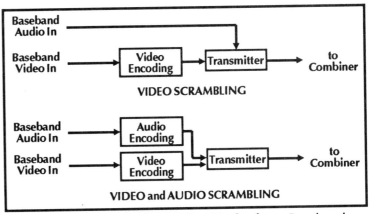

Figure 12-21. Baseband Encoding Methods. *Baseband systems can scramble each channel independently of all others. Thus some channels can be relayed in the clear. Either the video or audio signals or both can be scrambled as shown here.*

The Comband CT-1000 baseband system uses a combination of random horizontal sync offset and APL controlled video inversion to scramble the transmitted video. Six different scrambling modes controlled by a random algorithm or directly by the system operator are available. The sense of video inversion flips every time the APL changes more than 10 IRE over a 10 field period and the sync pulses are randomly offset by 80 IRE within a field. Encoding and decoding is accomplished by use of two "keys". The first key is a random number generator located within the encoder. The second is a code set by the system operator and is transmitted to the set-top converter with the wireless signal. Only the addressing computer knows this key. Both of these keys are used to generate the scrambling data that is inserted in lines 10 and 273 of the video signal. Each converter must periodically receive the transmitted key or an internal clock disables the unit. Channel maps and channel level assignment are relayed with this key and are repeated at least 30 times per minute.

This CT-1000 system is somewhat limited because the audio is not scrambled but masked. While the audio is muted by the set-top converter on an unauthorized channel, a cable-ready TV can decipher the audio information under standard operating conditions.

The Tocom HVP-III video processor scrambles the audio and video signals via random dynamic baseband video inversion and sync suppression. The data relayed with the wireless signal for use by converters in decoding this signal is encrypted and constantly changed to improve security.

The Comband analog compressed wireless system employs a spectral efficient signal format that allows two audio and video programs to be carried within the bandwidth of a single television channel (see Figure 12-22). In this process, the luminance and chrominance signals of the two channels are separated and the baseband information is relayed in a format different from that of a standard NTSC or PAL transmission. Video A contains the luminance information for video A, the chroma information for video A and B as well as a very narrow, positive-going horizontal sync pulse. Video B is simply the luminance information for program B. The chroma information on video A is time compressed by a factor of 10 to 1 and every alternate line of chroma for each program is omitted. These two Comband video programs are then routed to an IF modulator along with baseband audios for upconversion. Figure 12-23 shows the frequency layout and allows a comparison to standard NTSC format.

This Comband system securely scrambles both the audio and video signals. Channel A video appears severely torn and without color to a pirate viewer or a cable ready television receiver. Only an objectionable "buzz" can be heard in the audio on a television receiver. Channel B is totally scrambled since the picture carrier is offset in frequency by more than 4 MHz; the audio is also not available to a conventional TV receiver. The 3.58 MHz color subcarrier or "burst" is not transmitted in this Comband signal. This prevents television sets which obtain synchronization from the color burst from locking onto the program material.

The security of either baseband or Comband compressed signal encryption systems are generally furthermore augmented and complemented by addressability.

WIRELESS CABLE COMPONENTS

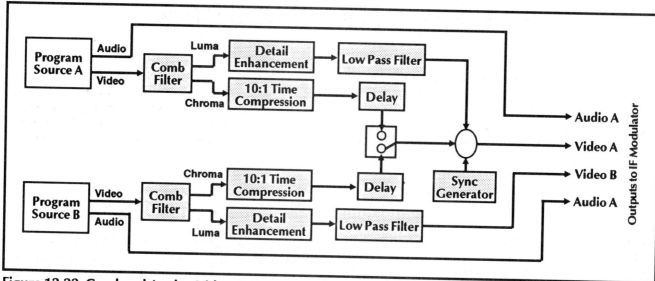

Figure 12-22. Comband Analog Video Compression Processor. *The Comband processor accepts inputs from two channel sources and outputs four signals, a mix of the input information (see text for more detailed description).*

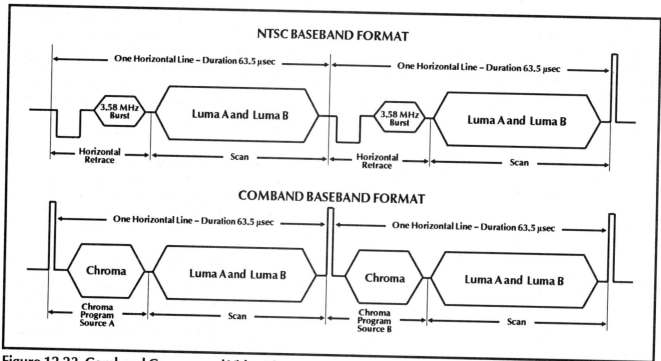

Figure 12-23. Comband Compressed Video Line versus Standard NTSC Video Line. *Two compressed channels fit into the time duration of just a single standard NTSC line. This top illustration is the standard NTSC baseband format and the bottom the Comband baseband format.*

C. OPERATIONS FACILITIES

Business office and warehouse requirements vary according to the size and complexity of the planned wireless system. Surveys of operators indicate that a system that serves 8,000 to 12,000 subscribers typically requires 3,500 feet^2 (325 m^2) of business office and warehouse space.

Television studio facilities are optional additions to most wireless cable systems. If feeds from 24-hour satellite networks are received, it is generally not necessary to invest in a studio. On the other hand, if one or more stand-alone program services or local advertising sales are envisioned, a studio with program integration and tape equipment would be required.

Telephone and data systems that support a wireless operation can range from sophisticated regional telephone centers serving a number of distant markets to a local telephone center with minimal call switching capacity and computer support. Telephone requirements should be structured around subscriber projections and, in general, should include four to six administrative lines, six to eight customer service lines, eight to ten lines for telephone sales and two to four lines for service dispatching and installation scheduling.

Subscriber billing and addressing systems are an integral component of a business office facility and, wherever possible, should be interfaced to ensure that accurate and efficient transactions occur. Typical computer billing systems nominally provide a subscriber data base encompassing buildings passed, address listings, subscriber account information, account receivables, marketing information and inventory control programs. As described earlier, the addressing system, via interaction with the billing system, controls the activity of all converters by authorizing or de-authorizing service to individual subscribers and generally offers additional features such as channel mapping, pay-per-view and converter inventory control.

The warehouse facility should have adequate space to store appropriate amounts of subscriber equipment such as antennas, converters, mounting masts, cable, block downconverters as well as miscellaneous installation materials. The warehouse, wherever possible, should be co-located with the business office as this provides the most economical approach to system operations. Additionally, if addressability is utilized, the warehouse should be within close proximity to the addressing computers in order to properly process subscriber converters prior to installation and upon their return to inventory.

Business Office Site

The wireless cable business office is the preferred location for the addressing computer system because all communications with the transmit site, the warehouse and the billing computer normally occur here. This site is optimal because it is manned and easily accessible for performing routine back-ups and maintenance. Communication with the billing computer is via a hard-wire RS-232 link if the two systems are co-located. Otherwise they are interfaced via modems and conditioned telephone lines if a separate billing system service bureau is to be established. Addressing data is sent to the transmit and warehouse sites via hard-wire baseband data if the they are co-located with the business office and by conditioned telephone line if the transmit site is at a distance of more than 150 meters or if the warehouse site is at a distance of more than 500 meters.

Customer service representatives are located at the business office. Interconnection problems with the computer systems are minimized if these computers and most terminals are co-located. One possible layout is illustrated in Figure 12-24).

Space and Facilities Requirements.

Adequate space within the business office should be designated as the "computer room" (see Figure 12-25). This area generally provides additional facilities for billing, addressing and telephone system computers. This room should have isolated and conditioned power, air conditioning and fire protection facilities. General specs for the computer room are:

- Each computer system requires power line conditioning which connection to a mains circuit (120 Vac/50 or 220 Vac/50 amp) through an isolating ground type receptacle.
- The room should be maintained at a temperature between 20 to 27°C as measured at

WIRELESS CABLE COMPONENTS

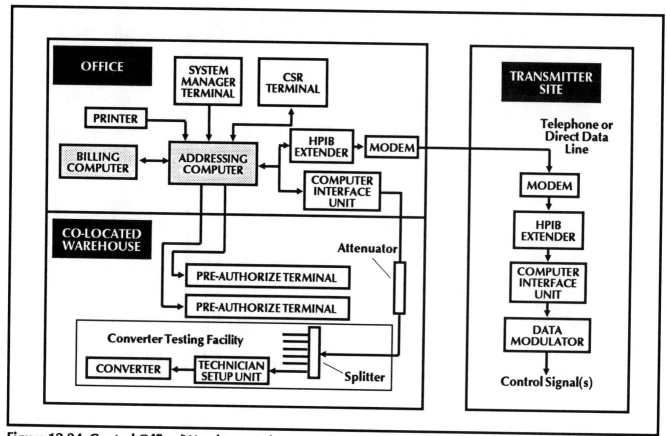

Figure 12-24. Central Office/Warehouse Schematic Layout. *This schematic illustrates one possible layout for the central office and warehouse facilities. Note that some types of converters can be authorized and set-up at the customer premise.*

the intake of the computer racks and a relative humidity ranging from 50% to 60% with no condensation. The computer equipment imposes an air conditioning load of approximately 6,000 BTUs/hour for each unit. This load is above and beyond the load presented by operations personnel, room lighting and heat transfer from external walls, windows, floors and ceiling of the computer room.

- The computer room should be fully protected from fire by the installation of a system such as a Halon fire protection system. This should additionally include ionization detectors and smoke/heat sensing units.

Billing Computer Interface.

The billing computer interface is an RS-232 ASCII line that relays data to and from the billing computer at up to 9600 baud and higher with or without modems. The billing computer is the master controller in the communication link. All transactions are initiated by this computer and a response is returned by

Figure 12-25. Computer Billing System. *This space has been set aside at the Corpus Christi wireless system central office for all the necessary components of a computer billing system. (Photo Steve Berkoff)*

the addressing computer. If the response indicates a data transmission error, the billing computer should re-transmit the data packet. If the response from the addressing computer indicates that the data was re-

254

ceived correctly but that the data content contains an error (i.e., converter serial number does not exist on an update transaction), the billing computer should notify the customer service representative and should log an error to the error reporting file. The billing computer cannot send a subsequent transaction until the previous transaction's response has been received from the addressing computer system.

Group transactions can also be supported to provide a rapid method to check the entire addressing computer data base. Group transactions are a collection of typically four individual transactions separated by optional group delimiters. In general, transactions over the communication link are provided to establish communications, log-off the billing computer as well as update, install, pre-program, disconnect or to reset a converter and compare converter record to billing system data.

Individual transactions are capable of supporting all features of the addressable converter. These generally include enabling and disabling the converter, changing authorized channels and enabling special events.

Addressing System Features

The addressing computer system is that part of a wireless system that manages and controls data and data transmissions pertaining to converters installed on subscribers' premises. Information in the addressing computer database can be modified by transactions that originate from the billing computer system or from a user interface subsystem which allows direct entry of data to the addressing computer. The user interface can only be used to modify converter data if the billing computer interface is shut down. The addressing computer system allows functions to be performed for subscriber service, tier management, reports and overall system management.

Communication to Transmit Site

Communication to the transmit site is via hardwire if the computer is located near the transmitters. Communications via modem and conditioned telephone lines or point-to-point microwave relay may be used if the business office is remotely located from the transmit site.

This communication path is used to define data to the transmitters that is relayed over-the-air to installed converters. During any bulk addressing time period, this communication link is used continuously to refresh all converters in the system.

D. RECEIVE SITES

At the subscriber location a directional receive antenna and downconverter assembly detects the wireless signal and converts it to standard VHF or cable television frequencies. The antenna is usually roof mounted on an elevated mast and aimed at the transmitter. It intercepts, focuses and detects the microwaves for conversion to an electrical signal. This signal travels over a short length of transmission line leading from the antenna driven element to the downconverter input, where it is amplified and downconverted to lower VHF or CATV frequencies. The downconverted signals are then relayed through RG-6 or RG-59 coax to a power inserter located inside the building and to the set-top converter connected directly to a television receiver (see Figure 12-26).

Reception locations within the wireless service area can include single family residences, multiple dwelling residences such as apartments and condominiums and bulk service facilities. These receive sites all share a common installation process because they all use a highly directional receiving antenna and downconverter assembly to receive wireless signals. The principle differences in receive site installations at the various types of locations throughout the system are related to antenna gain requirements, height and location placement, downconverter configuration and the method by which the downconverted signals are distributed to televisions within end user facilities.

WIRELESS CABLE COMPONENTS

In all cases, it is wise practice to conduct a signal survey to determine serviceability at a particular location before installation. Sometimes a change of two or three feet in antenna height or placement can cause several decibels change in available signal level. The site survey can be accomplished with an wireless receive antenna, block downconverter and a push-up fiberglass mast connected to a field strength meter or portable TV receiver. This procedure is explored in much more detail in Chapter 14.

Receive Antenna

The receive antenna must deliver consistent gain and a good directivity to minimize the detection of interference and unwanted signals arriving from directions other than the antenna boresight (see Figure 12-27, 12-28, 12-29, 12-30 and 12-31). The ratio of signals detected along the antenna boresight to those detected from 180° off boresight, i.e. from behind the antenna, is termed the "front- to-back ratio". A minimum of 20 dB front-to-back ratio is generally assumed to be acceptable.

Wireless receive antennas generally have gains ranging from 16 to 28 dBi although 12 dBi corner reflectors are used fairly frequently in close in situations. Antenna gain and thus size required at a given location is determined by system requirements, distance from the transmitter and signal distribution requirements at the receive site. The use of higher gain antennas allows reception of signals from a given transmitter at greater distances. Of course, this range improvement comes at higher receive site costs. Typical receive antenna specifications are listed in Table 12-4.

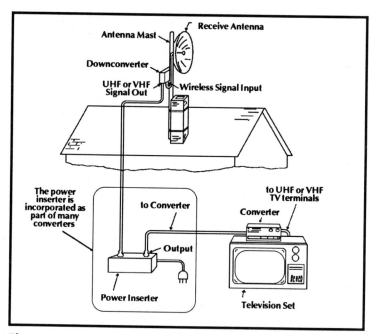

Figure 12-26. A Typical Wireless Cable Reception Site. *Signals are received by a small microwave antenna, downconverted, and relayed indoors to a television-top converter.*

TABLE 12-4. TYPICAL WIRELESS RECEIVE ANTENNA SPECS				
	Gain (dBi)			
	18	21	24	28
3 dB Beamwidth (degrees)	20	15	12	8
Front-to-Back Ratio (dB)	25	28	32	36
VSWR: 2.5 - 2.686 MHz	1.2:1	1.2:1	1.2:1	1.2:1
Nominal Impedance (ohms)	50	50	50	50
Cross Pol. Discrimination (dB)	25	25	25	25

WIRELESS CABLE COMPONENTS

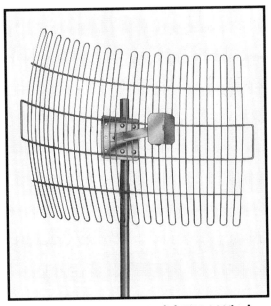

Figure 12-27. Lance Model 2524 Wireless Receive Antenna. *This welded metal wire antenna has a front-to-back ratio of 20 dB and a beamwidth of 20°. It has zinc and gold irridite standard protection and can be galvanized at request. (Courtesy of Lance Industries)*

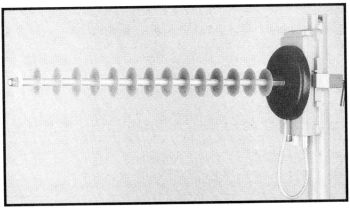

Figure 12-29. Yagi-408 Wireless Receive Antenna. *This Yagi antenna with integral downconverter is designed for close-in reception of wireless broadcasts.*

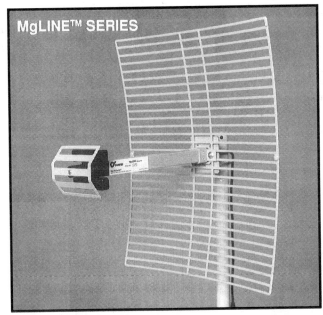

Figure 12-30. Conifer MgLine Series Wireless Receive Antenna. *This die cast, powder-coated magnesium antenna shown in the top photo has a built-in downconverter at the focus so only an F-connector and RG-6 coaxial cable is necessary for its installation. The lower illustration shows the polarized and cross-polarized reception patterns for this antenna. (Courtesy of Conifer Corporation*

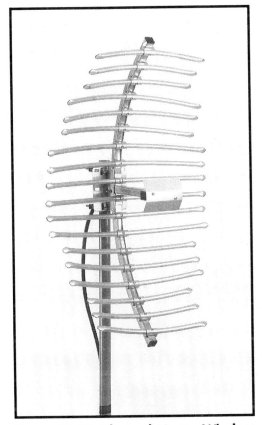

Figure 12-28. Channel Master Wireless Antenna. *(Courtesy of Channel Master)*

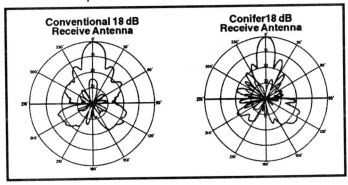

WIRELESS CABLE COMPONENTS

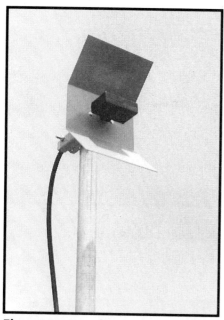

Figure 12-31. Channel Master Corner Reflector. *This wireless receive antenna is designed for use in strong signal areas. (Courtesy of Channel Master)*

Wireless systems can be designed to transmit signals on either horizontal or vertical polarizations. Similarly, receive antennas can be oriented to respond to either horizontally or vertically polarized signals, or both. A horizontally polarized receive antenna exhibits a varying degree of rejection or discrimination to interfering signals that are vertically polarized. The degree of rejection is typically between 20 to 30 dB, if the interfering signal is arriving at the same azimuth as the desired signal, to as low as 6 dB at other azimuths. The interference criteria generally employed conservatively adopts the minimal 6 dB polarization discrimination.

In some cases, signals of both polarizations are relayed from the central transmitter. A single antenna can be oriented at 45° to the direction of either polarization resulting in a 3 dB signal loss (see Figure 12-32).

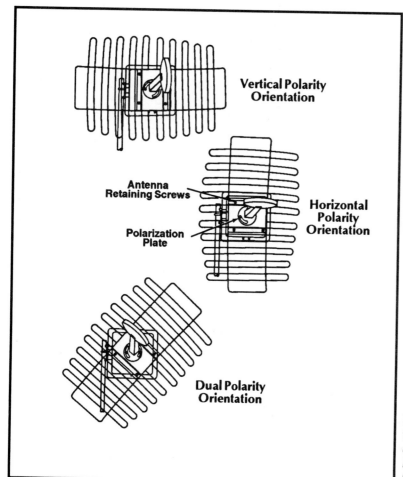

Figure 12-32. Antenna Polarization Orientations. *A wireless receive antenna can be oriented to receive vertical, horizontal or both polarity signals. (Courtesy of Texscan Corporation)*

Block Downconverter

The block downconverter is the heart of a wireless reception system (see Figures 12-33, 12-34 and 12-35). Units are available in gains typically ranging from 20 to 30 dB, a range of noise figures and various input frequency parameters as well as choices in output downconverted signal frequency bands. The selection of a downconverter should be based not only upon its noise figure and gain but also upon its signal handling capabilities. Some of the characteristics and requirement are discussed below and those of a state-of-the art downconverter are presented in Table 12-5. Of course, "state-of-the-art" characteristics are always evolving.

Noise Figure

The noise figure of a downconverter is a measure of the amount of noise added by this component to the received signal and is the major contributor of receive system noise. S/N ratio increases by 1 dB for each 1dB increase in noise figure (see Appendix B).

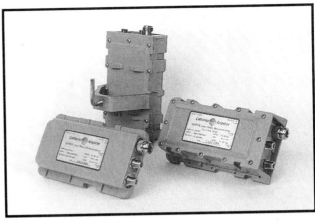

Figure 12-34. CalAmp Downconverter. *This photo of the slimline downconverters shows the clamp used to bolt the unit to a pole. These downconverters are available in a range of noise temperatures. (Courtesy of California Amplifier)*

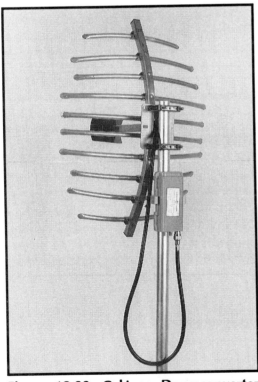

Figure 12-33. **CalAmp Downconverter Mounted on Mast.** *This downconverter is mounted on the mast close to the antenna dipole. The dipole is connected via a 50 ohm coax and N-connector. Two outputs are provided via F-connectors. One of these is a test point output. (Courtesy of California Amplifier)*

TABLE 12-5. DOWNCONVERTER SPECS
(Norsat 2500 Downconverter)

Gain	24 dB typical
Gain Flatness	1.5 dB pp in-channel
	3.0 dB pp in-band
Gain Stability	± 1.0 dB typical
Noise Figure	1.0 dB Typical
Input Frequency	2500 to 2686 MHz
Output Frequency	222 to 408 MHz
	662 to 848 MHz
Input VSWR	2.0:1 Max (1.5:1 typical)
Output VSWR	2.5:1 Max (1.5:1 typical)
Frequency Stability	± 50 kHz (−40° to 60°C)
L.O. Leakage	−60 Dbm max. @ RF port
L.O. Frequency	VHF: 2278 MHz
	UHF: 1838 MHz
IF Rejection at RF	−60 dB
Image Rejection	−50 dB
Intermod Distortion	−55 dB max
Cross Modulation	−60 dB
Spurious Outputs	−60 dB
Output Levels	± 5 dBm @ 1 dB comp.
Supply Voltage	+15 to +24 Vdc (IF cable)
Supply Current	250 mA Max (185 mA typ.)
Temperature Range	−40° to +60°C
Humidity Range	0 - 99% non-condensing
Input Impedance	50 ohms
Input Connector	"N" type
Output Impedance	75 ohms
Output Connector	"F" type
Test Points	20 dB below output
Dimensions	3.5 x 4.5 x 1 inches
Weight	0.8 pounds

WIRELESS CABLE COMPONENTS

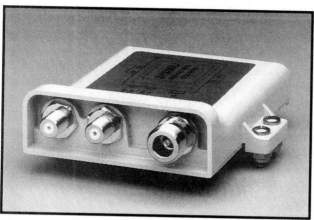

Figure 12-35. Norsat Downconverter. *The Norsat 2500 downconverter has a 24 dB gain, a noise figure of 2.0 dB, an input frequency of 2,500 to 2,686 MHz and two output frequency ranges: VHF 222 to 408 MHz and UHF 662 to 848 MHz. (Courtesy of Norsat International)*

Gain

Downconverter gain is the ratio of output to input signal level. A high gain unit serves both to compensate for losses in the drop cable and also to reduce the noise contribution of the television receiver. In some cases use of a low noise preamplifier may be warranted, although excessive gain can overload the set-top converter (see Figure 12-36).

Gain should be relatively constant across the frequency band of interest. A flat gain response produces consistent performance from the low to high frequency channels. Gain must also be flat within the frequency range of a single channel to maintain to correct relationship between visual, chroma and aural levels.

Dynamic Range and Signal Distortions

Selecting a good quality downconverter is crucial, especially in a multi-channel wireless system, in order to minimize excessive signal distortions caused by non-linear interactions among the signals that make up the various channels. This process, known as intermodulation, can become more pronounced as the number of channels increases. The range between the noise floor and the signal level that causes downconverter overloading is defined as the dynamic range. Signal distortions become more pronounced when input levels are near the top of the dynamic range, i.e. near the saturation point.

The National Cable Television Association (NCTA) specifies that distortion products, namely intermodulation, cross modulation as well as other distortions, should fall 57 dB below the signal level to be invisible in the video. A practical limit to distortion products of −50 dB should be maintained to achieve acceptable system performance. As a rule of thumb, a 20 dB gain downconverter should reduce third order intermodulation products to 50 dB below the signal when operating with two −20 dBm carriers. Cross-modulation, the superimposition of the carrier from an adjacent channel upon the desired signal, is generally not a problem if intermodulation products are well managed.

The number of channels that a downconverter can manage is related to the total amount of input power applied. Beyond a certain number of channels the amplifier can be overdriven. Theoretically, for each doubling in the number of input channels, the power of each channel must be backed off by 6 dB. Thus, for example, if four channels each having input powers of −21 dBm produce distortion products of 50 dB, then eight channels at −27 dBm and 16 channels at −33 dB would both produce the same amount of distortion. Since variations in antenna performance, transmitter power and signal

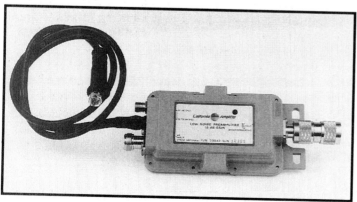

Figure 12-36. Low Noise Preamplifier. *Low noise preamplifiers may be used in fringe areas to increase signal levels to the downconverter. This unit has a 15 dB gain, a 1.5 dB noise figure and a RF range of 2.1 to 2.7 GHz. (Courtesy of California Amplifier)*

propagation can be experienced, and since the power levels at the transmitter are always maximized, it is very important to choose a downconverter with a wide dynamics range.

Interfering Signals

Unwanted microwave signals that fall within the frequency range of a downconverter may be downconverted to an adjacent television channel causing interference. Overloading may also result when a nearby microwave service transmits a strong signal close to the frequency band being used. This may cause noise and distortion in the desired signals being received.

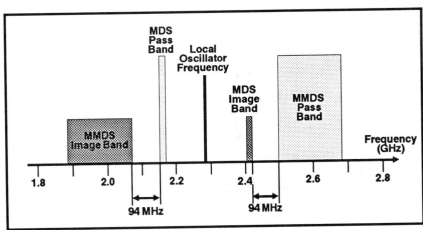

Figure 12-37. The Production of Image Frequencies. *Signals or noise in the MMDS image band can be translated by the LO into the MMDS pass band. Image rejection filters can largely eliminate the effect of this image band on the desired signal. However, if both MMDS and MDS bands are broadcasted the situation is somewhat more complicated as explained in the text.*

In an urban environment interference from a variety of sources including amateur radio, radar and direct pickup of UHF or VHF broadcasts and other intermediate frequency signals is more likely than in a rural area. A wireless cable system is not a "closed" system like a CATV distribution network. In addition, urban wireless systems often must compete with existing cable systems and therefore must have a wide channel menu. More channels translates into more potential for overloading downconverters. These considerations suggest that a high dynamic range downconverter is crucial in a densely populated urban market.

Image Rejection

The wireless IF signal is downconverted by mixing it with a fixed signal from a local oscillator (LO). This process also produces the same IF band of frequencies from combination of the LO signal with an "image band" of frequencies. The image and IF pass bands are also equally spaced on both sides of the LO frequency.

When signals in just the 2.5 to 2.7 GHz range (MMDS/ITFS) or another similar narrow range are transmitted, a single band downconverter can be used. In this case, image rejection is simple because the image band is relatively far from the main signal band. It can be removed with a band pass filter that passes just the desired band or an image band trap. If both MMDS/ITFS and MDS bands are broadcasted as in Figure 12-37, both image bands fall within 94 MHz or the main signal and image rejection is not as simple. A dual band downconverter that has 10 dB protection against image signals is sufficient if no other signals are present in the image band. However, a 30 to 40 dB of image rejection is important if an interfering signal is present in the image band.

The Effect of Oscillator Performance

Downconverter performance is effected by the quality and type of LO employed. A perfect LO would generate a signal of just one frequency and no others. In the real world, spurious signals as well as harmonics of the LO frequency are fed into the mixer along with the desired signal. A LO with a stable dielectric resonant oscillator that generates a single frequency is usually the best option. Models that multiply the frequency produced to the desired output have more potential to generate harmonics that themselves can produce intermodulation products and distortions. Even mechanical vibrations, known as microphonics, that result from the effect of rain drops, hail or winds can cause the oscillator to "jitter" and generate unwanted signals.

In PAL or SECAM broadcast systems where phase noise can be a concern, downconverter LOs must generate "clean" signals. Phase locked loop oscillators often generate poor phase noise characteristics that result in video flashing, streaking or other color distortions.

WIRELESS CABLE COMPONENTS

Note that downconverters can generate other types of spurious responses that can be delivered to the mixer and amplified along with the input signal. Quality units will produce a minimum amount of spurious products.

Return Loss

Both the input and output impedance must be matched to that of the input and output cables. Since the match is never perfect, there will always be a small component of reflected signal. When a reflection from the cable/set-top converter reaches the downconverter output, a portion would then be re-reflected. If the output return loss is too great this would appear as a ghost on the television picture. Return loss is measured by voltage standing wave ratio, the VSWR.

Environmental Effects on Performance

A downconverter operates in a harsh environment that includes wind, snow, rain, hail or possibly salt corrosion. A top quality unit will have reasonable immunity to lightning damage and cannot allow intrusion of any form of water. It must perform consistently over the years up to its original specifications.

Downconverter Designs

Two general block downconverter designs are outlined in Figure 12-38. Placing an RF amplifier in front of the RF filter has advantages: the lowest noise figure possible can be attained, a lower cost RF filter can be used and the overall housing size can be reduced. However, the makes the RF amp susceptible to overloading by out-of-band RF signals, both IF and image frequencies are amplified before effective RF filtering is effected, the bandwidth of the image notch is not wide enough to cover all frequencies that are required and the unit is more susceptible to lightning damage. Alternatively, placing the RF filter at the input in front of the amplifier stage avoid overloading of the amplifier, attenuates IF and image frequencies before amplification, permits high levels of image and IF rejection and lowers the possibility of lightning damage. However, this design makes it more difficult to attain a lower noise figure. Clearly, a downconverter that inserts its RF filter in front of the RF amp would be preferred compared to one with the opposite arrangement but with an identical noise figure.

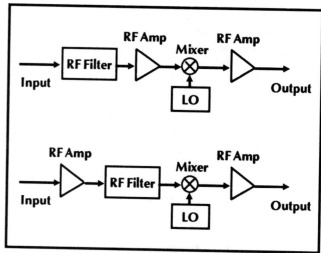

Figure 12-38. Two Block Downconverter Designs. *The placement of the RF filter and RF amplifier has consequences in downconverter performance. These are described in the text.*

Set-Top Converter

The wireless set-top converter is the customer interface. It provides signal security and control as well as a number of other functions that are transparent to end-users. These including channel mapping, automatic tuning to barker channels and downloading of system parameters. Other features such as parental control, PPV and channel selection are directly controlled by subscribers. Most modern converters also incorporate a tuner with two inputs: one for off-air UHF/VHF signals and the other for wireless signals. The correct input is then automatically selected by subscriber choice of channel number. Converters also usually pass MTS stereo to MTS-capable television sets and VCRs.

Most modern wireless converters feature a handheld remote that control automatic selection between off-air and wireless broadcasts, channel tuning, on/off, volume control as well as other remote control functions. For example, the Tocom 5507-MU includes volume control, last channel recall, mute, favorite channel selection and local parental access control while the CT-1000 optional remote provides volume control, mute, direct channel and scan up/down tuning, last channel recall, favorite channel memory and parental control (see Figures 12-39 and 12-40).

WIRELESS CABLE COMPONENTS

Figure 12-39. Tocom 5507-MU Set-Top Converter. *This remote controlled converter has a VHF/UHF tuner with inputs for both wireless and off-air signals. It covers the 54 to 806 MHz range for NTSC, PAL M and PAL N signals and 47 to 862 MHz for PAL B/G signals. It is VCR- ready and IPPV capable and employs random dynamic baseband scrambling. The ITM-111 telephone modem can be incorporated into this set-top to support store-and-forward IPPV ordering. Information about the purchase is stored in the converter for later relay via phone line to the headend. (Courtesy of Jerrold Communications)*

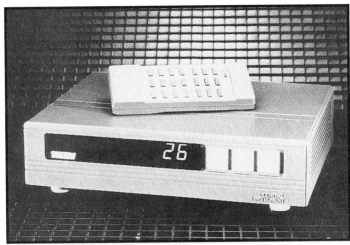

Pay-Per-View Capability

Some converters feature internal telephone modems so subscribers can have immediate access to interactive entertainment services such as pay-per-view (PPV) or impulse PPV (IPPV).

Hotel movie vendors have known for some time that IPPV results in buy rates typically twice that of the traditional order entry method. However, until recently impulse PPV required a two-way RF communication path to implement the order entry process. Wireless cable, traditionally a one way transmission path, has only recently started to utilize store and forward technology to create the necessary two-way path and provide subscribers with impulse purchasing.

A modern set-top converter can be pre-programmed with a finite limit of PPV purchases that are available to the subscriber. This data is fed into the set-top through the forward transmission paths of the microwave broadcast itself, with the signal usually embedded in the vertical blanking interval of the standard television signal. This pre-set purchase level is determined by the wireless operators credit polices and should reflect a realistic ability for payment by the subscriber.

The subscriber can then purchase real-time IPPV events with the resulting purchase is then stored in the set-top memory. As each purchase is made via the impulse buttons on the remote control or the set-top, the activity is recorded by its internal memory. At a pre-determined time, usually in the early morning hours, the converter then utilizes its internal modem to return a call to the subscriber control or the billing computer center and to download the purchase information for that customer into the billing system.

IPPV is thus made available to the wireless cable subscriber without the need to construct a real-time return line from each subscriber outlet. This technique, known as store and forward, is akin to the VideoPal system developed by General Instruments for use in home satellite PPV entertainment.

In some wireless cable systems, this telephone return path also allows the set-top to be utilized in an interactive mode between the subscriber and the wireless operator. Advanced functions such as accounts review, independent video messaging systems and upcoming events are available.

Figure 12-40. Zenith Z-TAC Converter. *This is an addressable baseband decoder that uses dynamic sync suppression and active video inversion combined with optional audio scrambling. Each decoder has a unique address embedded into permanent memory that can be addressed by the control computer. Multi-tiered, addressable control of all basic and premium services as well as PPV is possible. (Courtesy of Zenith Cable Products)*

WIRELESS CABLE COMPONENTS

Technical Demands in a Wireless versus a Wired Cable Environment

Set-tops must function in a much less forgiving environment than standard cable converters that are connected to a wired network (see Figure 12-41). In a wired distribution system, most factors that affect signal level and S/N ratio such as line loss, amplifier tilt or return loss are predictable. Wireless converters also must be capable of managing similar static variations in signal level including free space path loss as well as stationary blockages and multipaths. However, they must also perform under the influence of dynamic variations, namely changes in signal level that occur randomly and vary with time. Dynamic variations stem from a number of sources including vibrations in transmission equipment resulting from wind or mechanical disturbances, objects moving in the path between the transmit, and receive antennas as well as objects moving within the Fresnel zone just on the edges of the direct line-of-sight path. Such dynamic factors can typically reduce signal power by 2 to 4 dB with peaks up to 10 dB. Variations generally occur at rates of 2 to 10 Hertz.

Converters in a wireless system must also contend with an increase in signal distortions that results from the need for a high power microwave power amplifier at the transmit site. An RF wired system generally operates at substantially lower powers. In addition, the inclusion of both a transmitter and downconverter in the broadcast circuit introduces extra frequency errors, typically on the order of ±100 kHz.

In most countries, wireless operators have access to a more restricted operational bandwidth than do wired distribution systems. As a result, wireless set-top converters must be able to detect data that is transmitted within the channel bandwidth. This effective bandwidth can also be increased by including both VHF and UHF broadcasts, a design that is facilitated by converters that can tune to these signals. In addition, in the United States only 13 of the 33 wireless channels are available for full time use by wireless operators. The remaining 20 can be time-shared with educational services. Consequently, converters must have the ability to share bandwidth, as process known as channel mapping.

Converter Design Considerations

Wireless converters must be designed to meet the stringent environmental demands that were discussed in the previous section. These impact three important areas: RF input, scrambling system and addressing system requirements.

RF Input Requirements

An RF tuner must have an unusually wide dynamic range given the wide variations in signal levels within the market area of a wireless transmitter as well as dynamic variations in level that can be experienced. For example, as a result of free space path losses alone the received power level decreases by 6 dB when the distance from the transmitter is doubled. Signal levels being applied to the input of a downconverter within a typical wireless system can easily vary from –10 dBmV to +24 dBmV, not accounting for dynamic variations. This upper level ap-

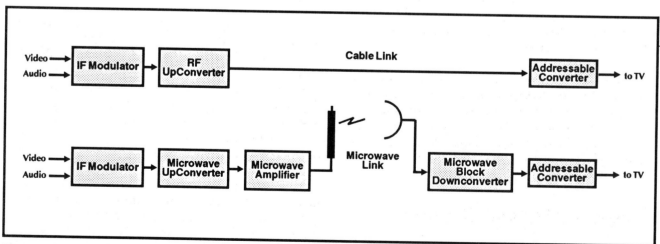

Figure 12-41. Differences between Wireless and Wired Cable Environment. *A wireless broadcast system must contend with a more variable environment that a typical SMATV or CATV system. The wireless system requires use of a high power amplifier and transmit antenna at the broadcast site and a receive antenna and block downconverter at the customer site.*

plied to a set-top is usually limited to +24 dBmV by the dynamic range of most downconverters.

The RF tuners section of a set-top must not only manage variations in signal strength but must do so in an environment where many adjacent channels may be present. Such operational conditions which could potentially generate unacceptable levels of distortion products mandate use of a high quality design. In addition, the automatic gain control (AGC) system must handle changes in power levels while minimizing distortions.

Given the potential for as much as 100 kHz variation in input frequency, a set-top must have an effective automatic frequency control (AFC) system to have the ability to lock on and fine tune the wireless channels.

Scrambling System Requirements

The descrambler built into a wireless converter must be able to manage the increased levels of non-linear distortions and dynamic changes in signal level without a loss in picture synchronization or an increase in noise. This is a challenging design goal given that in many scrambling systems the majority of sync pulses, a useful measure of signal power level, are suppressed.

Addressing System Requirements

Given the limited bandwidth of wireless broadcast system, the data that controls set-top addressing as well as channel mapping must be relayed in-band. A converter must be able to decipher this data with a minimum errors in an environment where signal levels are fluctuating, S/N ratios may be low and multipath conditions may occur. This is usually accomplished with FM or FSK modulation techniques and by inserting the data on one or more channels.

E. SINGLE and MULTI-UNIT RECEIVE-SITE CONFIGURATIONS

Single Family Residences

The single family receive site (see Figure 12-26) consists of the equipment described above plus provisions for receiving and combining the local VHF and/or UHF channels. As previously discussed, most modern converters accept both local channel and wireless inputs thereby avoiding the inconvenience of an A/B selector switch.

In general, the received signal level should in most cases be adequate to provide service to two or three TV sets within the residence. Where multiple additional sets are requested, a low gain (10 dB) distribution amplifier may be required to provide an adequate signal power to all television sets.

Multiple Dwelling Residences

In apartment buildings, condos or hotels the primary requirement is provision of an internal cable distribution system serving all units within the multiple dwelling residence. A cost effective approach for the wireless cable operator is, wherever possible, to utilize an existing master antenna TV system in order to distribute the wireless signal to subscribers. Where MATV systems do not exist or cannot be used by the operator, an internal or external wired system must be constructed at the user site.

Several low-cost and effective wireless SMATV headend designs can be used (see Figures 12-42, 12-43, 12-44, 12-45 and 12-46). These have substantially lower costs than those that incorporate reception and processing of satellite television signals and therefore allow multi-unit wireless / SMATV systems to be quite competitive. In addition, multi-family wireless installation are also generally about 40% less costly than single family sites.

Subscribers in multi-unit situations are generally billed according to one of two methods of: discreet service and bulk packaging. The primary difference between discreet service and bulk delivery systems is in the way service is obtained from the wireless operator. In discreet service systems, the operator provides the wireless channels in combination with local VHF and UHF channels and inserts them into

WIRELESS CABLE COMPONENTS

the central distribution system. Any resident desiring to purchase the wireless service contacts the operator directly and thus subscribes to the service. The installation required for this type of subscriber is limited to the connecting appropriate converter(s) to a subscriber's television.

In bulk service delivery, the owner of the multi-dwelling facility (such as hotels, motels, etc.) would purchase some or all of the wireless programming and deliver it to all residences within the complex. With this method, all purchased channels would be descrambled at the building central processing point for insertion into the distribution system.

Figure 12-42. A Multi-Receive Antenna Installation. *Three separate antennas and downconverters here feed three individual units. This could be avoided by using the design shown in Figure 12-43. (Photo Steve Berkoff)*

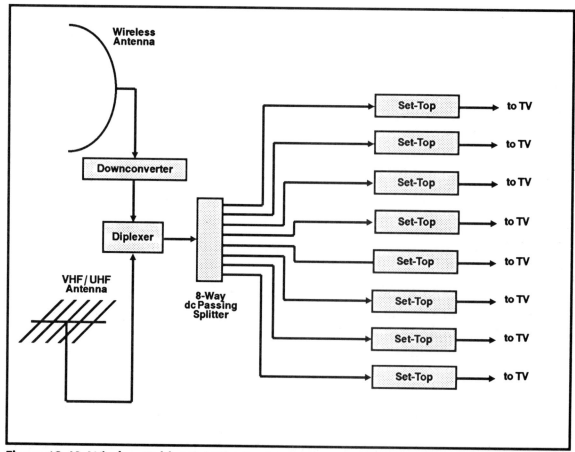

Figure 12-43. Wireless Cable Mini-Distribution System. *A low cost wireless distribution system can be created by use of an amplifying splitter that passes dc power of all ports. This device allows any one of the connected converters to power the downconverter and avoids the problems associated with installing wireless in smaller multi-unit systems as illustrated in Figure 12-41.*

WIRELESS CABLE COMPONENTS

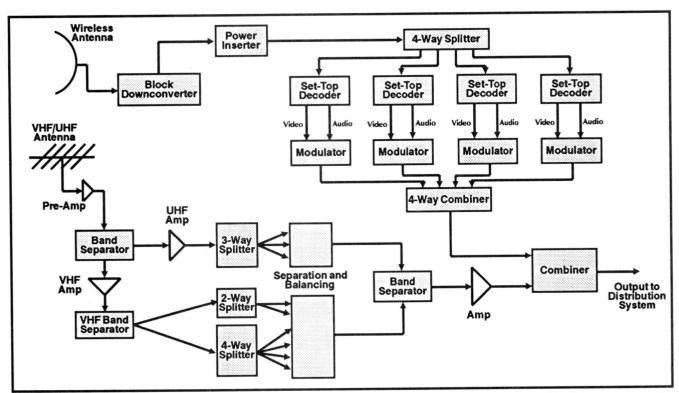

Figure 12-44. Four-Channel Wireless / Nine-Channel Off-Air Bulk Billing Receive Site. *This illustrates a simple, low-cost wireless headend that feeds a small distribution system. Four wireless set-top decoders feed their signals into four adjacent channel modulators. The wireless signals are combined with three UHF and six VHF off-air channels. The output signal passes through the distribution system to standard cable set-top converters.*

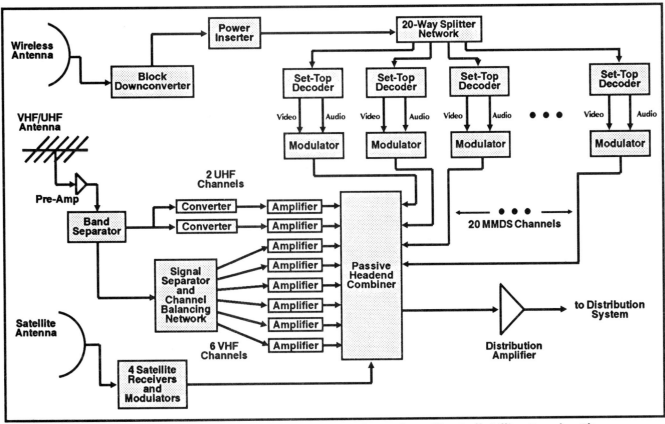

Figure 12-45. 20-Channel Wireless / 8-Channel Off-Air/4-Channel Satellite Bulk Billing Receive Site.
This illustrates a simple, low-cost wireless headend that feeds a small distribution system. Four wireless set-top decoders feed their signals into four adjacent channel modulators. The wireless signals are combined with two UHF and six VHF off-air channels. The output signal passes through the distribution system to standard cable set-top

WIRELESS CABLE COMPONENTS

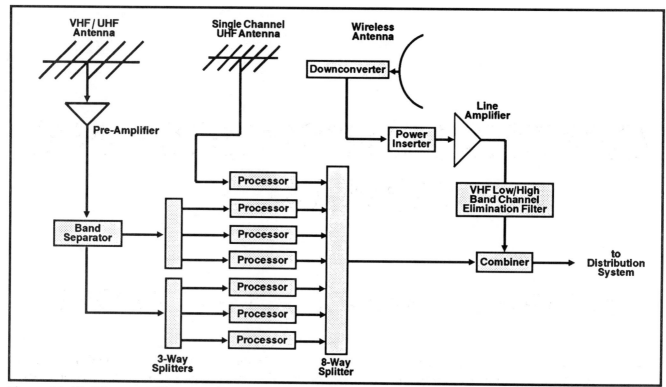

Figure 12-45. Discrete Billing Wireless Headend with One-Channel UHF and Six-Channel VHF. *This headend is designed to receive both wireless and off-air channels. In this case the wireless channels are passed to a set-top converter in each subscriber's residence. Billing is based on an individual subscriber basis. Any customer not wishing to receive the wireless channels simply does not receive a set-top. Separate UHF and VHF antennas are incorporated into this system.*

F. VIEWER QUALITY ASSESSMENT and S/N RATIO

The relationship between viewer quality assessments and S/N ratio was developed by the landmark 1959 TASO (Technical Allocation Study Organization) studies as well as the National Cable Television Association study in the United States. These results, which have been discussed earlier in Chapter 4, have been fundamental to the development of the NTSC television system used in North American and other countries. They are considered to be directly applicable to both PAL and SECAM formatted broadcasts.

It is generally accepted that a 45 dB S/N ratio provides an excellent quality signal in the opinion of a very high percentage of viewers. In terms of modern, urban cable television transmission in North America, it represents the very best that coaxial cable technology generally offers, even when using feedforward amplifiers and high quality headends. For example, the standard employed by the Canadian Department of Communications for cable systems is an S/N ratio of a 40 dB minimum using identical measurement references.

CHAPTER 13. WIRELESS SIGNAL COVERAGE

A. SIGNAL COVERAGE and LIMITATIONS

Wireless cable is a "line-of-sight" broadcast service generally limited in range to approximately 50 km from the transmitting site. Reception of these signals requires a relatively unobstructed transmission path because microwaves can be partially absorbed or attenuated by trees, buildings, hills, dense vegetation or other obstructions. Consequently, a installation completed in winter may suffer from picture deterioration in the spring when the trees in the path between the transmit and receive antennas grow new leaves.

Wireless pictures may also suffer from "ghosting". Ghosts result when multiple signals arrive at the receive antenna within microseconds of each other. One signal arrives directly from the transmitter, as intended, the other after having bounced off nearby buildings, structures or hills. Ghost may be eliminated by moving the antenna to a different location or by using a better quality dish or both. If a larger receive antenna with a more narrow beamwidth is installed, it is more capable of ignoring extraneous off-axis signals.

Transmissions from other sources can also interfere with reception of wireless cable signals. Amateur radio transmitters or other point-to-point microwave links operating in or close to the wireless band may interfere with reception, especially when poor quality downconverters are employed. Even microwave ovens that operate in the 2,450 ±50 MHz band, can also interfere when low quality downconverters are installed.

A preliminary analysis of the signal propagation potential within a given area can be accomplished by plotting the wireless signal contour on a topographical map to identify problem areas (see Figure 13-1). If significant portions of the market cannot be served, a change from the proposed or existing transmitting site or use of a repeater can be considered. Following this preliminary analysis, the next step is to conduct signal surveys throughout the proposed service area to confirm and/or identify areas of low signal power. This process is both beneficial and economical to the operation of a system because technicians would not be dispatched to poor or zero signal areas until "beam benders" or local cabling were installed.

The performance of a wireless system follows the basic laws of physics. Mathematical equations can be used to predict the signal power received at downconverter inputs. System performance measurements outlined in more detail in Appendix B. The S/N ratio can be calculated from factors such as:

- transmitter power and frequency
- power loss in components such as the feedline, combiner and diplexer
- transmitter antenna gain, height and location
- free space path loss in transmission between the source and end user
- receive antenna gain
- receiver cable loss
- downconverter noise figure
- ambient noise at receiver
- signal bandwidth

WIRELESS SIGNAL COVERAGE

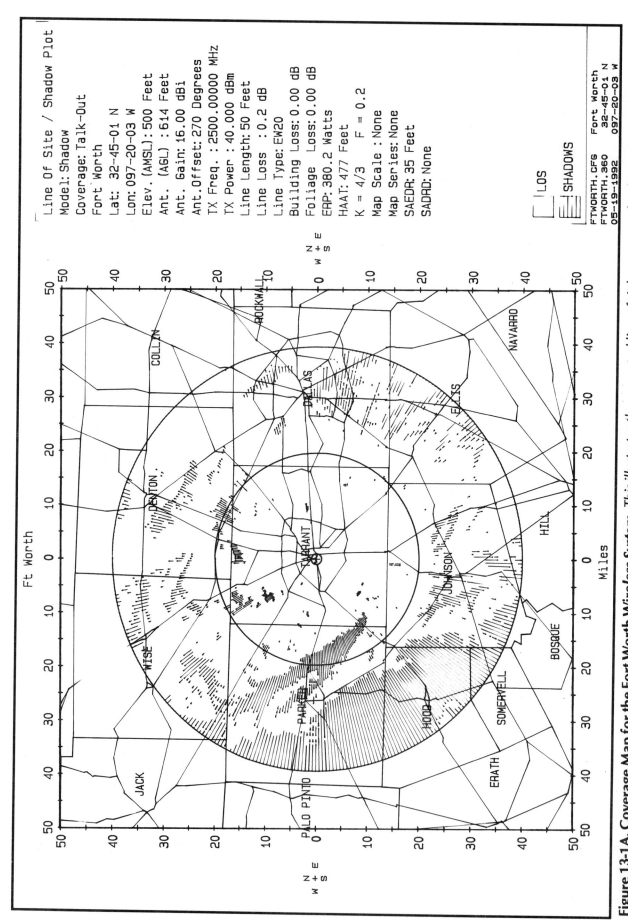

Figure 13-1A. Coverage Map for the Fort Worth Wireless System. *This illustrates the expected line of sight coverage for this Ft. Worth, Texas market area. 360 radial lines are projected from the transmit site. All lines shown in black are shaded areas; the white or opened spaces represent clear areas for reception. The counties and main roads are also shown to aid in identifying specific areas. (Courtesy of Chris Holman, AWS)*

WIRELESS SIGNAL COVERAGE

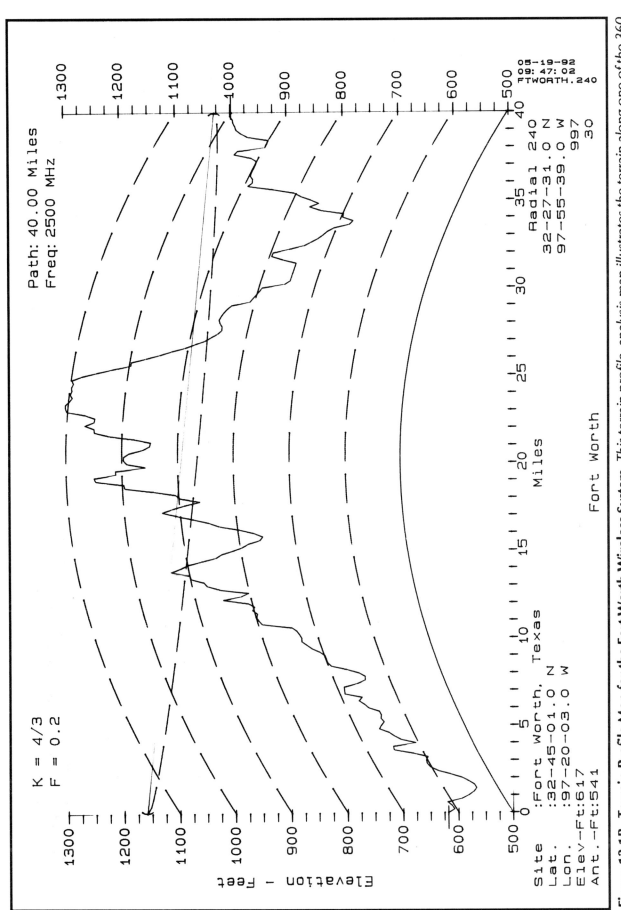

Figure 13-1B. Terrain Profile Map for the Fort Worth Wireless System. *This terrain profile analysis map illustrates the terrain along one of the 360 radials, #240 at 240 degrees, shown in Figure 13-1A. The scale on the left shows the elevation of the transmitter; the one on the right the receive antenna. In this case the transmitter is at a 1,170 foot elevation. A straight line drawn to the receive antenna, here at 40 miles and a 1,020 foot elevation, clearly shows the intervening obstacles. The curved dashed line shows the required Fresnel zone clearance. (Courtesy of Chris Holman, AWS)*

WIRELESS SIGNAL COVERAGE

Alternatively the size of the receive antenna and downconverter noise figure can be determined as a function of the required S/N ratio and other system variables. Of course, the validity of these results is dependent upon the accuracy of the information used in these calculations.

Within a given wireless system, signal level is a function of the distance from the transmitter to a receive site, the terrain between these two points, receive antenna gain and downconverter noise figure. For an existing system there are, for all practical purposes, usually only two variables that can be varied in determining received S/N ratios: downconverter noise figure and receive antenna size. Once the output S/N ratio has been chosen, the appropriate combination of antenna and downconverter to achieve this result can be selected. Optimally the S/N ratio should exceed 46 dB at the downconverter output. In most cases throughout a service area, attaining this signal-to-noise ratio requires a received signal level of 0 to +10 dBmV, approximately 1.0 to 3.1 millivolts. The variation of picture quality with S/N ratio is discussed later in this Chapter.

There is a common misperception that the higher the downconverter output signal level, the better the picture quality. In reality, downconverter noise figure and received signal level at the downconverter input determine the S/N ratio that sets the ultimate limit on picture quality. A higher gain downconverter generates more signal to drive the cable and the equipment between the converter and TV. However, unduly high signal levels can overdrive the output stage(s) in a downconverter and worsen picture quality.

B. COVERAGE PREDICTIONS

Service Area Coverage

The first step in an evaluation of a wireless cable service area is to construct a shadow zone map – a computer generated projection of the line of site reception in the proposed service area. Preliminary calculations, that are based on transmit tower height, receive site height and terrain elevations, should indicate the transmitter power that is adequate to provide coverage for the intended market. The shadow zone map highlights areas where reception will most likely be blocked by intervening terrain but does not identify potential blockage due to smaller structures such as trees and similar unpredictable obstructions. If the proposed service area has a high concentration of tall trees, this should be factored into estimates of overall coverage area. A rough assessment of topographical maps should also suggest the number of beam benders or, if permitted, repeaters that would be necessary to meet coverage goals. This preliminary conclusion will later be verified by the detailed site design process.

General Coverage Calculations

The theoretical maximum distance that a signal can travel to achieve a given S/N ratio can be calculated for each combination of transmitter power, transmitting antenna gain, receive downconverter noise figure and receive antenna gain.

The theoretical maximum path length versus S/N ratio for two transmitter powers, 100 and 50 watts, is listed in Table 13-1. Among the assumptions made in arriving at these figures are:

- Cable loss to transmitting antenna = 3 dB
- Transmit antenna gain = 15 dBi
- No loss in cable at receive dish
- Downconverter/antenna noise figure 7.5 dB
- Operating frequency = 2.59 GHz

However, path lengths in excess of 80 kilometers are rarely or never achieved for a number of reasons. First, the curvature of the earth limits the range of a transmitting antenna. Even if the transmitter were sited at 330 meters above ground level, subscribers beyond 80 kilometers would be below the horizon relative to the transmitter. Second, factors such as atmospheric absorption of signals and losses outlined below also limit range.

For comparison purpose, Figure 13-2 shows the expected coverage obtained using 15, 18 and 21 dBi gain receive antennas receiving signals from a 10 watt transmitter. At 45 dB S/N ratio picture quality should be excellent; at 40 dB S/N ratio the picture is marginally degraded and visible noise is just begin-

TABLE 13-1. MAXIMUM THEORETICAL PATH LENGTH (km) versus RECEIVE ANTENNA GAIN and S/N RATIO

S/N Ratio	Transmitter Power = 100 watts Receive Antenna Gain (dBi)			Transmitter Power = 50 watts Receive Antenna Gain (dBi)		
	24	21	18	24	21	18
35	303	214	150	189	134	95
36	270	191	135	168	119	84
37	240	170	120	150	106	75
38	214	152	107	134	95	67
39	191	135	96	119	84	60
40	170	120	85	106	75	53
41	152	107	76	95	67	47
42	135	96	68	84	60	42
43	120	85	60	75	53	38
44	107	76	54	67	47	34
45	96	68	48	60	42	30
46	85	60	43	53	38	27
47	76	54	38	47	33	24
48	68	48	34	42	30	21
49	60	43	30	38	27	19
50	54	38	27	33	23	17
51	48	34	24	30	21	15
52	43	30	21	27	19	13
53	38	27	19	23	17	12
54	34	24	17	21	15	11
55	30	21	15	19	13	10
56	27	19	14	17	12	8
57	24	17	12	15	11	8
58	21	15	11	13	10	7
59	19	14	10	12	8	6
60	17	12	9	11	8	5

ning to be perceptible. This degradation would not be annoying to most viewers.

An additional "grazing" loss indicates the reduction in coverage resulting from specific path losses over and above normal free space path losses. These additional losses usually result from insufficient path clearance over hills, trees or other obstacles such as adjacent buildings. Note that although free space path losses can be counteracted by increasing antenna size and receiver amplification, noise introduced externally and within electronic components eventually becomes a limiting factor.

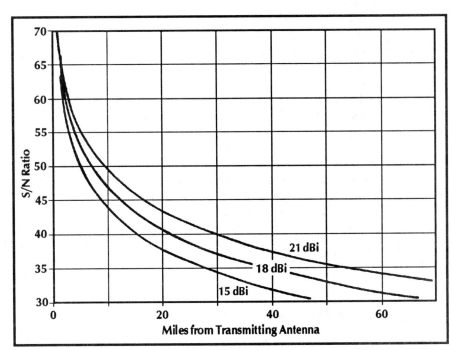

Figure 13-2. S/N Ratio versus Distance from Transmitting Antenna. *This graph shows how S/N ratio decreases with distance from a 10 watt transmitting antenna. Most wireless operator attempt to provide at least a 46 dB S/N to all subscribers.*

WIRELESS SIGNAL COVERAGE

TABLE 13-2. WIRELESS COVERAGE PREDICTION
(distances in kilometers)

Grazing Losses (dB)	Effective Radiated Power (dBw)					
	6 dBw		9 dBw		12 dBw	
	A	B	A	B	A	B
0	3.5	6.2	5.0	8.8	7.0	12.5
3	2.5	4.4	4.5	6.2	5.0	8.8
6	1.8	2.8	2.5	4.4	3.5	6.2
9	1.0	2.2	1.8	3.1	2.5	4.4

NOTES
A is 40 dBm contour; B is 45 dBm contour
6, 9 and 12 dBw equivalent to 4, 8 and 16 w transmitter power

Even when a subscriber can "see" the transmitter there must be more than just marginal clearances over any obstructions or "Fresnel" losses occur. Due to a characteristic of the microwave radiation, a minimum distance, known as the path clearance or Fresnel Zone clearance, must always exist between the boresight of the radio beam and any obstruction if additional path losses are to be avoided.

Table 13-2 shows the reduction in coverage resulting from path losses over and above normal free space path losses. For example, a clearance of 18 meters must be maintained at mid-path on a 30 kilometer path, if the free space path condition is to be realized. If the actual clearance is lower, supplementary losses will be incurred even if the path remains unobstructed. The additional loss may be as low as 1 dB or as high as 20 dB.

Therefore, a conservative approach should be taken in planning the path of an wireless signal. Path calculations should include an allowance for possible additional losses due to grazing. A 6 dB margin or grazing factor should be included in all wireless coverage predictions.

In summary, the coverage area of a wireless system depends upon four variables. First, a clear line of sight between the transmitter and receiving antenna must exist. This mandates high placement of the transmit antenna. Second as transmitter power increases, so does range. At the subscriber site, choosing a low noise downconverter and a high gain antenna also effectively increases range. The economic tradeoffs between these strategies depend upon the expected number of subscribers and the housing demographics. For example, if the subscriber count will be high, a sensible strategy would be to increase transmitter power and height. However, if most subscribers lived in multi-unit housing it might be feasible to use a low transmitter power in conjunction with high gain receive antennas and SMATV distribution systems.

C. EXTENDED RECEPTION TECHNIQUES

Blocked Areas

Reception of wireless signals can be totally blocked by a high rise building, hill, mountain or other physical obstruction. When this occurs, a signal cannot be received from the primary transmission source. Correcting this situation can require significant changes to the receiving configuration.

Shadow Areas

The effect on reception in a "shadow" area is quite visible. A partial blockage can occur for a comparatively small percentage of time when unusual weather conditions exist. Blockages may also occur randomly from point to point within an area. For example, a region of small rolling hills is a shadow area because while reception would be excellent at the top of the hills, it disappears in the intervening valleys. In such shadow areas raising the receiving antenna support mast or even moving the antenna a relatively small distance may markedly improve signal reception. Alternatively locating the transmitting antenna on a nearby mountain would overcome shadowing. In this sense, shadowed areas differ from blocked areas.

Localized Blocked and Shadow Areas

Within the service contour of a wireless transmitter a significant percentage of standard receive antenna locations might be either blocked or shadowed. Since it is not feasible or necessary to precisely identify such localized areas in advance, these are not generally noted on service contour maps. However, major blockages or shadow areas that might affect comparatively large areas are identified during a detailed engineering service contour analysis. This process includes field visits, multi-radial computer analysis and service area topographical digitization calculations.

Because of the line-of-sight nature of wireless microwave signals, the presence of a single building, house or fully-leafed tree can create a blockage at an isolated location. While such areas meet the definition of being blocked or shadowed, it is usually possible to offer service by using one or more augmented or extended reception techniques. A summary of such reception techniques that can be used to eliminate blockages and/or shadow areas are presented below.

Repeaters or Beambenders©

Repeaters can be utilized to provide multichannel television service to blocked or screened areas. A beambender (a copyright term of California Ampifier) or booster is a high gain, low power, broadband amplifier capable of receiving a wireless signal from the regional transmitter, amplifying it to a level in the neighborhood of 5 dBm per carrier, and then re-transmitting the multichannel signal via a directional antenna into a well-blocked reception area (see Figures 13-3, 13-4 and 13-5). The blocked area should have more than 60 dB isolation from the main transmitter signal. The beambender antenna height is normally kept low to control pattern distri-

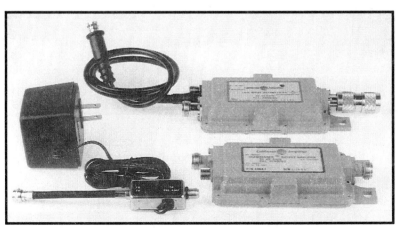

Figure 13-3. A Beambender© Kit. *These components include a low noise preamplifier (at top), a power transformer for inserting power into the line and an output amplifier (at bottom). Reception, amplification and retransmission are all accomplished in the 2.1 to 2.7 GHz microwave range. (Courtesy of California Amplifier)*

WIRELESS SIGNAL COVERAGE

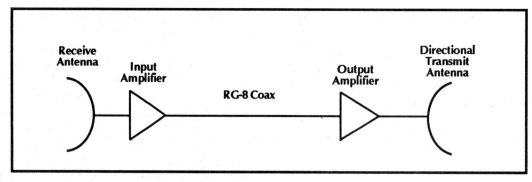

Figure 13-4. Beambender© Technical Configuration. *A beambender typically employs a receive antenna of 12 to 30 dB gain, a low noise preamplifier having a gain ranging from 10 to 30 dB, a short run of RG-8 coax, an output amplifier of approximately 35 dB gain and a transmit antenna of 20 to 30 dB gain. When the input signal is sufficiently strong, the input preamp may not be necessary.*

bution and eliminate possible interference with the central, higher power wireless transmitter.

Since the boosted signals are carried "on channel," they must be well isolated from the main signal. There cannot be any significant overlap of coverage between the beam bender and the regional transmitter without some potential for interference. In those cases where the blockage itself provides sufficient isolation, the beam bender has no opportunity of interfering signals from with the main transmitter. In areas of partial blockage or screening, the repeater can be aimed towards the main transmitter so that receive antennas detect only the repeater signals. Alternatively, signals can be re-broadcasted on a polarity orthogonal to or 90° rotated to that of the original signal. Note that in the United States the FCC does not allow changing polarities when using beambenders. Table 13-3 outlines one example of the minimum separation required for a transmitter on a 200 meter tower operating at 32 dBw (a 50 watt power radiated by an antenna having a 15 dBi gain) to have an undesired field strength of less than −20 dB at any location within its service area.

Repeaters are also used in situations where extension of service is required in a populated area, just outside or at the edge of the range of a central transmitter. Note that while beam benders can be installed to fill in blocked areas in any wireless system, in some countries including the United States, repeaters to extend coverage areas are not permitted.

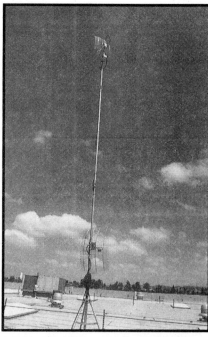

Figure 13-5. A Beambender© System. *These two photos show a beambender that is used to re-transmit signals into a shadow area. The lower horizontally polarized antenna receives the incoming signal while the upper one re-broadcasts vertically polarized signals into the shadowed area. The transmit antenna is oriented at 90° relative to the receive antenna to minimize co-channel reception problems. The electronic components are located under the weather-proof covering. (Photo Steve Berkoff)*

WIRELESS SIGNAL COVERAGE

Receiving Antenna Enhancements

Given that wireless receive antennas are small and lightweight, they can be easily and inexpensively elevated in situations where there is an isolated blockage due to a building or tree. Relocation of the receiving antenna to another location often may solve a blockage problem. When necessary a 3.2 cm O.D. mast can be employed to raise the antenna an additional 10 to 15 meters.

Local Cable Distribution Systems

Wireless cable is primarily intended to be a means of delivering multichannel television to residential households at a capital cost well below that of coaxial cable or fiber optics distribution systems. However, there are locations within a wireless cable service area where the local terrain or other factors make establishment of a local cable distribution system the most cost-effective means of providing reception and distribution. For example, a wireless cable signal can easily be received at one building and subsequently relayed by self-supporting messengered cable installed aerially to one or more adjacent structures. In such a case, just one receiving antenna and block downconverter would be utilized.

Higher Gain Receive Antennas

While a higher receive antenna gain cannot compensate for a blockage or for an unlucky location within a shadowed area, higher gain can be utilized to extend the range of reception significantly. Standard receive antennas typically range from 18 to 24 dBi gain; occasionally 28 dBi gain units are used. The difference between 24 and 28 dBi is equivalent to more than doubling the size of a parabolic antenna, or conversely, more than doubling the transmit power of the transmitter. Higher gain antennas can be effectively employed by buildings in a lower signal environment or in cases where more than one building would be served by a single antenna. This ensures reception of a sufficiently high output S/N ratio required to drive a small SMATV system. Larger diameter, higher gain antennas have higher forward gain, higher front-to-back ratio and reduced side lobes and can therefore be used to reduce interference while improving received S/N ratio.

Another option to using higher gain receive antennas in marginal and fringe areas is use of a preamp installed between the antenna and downconverter. This device provides extra gain increasing signal levels relayed to the downconverter but does not increase the S/N ratio and also adds some additional distortions.

These extended reception techniques can be used to serve blocked or shadowed areas. One or more economically viable methods to augment reception are generally available to a wireless operator. Nevertheless, there are a limited number of situations where reception can be obtain only by direct cabling from adjacent trouble-free areas. Of course, if buried cable construction is required for signal distribution, costs will be higher.

TABLE 13-3. MINIMUM ANTENNA SEPARATION
(50 watt Transmitter with 15 dBi Antenna Gain on 200 meter Tower)
(Undesired field strength less than −20 dB at any location)

Channel Type	Polarization	Offset	Minimum Separation (km)
Co-channel	Same	No	134
Co-channel	Same	Yes	111
Co-channel	Orthogonal	No	126
Co-channel	Orthogonal	Yes	104
Adjacent	Same	N/A	58
Adjacent	Orthogonal	N/A	54

© RONARD INDUSTRIES, INC. 6-70

RONARD

**Fax: (219) 872-6681
US & Canada
1-800-TRIPOD-4**

ANTENNA INSTALLATION HARDWARE

ENGINEERED FOR THE WIRELESS CABLE INDUSTRY

TRIPODS
2' - 3' - 5' - 10' — AVAILABLE WITH BOLT ON FEET FOR EASY SERVICING & REMOVAL WITHOUT DISTURBING ROOF
HIGH WIND LOAD TRI-MOUNTS — 5' - 10'

Y-TYPE CHIMNEY MOUNTS — WALL MOUNTS
4" - 6" - 8" - 12" - 18" - 24"

ADJUSTABLE EAVE - EAVE MOUNTS
RONARD EXCLUSIVE DESIGN
ADJUSTS FROM 45" - 60"

EXTRA HEAVY DUTY T.V. MASTING
HELPS ELIMINATE PICTURE FLUTTER
GUY WIRE
LAG SCREWS - GUY RINGS - GUY HOOKS - TURNBUCKLES

RONARD INDUSTRIES, INC.

P.O. Box 708 • Michigan City, Indiana 46360
US & CANADA 1-800-TRIPOD-4 • Fax (219) 872-6681

CHAPTER 14. RECEIVE-SITE INSTALLATION

One of the crucial ingredients to a successful wireless cable system is an excellent customer installation. Top quality workmanship is essential for two reasons. Each installer directly represents the wireless operator. Signal reception quality also depends upon the installation methods used. Furthermore, the technical nature of a wireless broadcast system requires that an operator make a substantial investment at end-user locations and installation funds should not be squandered.

The detailed methods to be followed to insure that equipment is properly installed and customers are well served are outlined in this chapter. All installation personnel should be familiar with the practices outlined below.

A. INSTALLATION PLANNING and SITE SURVEY

Each installation should be well planned before crews arrive at the site. This strategy minimizes both the amount of time spent in each customer's home and the inevitable disruption to the customer. Such planning includes determining the location of the receive antenna, choosing the cable entry as close as possible to the television set, minimizing the amount of cable used inside the customers home and pre-cutting all jumpers prior to installation.

Arrival at the Customer Site

Upon arrival at the customers home, the installer should first park the company vehicle in front of or as close to the building as possible but definitely not in the customer's driveway. Operator supplied safety cones should be placed two feet in front and back on the outside street corners of the vehicle. Always be aware of the traffic around this vehicle, especially when removing or replacing equipment and materials in the van.

The installer should then introduce himself, show a company photo ID to the customer and inform this customer of the intent to install the wireless television reception system. The next step is a detailed site survey to plan for the installation.

Site Survey

Although all sites would have been previously studied by a signal survey crew, a detailed, on-site survey is required prior to beginning any particular installation. This not only allows a technician to plan the job, but also results in a paper record of installation details. During the survey, all meter readings, components installed as well as the mast height should be recorded on the installation work order. The installer should also create a simple drawing of the antenna mast, guy wires and ground system placement on the back of the work order.

RECEIVE-SITE INSTALLATION

The first step of the site survey is to verify the signal level at the job site using a test kit that includes a lightweight, telescoping fiberglass push-up mast with attached wireless antenna and downconverter. The mast should have 5-foot sections that allow it to be raised as high as the allowable local height limit; the antenna gain should be an average of the antenna gains to be used in the system. In addition, either a battery-pack or extension cord should be used as a power source. When a battery pack is used, it is important to keep it fully charged because a low charge can cause a faulty reading. This test assembly should be supplied by the wireless operator. Under no circumstances should an installer use a metal push-up mast for survey work. If it were to topple during the survey and contact a power line, serious injury or even death could result.

This survey mast should be carried to the roof and set at points where the permanent mast could and might be installed. The required mast height can be found by first connecting the battery power supply and signal level meter and then by raising the mast in five foot increments while checking for the desired signal level. All points from roof level to the system height limitation should be checked.

While most set-top converters must have inputs in the 0 to 5 dBmV range for optimal performance, the most satisfactory method to ensure adequate performance is to achieve a S/N ratio of a minimum of 40 dB to 45 dB. S/N ratio can be calculated from downconverter (DC) gain in dB, downconverter noise figure also in dB and the signal level in dBmV read by a signal level meter as follows:

$$\text{S/N Ratio} = 59 - \text{DC Gain} - \text{DC Noise Figure} + \text{Signal Level}$$

The number 59 used in this equation is a constant relating to a quantity known as the noise floor. Also, once a standard test kit has been established, the DC gain and noise figure will be known quantities and calculating S/N ratio becomes a simple matter of adding one number to the signal level reading.

When reading the signal level an installer must be aware of physical factors that could affect system performance. Microwave signals are highly susceptible to absorption by trees and other foliage. Acceptable signal readings taken in winter behind trees that have lost their leaves may be reduced by as much as 12 dB in spring when they bloom. A wireless antenna must therefore not view the transmitter from behind a tree when a survey is performed in winter. Distant trees can also cause reception problems. While they may not be visible to an installer, but may still lie in the transmission path. If a technician notices signal fluctuations exceeding 2 dB during a survey, this may be an indication of distant foliage problems. Raising the mast an additional 10 feet or more may correct for such fluctuations.

Once a satisfactory reading has been obtained, raise the antenna an additional 5 feet to determine if signal level increases significantly. Signal gains exceeding 3 dB and therefore 3 dB increases in S/N ratio may be achieved by raising the mast by only 10 feet. Installing the receive antenna somewhat higher than required creates a margin for error in any problems are encountered at some point in the future.

If a signal level that is sufficient cannot be obtained, an installer should contact the central dispatch office to discuss the problem before notifying the customer. One solution may be use of a larger receive antenna. If a mast height greater than 35 feet is required, the installer should also contact dispatch to request additional manpower.

After verifying mast height and location, the placement of guy wires, the ground system, down-leads and interior cable runs should be determined. All pertinent data must be recorded in the appropriate section on the work order.

Customer Approval

Upon completion of the site survey, the customer must approve the proposed installation before the job begins. To this end, the installer must outline the installation to the customer and have the customer indicate his approval by signing the work order.

The technician should refer to the design sheet that has been prepared and walk the customer around the home explaining how the installation will be done. The customer should understand where the antenna will be installed, its height above the roof as well as the placement of all guy wires, grounding wires and down-leads into the home. The installer must show the customer where the entrance point will be drilled for the down-lead and must explain that a wall plate will be provided on the interior wall and that all openings will be weatherproofed against water leakage.

The installation process should begin only after the customer has approved by signing the design sheet.

B. MAST, ANTENNA and DOWNCONVERTER INSTALLATION

An antenna and its supporting mast must be be securely installed in a vertical orientation.

The first step in installing the mast and antenna system is to attach the baseplate to the customer's roof. If the mast is relatively short, ten feet or less, a standard roof mount baseplate, wall, saddle peak or chimney mount system can be used to hold it in place (see Figures 14-1 and 14-2). All installations using a mast of height ten feet or greater, with the exception of chimney mounts, must be secured with guy-wires in addition to any base mount structure that is utilized to attach it to the roof.

When attaching the baseplate to the roof, first line up the plate in its predetermined location and then drill pilot holes into the roof structure to accept the lag bolts for the base mount. All pilot holes must be drilled into rafters or other solid roof supporting structures. In no circumstance may antenna bases, down-guys or any other supporting structure be anchored into plywood or any other thin roofing structure. In addition, holes should never be drilled into a flat roof, as they can form a collection point for standing water to enter the customer's home. Antennas and masts must be side mounted on buildings with flat roofs.

Once pilot holes have been drilled, the weatherproof integrity of the mount system can be ensured by first filling all pilot holes with an approved roof sealant. Then a pitch pad should be placed above the pilot holes. The base is secured into the pilot holes with lag bolts. After the lag bolts are installed, additional roof sealer should be used to cover the lag bolts and the base plate structure to ensure a weatherproof enclosure. A similar procedure should also used for all guy-wire anchor points.

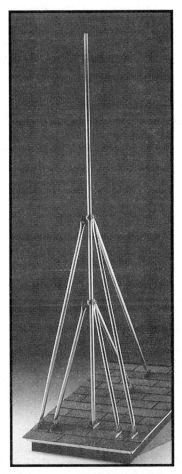

Figure 14-2. Universal Tripod Mount. *This mount universal mount conforms to any roof configuration and has 3-collar adjustment screws. It is available with 3-foot leg and a 5-foot mast or 5-foot legs and a 10-foot mast. (Courtesy of Ronard Industries, Inc.)*

Figure 14-1. Eave Mount. *This mount can be adjusted from 45 to 60 inches. A gusseted construction is used to accept a push-up mast. (Courtesy of Ronard Industries, Inc.)*

RECEIVE-SITE INSTALLATION

This base system can be used for tilting the mast up to a vertical position. In the case of a simple roof base, this is accomplished via the pivot that is incorporated into the base structure. The process of installing a tripod roof mount is outlined below in Figure 14-3.

When a tripod base is used to support the mast, it is important that the base is aligned so that two of the legs can be used as a pivot and the third as an anchor to erect the mast. Once the base has been secured, it can be laid on its side and so the mast can be inserted. The antenna and downconverter can then be secured to the top of the mast, or mast section in the case of a push-up mast. To secure it and ensure that the antenna will not rotate in the wind, drill a small hole through the antenna mount into one side of the mast. Insert a self tapping screw into this hole and cover it with sealant.

At this time prepare two downconverter leads one that will serve as a test point and the other as the customer's downlead and attach these to the downconverter. The test point lead should be coiled twice at the bottom of the mast and secured with tie wraps. Then fit a 75-ohm terminator or an F-81 barrel fitting to the end of the test lead. All of these connections, as well as the end connector leading from the antenna to the downconverter input, must be greased and weatherproofed using Electroseal tape or some other type of weatherproofing material. Make sure that the downconverter is oriented so that the F-connectors face downwards. This will prevent rain and moisture from other sources from entering and accumulating in the downconverter.

It is important to use only UV resistant materials and metal hardware when installing downconverters and attaching outdoor cables. Ordinary plastic wire ties or electrical tape will weaken and become brittle after just a few months of exposure to the elements.

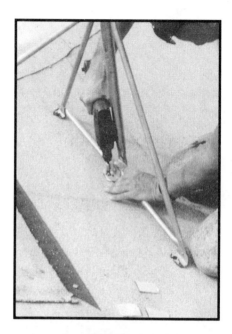

Figure 14-3. Installation of a Jiffy Mount. *Installation of a Jiffy mount, a short, self-contained tripod mast and base that is used on customer roofs, is shown here. In the top left photo the baseplate is attached. At the top right and bottom left the third leg is installed. The bottom right photo shows the completed installation with a 12 dB corner reflector mounted on top. (Photo by Steve Berkoff)*

RECEIVE-SITE INSTALLATION

Once the receive antenna and the downconverter have been secured to the mast and their leads have been attached, the tripod base structure can be tilted up and the third leg secured. When a simple base mount is used the guy-wires can be attached at this time.

When installing a push-up type of mast, raise one section at a time. As each section reaches is extended to its limit, insert the supplied cotter key into the side of the mast and secure that section. It is imperative that this key be placed in the mast to prevent it from spinning under a wind load once the installation has been completed.

With a push-up mast installation, a three foot tripod base is required for all assemblies up to 20 feet in height. A five foot tripod base is recommended for all installations up to 35 feet in height; a ten foot tripod based mount is recommended for push-up masts exceeding 35 feet (see Figure 14-4).

Just one person should be able to install any push-up mast on a five foot tripod base up to 35 feet in height. For masts exceeding this height, it is recommended than an additional helper be available to help secure and steady the guy-wires as the mast is raised into place.

Guy-Wires

After the push-up mast has been fully extended, the guy-wire system must be anchored to prevent the mast from bending in the wind. Guy-wires should be installed on each unprotected mast section that exceeds 10 foot in length. An unprotected section is one that is not secured by a wall mount, tripod, bipod, jiffy mast or trimast system.

Figure 14-4. 50-Foot Push Up Mount. *This 10-foot tripod base with an extremely high 50-foot mast section supports an antenna that is oriented to receive horizontally polarized signals. This set-up is relatively expensive, $100 or more for just the installation hardware. It would be used in deep-fringe applications or in areas with severe blockage by terrain. (Photo by Steve Berkoff)*

If at all possible all guy-wire anchors should be set into the side of the rafters overhanging the roof structure, instead of screwing the guy anchors through the roof membrane into the rafters. This strategy provides a secure grip for the guy anchors and minimizes their chance of being pulled out of the roof structure by the action of wind loads on the mast system. However, do not allow the wires to rub against the side of the house or roof structure. When setting guy anchors directly onto a roof, drill pilot holes and fill them with an approved roof sealant prior to securing the guy anchors into the these holes.

All guyed section masts up to 35 feet in height should have three guy-wire sections per ten foot length spaced evenly at 120 degrees from the other two (see Figure 14-5). Two of these guy-wires should be installed with a wire tightening device, such as a turnbuckle (see Figure 14-6). Guy-wires should be placed on every ten foot section of the mast and at its top. All masts exceeding 35 feet in height should have four guy-wires per section, spaced at 90 degrees from each other (see Figure 14-5 and 14-7). At least two turnbuckles are required on three guy mounts; three turnbuckles are specified for four guy mounts.

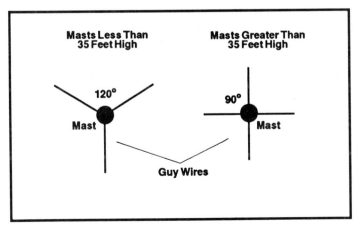

Figure 14-5. Guy-Wire Spacing. *Mast up to 35 feet in height should have three guy wires oriented at 120° from each other as shown in the left. Mast exceeding 35 feet in height should have four guy wires at 90° from each other as shown on the right..*

RECEIVE-SITE INSTALLATION

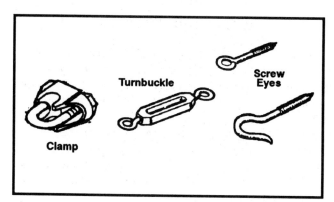

Figure 14-6. Guy Wire Hardware. *A turnbuckle, clamp and screw eyes are illustrated here*

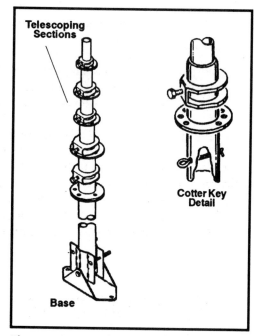

Figure 14-7. Push-Up Mast. *This figure illustrates how the telescoping sections of a push-up mast fit together as well as the cotter key detail at the base of this mast.*

Guy-wires should be run through the turnbuckles or roof hooks with two turns of the wire around the hook or turnbuckle loop. The guy-wire is then close wrapped against the loop for a minimum of ten turns around the down guy. It should then be covered with sealant tape to prevent corrosion. All turnbuckles should be secured to the rafter hook through direct mechanical connection without the use of guy-wires (see Figure 14-8).

As the mast base is being pushed up into place, all down leads should be secured to the mast with black wire ties spaced every 16 inches. No tape or other fastening structure should be used for this purpose.

In some instances, one or more guy-wires must be secured directly to the ground, rather than to a roof structure. In this case, a screw anchor is utilized to screw into the earth around the customer's home and secure the guy-wire to it. In this case, an installer must use the supplied yellow PVC caution tubes over the guy wires at the ground level. These tubes should extend upward to a height of approximately eight feet from the ground anchor over every guy-wire attached to a ground anchor. These serve to prevent the customer or any other person from accidentally running into the guy-wires and injuring themselves.

Antenna Polarity Orientation

Wireless transmission antennas broadcast either vertically or horizontally polarized signals. The polarity of the receive antenna must match that of the transmitter. Failure to do so can result in signal loss of up to 20 dB or more. Almost all wireless antennas can be installed to detect either sense of polarization. The orientation of the dipole mounted at the focus of the receive antenna is the determining fac-

Figure 14-8. Guy-Wire and Turnbuckle. *Turnbuckles should be directly connected to the screw eye hooks in the roof and then secured with a wire wrapped through and between them as illustrated here to prevent them from unraveling under wind loads. Note that the screw eye hooks are waterproofed. (Photo by Steve Berkoff)*

RECEIVE-SITE INSTALLATION

tor. Horizontally polarized signals can be received by "horizontally mounted" antennas; vertically polarized signals by "vertically mounted" antennas (see Figure 12-32).

In some cases, when both senses of polarizations are transmitted the receive antenna must be mounted at a 45° angle to receive signals of both polarities. However, when a dual-polarity orientation is used, signals are decreased in power by 3 dB.

Antenna Tuning

After the mast is installed and secured, a signal level meter can be used to tune the wireless antenna for maximal signal strength. This meter should be connected to the base of the antenna down-lead. During this procedure a separate power supply must be connected to the downconverter.

Rotate the antenna structure about the vertical axis until the signal meter indicates a peak reading on the channel being measured. When the signal level reaches a peak, secure the mast by tightening all of the bolts on the tripod or other base structure. It is important that all bolts be tightened and lug nuts secured to prevent the mast from rotating in the wind.

Off-Air Antennas

When an off-air antenna is used to receive off-air channels, it should be secured onto the mast structure after the wireless antenna has been tuned and secured in place. With the use of a step-ladder install this antenna as high as possible on the mast structure and then secure it to the mast, but do not tighten the bolts at this point. Next prepare and connect a down-lead to the coaxial connection on the antenna and weatherproof this attachment with an approved weather sealant.

First, connect a signal level meter to the down-lead and tune to a known off-air channel that will be received. Rotate the off-air antenna to maximize signal strength. When this orientation has been found, tighten all bolts down to prevent any further movement. As is the case with the wireless antenna, a small self taping screw installed through the mount and mast is used to prevent turning in the wind. The down-lead for this antenna should also be secured to the mast with wire ties located every 16 inches.

When a diplexer is employed to combine wireless and off-air signals, install it at the base of the mast with the same weatherproofing treatment used for all exterior fittings and connect it into the grounding system. When dual cable runs are installed, a dual ground block should be installed at the bottom of the mast and connected to the ground system.

C. GROUNDING and WEATHERPROOFING

Grounding

Once both wireless and off-air antenna and down guys have been installed, the grounding system for the mast and antennas must be prepared. While the establishment of a grounding system provides protection to subscriber electronic components, is will not prevent damage in the event of a direct hit. The best that can be expected is that the electrical charge would be carried to ground, not into the customer's home. In addition, a well-installed grounding system can prevent damage to components in the event of an electrical discharge from other faulty equipment within the home or from accidental contact with power conductors.

The mast, base and the down-leads from the receive antenna must be properly grounded. In addition, the run from the downconverter test point to the base of the mast should be terminated with an F-81 barrel and a 75 ohm terminator to close this conductive path. This affords the maximum level of protection against both exterior and interior electrical discharges. During storms and periods of high winds static charges can build up on the mast and guy wires. In a thunder storm this static build-up can act as an attractor and increase the risk of a lightning strike. A good ground path from the mast system can dissipate such static charges as they develop and thus reduce the risk of a hit.

RECEIVE-SITE INSTALLATION

The U.S. National Electric Code (NEC) requires that all coaxial cables potentially exposed to lightening, accidental contact with lightening-carrying conductors or power conductors operating a voltage that exceeds 300 volts must have their outer connective shield grounded as close to the entrance to the building premises as possible.

The ground point can be created by use of a ground block. This block can either be mounted at the entrance to the building or, in many cases, mounted at the base of the wireless antenna mast. Note that this latter strategy is not in accordance with NEC code.

The ground block should be connected to the ground point via a stranded or solid copper wire no smaller than #14 gauge but preferably #6 gauge that has a current carrying capacity at least as great as that of the coaxial cable conductor. The grounding conductor should be run from the ground block to the ground point in as straight a line as possible. Care should be taken to avoid any sharp bends in this line that can create an interruption in ground potential. Ground leads should be limited to at most 150 feet in length.

Existing ground points that can be utilized in many circumstances include the cold water pipes, the electrical service wire conduit or the power system ground (see Figure 14-9). In many areas, an existing ground system is available on new construction, as required by local electrical codes.

If no ground point exists, one must be created by driving a ground rod into the earth. The length of the ground rod should be determined by local soil conditions. When the wireless antenna is connected to a ground rod, NEC codes recommend using a jumper no smaller than #6 copper to connect the ground block to the ground rod. Keep the length of the wire between the ground rod and the grounding block as short as possible and avoid any sharp bends in the grounding path.

Weatherproofing

All connectors, N- and F-connector fittings, test points on the exterior of the home and penetrations into the home require weatherproofing. The damage caused by water leaks and the subsequent corrosion can be substantial.

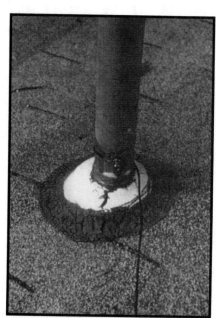

Figure 14-9. Grounding to an Existing Electrical Service Conduit Riser. *The left photo shows a ground clamp attached to the base of the antenna mount leg. The middle photo shows the grounding wire attached to the existing electrical service conduit riser. The photo at the right shows an expanded view of the riser. (Photo by Steve Berkoff)*

N-connectors installed on coax running from the antenna dipole to the downconverter must be weatherproofed to eliminate the possibility of ingress at these points. Silicone grease should be applied to the threads of the male portion of the N-connector before connecting it to the downconverter. Then ply-seal tape should be wrapped around the connection and over a portion of cable.

All exterior F-connectors require weatherproofing as well. When preparing the fittings for the F-connector, place a air-shrink boot on the cable prior to installation of the fitting. When connecting the fitting to an F-81 connection point, place a thin layer of teflon grease on the threads of the F-81 connector. Then thread the connector onto this point and tighten it, finger tight plus one quarter turn with the supplied torque wrench. The air-shrink boot is then placed over the fitting and allowed to set.

Weatherproofing practices should also be applied to all other potentially exposed electrical contacts. The grounding system at its connector points should be covered with a thin layer of teflon grease as well to help prolong the life of these connections and minimize corrosion due to ingress of moisture. Service technicians must also remember the importance of weatherproofing when servicing a system and to redo all weatherproof points that were opened for repair or testing purposes.

Cable entrances into the customer's home also provide a point for water ingress and require special attention. Wherever a cable enters via a hole drilled into a customer's home, the hole must be secured against water entry and potential damage. All points of entry should be fitted with cable plugs through which coax enters the customer's home. This plug should be secured and weatherproofed to the wall by RTV sealant. A wall plate and F-81 fitting is to be installed on the interior wall. In addition, the cable entry itself should have a 6" drip loop formed so that any water running down the cable will drip off the bottom of the loop and not into the customer's home (see Figure 14-10).

D. CABLE and CONVERTER INSTALLATION

Once the antenna, downconverter and mast have been installed, the next step is to run an RG-6 coaxial cable from the antenna into the customer's home.

Aesthetic considerations dictate that this wire should be concealed as fully as possible. Customers do not want to see a spiders web of cables hanging all over their homes To this end, the cable should be placed on a side or back wall and should never be routed down a front wall of a customer's home. If at all possible, the wire should follow the lines and contours of the house as well as facing structures that can serve to conceal its presence. Hide wires under eaves, along drain spouts and decorative trim. Run this line down to the entrance point as close as possible to the customer's TV set. All cable runs should be either vertical or horizontal. In no case should a wire be run diagonally from one point to the next. Given these constraints, keep the cable run as short as possible to reduce signal attenuation. Also, make sure that the cable bending radius is at least 10 times the diameter of the cable. Sharp kinks could altering its impedance and result in some signal reflections and additional attenuation.

Use plastic cable clips every 18 inches to secure the cable on wood and brick structures and metal clips on trailer sidings and other aluminum structures with the same 18 inch spacing. All clips installed brick walls should be attached only to the mortar, never into the brick itself (see Figure 14-10). All cable runs on stucco walls must be attached with metal clips and anchors drilled into the stucco. Under no circumstance should staples be used when installing coax because of their potential to damage the drop cable.

If the receive antenna is mounted on a garage or building not attached to the home, a portion of the drop may be routed underground. In this case, flooded cable that is designed for underground use can be installed at a minimum depth of 6 to 12 inches underground.

Mount all ground blocks, splitters and test points at outside locations that are convenient for service. Securely anchor all ground blocks and splitters with the supplied hardware. Do not use tie wraps to secure ground blocks to mast bases.

RECEIVE-SITE INSTALLATION

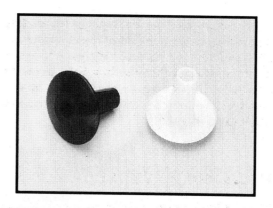

Figure 14-10. Cable Entry into Home. *This cable is routed along a brick wall. It is placed below the entry point to allow for creation of a drip loop to prevent water from entering the home. A wall plug shown in the top photo is inserted at the point of entry and three Roka clips hold the coax in place. These Roka clips are attached to the mortar, not to the brick. (Photo by Steve Berkoff)*

Drilling the Cable Entry

It is important to be aware of the potential damage that can be caused when drilling the cable entrance hole through a customer's wall. All holes should always be drilled from the inside out, only after verifying that structures such as gas meters, water pipes, or electrical conduit that could be damaged are not on the outside wall in the path of the drill. When drilling of the hole through the customer's wall is complete, it is important to clean up all shavings and dust that have been created.

Install a wall plug with a drip loop and then weatherproof the outside entry point (see Figure 14-10). The cable should tie into a a wall plate on the interior wall of the home (see Figure 14-12). Once the interior wall plate has been installed run a length of coax from this plate to the back of the converter located at the television set. In those cases where this run exceeds 6 feet, secure the cable by either attaching it to the baseboard with the required cable clips or by installing it under the carpet between the tack strip and the wall. This minimizes the possibility that the customer might trip over any loose wire and subsequently either injure himself or herself or damage the converter box.

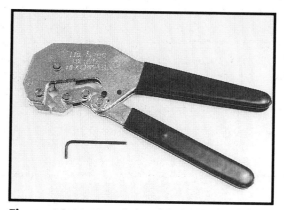

Figure 14-11. Coaxial Cable Crimp Tool. *This tool is used to crimp RG-6 and RG-59 coaxial cable.*

Fitting and Connectors

The majority of all cable system problems result from poorly made or corroded fittings and connectors. The use of proper tools and techniques can go a long way towards insuring the quality of every fitting in the system (see Figure 14-11)

When installing coaxial fittings, insure that no shield wires protrude past the ferrule case. Also make sure that the shield wire is not wrapped around the center conductor and that the center dielectric is installed flush with the inside of the fitting. Crimp the fitting with the appropriately sized tool and use the supplied torque wrench to tighten the fitting. It is wise practice to closely inspection each fitting as it is to eliminate future problems.

RECEIVE-SITE INSTALLATION

Figure 14-12. A Coax Wall Plate. *The coaxial cable that runs to the converter is screwed into this wall plate fitting.*

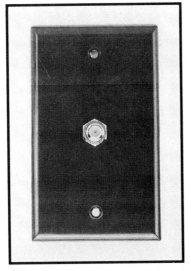

When running wire inside the customer's home, use best efforts to hide the wire from view. In no instance, should the wire be routed around doorways or across hallways or thresholds. Cables should not be installed under carpets where the customer can walk on the cable, for example, across the threshold of a doorway or hallway.

Connection to the Television Set

Connect the incoming cable to the set-top converter at the customer's TV set. Then connect the output of the set-top to the input of the television set. At customer request, the output can be routed through a VCR input. Then the output of the VCR can be connected into the TV set. Various configurations can be used to install a converter and VCR.

Once these connections have been completed, have the customer turn on the TV and VCR and verify their operation. The job supervisor should be advised of any problem with the customer's equipment before proceeding with the installation.

All wires behind the customer's set should be secured with wire ties in a neat and orderly fashion. Then plug the television into the back of the set-top box and the set-top box to a power receptacle.

After installing the converter, fine tune verify its operation. If a VCR is hooked up to the set-top, ensure that it is able to tune to various channels and is operating properly. At this point, document the location and operation of all set-top converters on a form that should be supplied by the system operator. Finally, close or open any doors or drapes who position may have been adjusted or changed during the course of the installation.

Under Carpet Installations

When installing cables under a carpets, use a pair of needle nose pliers to gently pull the carpet away from the wall and exposed the space between the tack strip and the wall (see Figure 14-13). Take care not to pull the carpet off of the tack strip. Then insert the cable into this space and push the carpet back over it and under the base board with the use of a wide putty knife.

Before attempting any wall or any under-carpet installation, thoroughly discuss the method to be used with the customer and clearly explain the potential damage that might occur to the carpet, especially on older and thread-worn rugs. Have the customer sign approval for under-carpet installation on the work order before beginning the job. Call the supervisor immediately if any damage occurs.

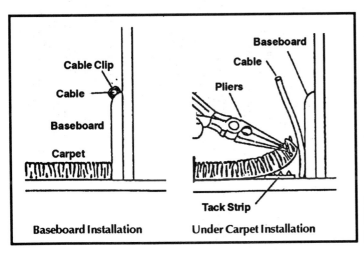

Figure 14-13. Baseboard and Under-Carpet Installation. *The details of both baseboard and under-carpet installation are illustrated in this diagram.*

289

RECEIVE-SITE INSTALLATION

If impulse pay-per-view technology is used, the next step is to hook-up the phone modem connection as outlined below in the telephone installation section.

Telephone Connections

Many of the set-top units designed for impulse pay-per-view operation require a two-way communications interface. This converter interface is via a built-in modem connected to a standard phone jack. The store-and-forward technology allows a customer to purchase movies and other one-time special events without having to call in an order and deal with an operator at the wireless business office.

A major inconvenience is often encountered when installing two-way interactive telephone connected systems. A telephone jack may not be located besides or near the TV outlet. This is rarely the case in most living rooms or bedrooms. In general, it is necessary to run a dedicated phone line from the closest telephone jack or junction point to an outlet behind the television set.

The two most common methods used to install such a telephone line is to either run it around the baseboard or to place it under the carpet. In many cases, a combination of these two techniques are required. Installing a telephone line is similar to running a standard coaxial cable. All routes and materials to be used must be anticipated. Of course, the location of the television set and converter must first be determined.

Once the route has been selected, measure the length of phone line needed and run the cable. Then prepare the connectors and install them at both ends. Finally, attach the line to the baseboard and around the door frames using either a staple gun or insulated cable clips. Once the line has been installed, plug one end into the converter and the other into the phone jack.

Most modern telephone equipment uses modular jacks identified as RJ11, RF14 and RF25 connectors. These are attached with a special purpose crimp tool that is available from many electronic suppliers such as Radio Shack.

To install telephone connectors, cut the ends of the line at right angles. Avoid any diagonal cuts. Insert the cable into the stripper section of the attachment tool and squeeze down on the handle while pulling out the cable. The action removes the correct length of insulation and leaves four leads black, red, green, and yellow ready to be connected to the modular plug. Then insert the plug into the fitting tool and the leads into the plug. Make sure the leads are correctly oriented in the connector. Squeeze the tool shut to secure the connector.

In some installations, a customer may request that the telephone line not be visible and not be routed around the baseboard and door frames. In such circumstances, if possible, install the line under a carpet. In many cases, an under-carpet installation will be faster and neater than a baseboard type installation. However, there is always the risk of damaging the carpet, especially when pulling older carpets away from the wall or baseboard tack strips. This potential for damage should be thoroughly explained to the customer before proceeding with the procedure.

During carpet installation, the carpet installer generally leaves a space of typically 1 to 2 inches between the tack strip that is used to secure to the carpet to the floor and the baseboard on the bottom of the wall. This is the space used to hide the telephone cable. Use a needle-nose pliers to pull the carpet up and away from the baseboard while taking care not to remove it from the tack-strip itself. Then lay the telephone cable into this empty space, push the carpet back over it and secure it under the baseboard using a wide-blade screw driver or putty knife. Then plug the telephone cable into the converter and into the exiting telephone jack in the same fashion as in a baseboard installation.

A duplex or two-way phone jack adapter can be used at the converter telephone jack. This has the advantage of allowing connection of both the new set-top line and the customer's existing telephone into the same jack.

Once the converter has been installed and is tied into the telephone line, as a final check verify converter operation using the protocol established by the set-top manufacture.

E. CUSTOMER EDUCATION

After completing the installation, both the installer and customer should verify converter operation and check the quality of all channels on the system. This information should be recorded on the work-order.

At this point the customer should be instructed in the operation of the converter. This should consist of a demonstration of the following. In all cases the installer should have the customer repeat the action to verify the instruction is effective.

- Use of the remote control to turn the set-top on and off.
- Use of the remote control for changing channels.
- Use of the remote control for changing volume levels on each individual channel.
- Use of the remote control to order an impulse pay-per-view event. One event should be ordered at this time for free viewing by the customer. After the customer demonstrates knowledge of this function, the television should be left tuned to this impulse function. This encourages the customer to order pay-per-view events that are available at later dates.

When the installation and demonstration are completed, any additional questions should be answered. It is crucial to spend the time necessary to answer any and all questions because service calls generated as a result of poor instruction by the installer are generally charged to the operator.

Additional Outlets

In some cases, a customer may require an additional outlet or outlets for televisions set(s) to be located in another portion of the home. These can be installed by splitting the down-lead into two or more components using a power passing splitter. Mount the splitter close to the ground block which is located near the entrance point of the main TV leads. Then connect an additional cable to this splitter point and runs this cable around the house to the second outlet using the prescribed clips or fasteners to secure the cable.

At the second cable entrance, drill a hole and install the cable with the accompanying wall plugs and drip loops as specified for the primary outlet installation. Then install the cable at the second outlet to either the baseboard or under the carpets in the space between the tack strip and the wall. Place the set top converter on the customer's set and connect it to this lead.

Clean-Up

After completion of all installation functions, it is the responsibility of the installer to clean-up the work area. All scrap cables, down guys and other installation materials must be removed. Any dirt or rock excavated for underground cable placement must be restored and cleaned-up. All debris caused by drilling through walls must be vacuumed. All cable wires and scraps created while preparing fittings must be removed.

The installer must also use a portable mini-vac to clean up small debris around the work area. As a final touch, wipe the face of the customer's TV screen with a soft rag to remove any accumulated dust. Following this procedure only on picture tubes, and never on a big screen TV as it may cause damage.

The job site must be left in the same condition in which it was found. It is the installer's responsibility to clean-up all trash and scrap created during the installation process.

Customer Sign-Off

When everything has been completed, have the customer sign the installation order. Then installer should call in the completed job to the dispatch center and verify that the next customer is home and available for installation.

RECEIVE-SITE INSTALLATION

F. SERVICE PROCEDURES and SAFETY

Although a wireless cable system is much more dependable than traditional CATV technology because a distribution network is not required, it does require occasional service. Some of the common service problems encountered in a wireless operation and their remedies are outlined here.

When a technician arrives at a customers home, it is important to remember that call has been made because a problem exists and that the customer is probably not happy with the service or the company at this moment. The technician must reassure the customer that the problem will be remedied and that the company is sorry for any resulting inconvenience. It is imperative that "people skills" be used to minimize the customer's aggravation at having service interrupted.

On every service call, each technician must completely check the condition of the installation including:

- Verify guy wire tension is sufficient
- Check that the mast is in a vertical orientation
- Verify the receive antenna is aligned to receive the correct sense of signal polarity
- Look for rust on mast system
- Insure that all down-lead attachments are in place and secure
- Inspect all weatherproofing
- Check the ground system
- Verify that all exterior cable runs are still attached at all clip points and that the wall plug is still in place
- Insure that all interior runs are still fastened down and secured

All of these items must be inspected on every service call and checked-off on the service inspection record. All additional service items encountered and/or repaired during the service call must also be documented on the service record. After completion of a service call, the technician must have the customer sign-off on the service record. As is the case with all installations, the service technician must call the dispatch office after every call, especially when customers do not keep appointments or when other problems arise.

Common Problems

The most common service problems encountered are:

Poor Fittings

Poor or corroded fittings are the most common cause of system failure. Improper installation techniques and poor weatherproofing are the most common source of fitting failure. Improperly operating fittings can cause problems ranging from a total system outage to low signal levels, ghosting and noise in the picture.

Whenever a service technician opens up an exterior fitting, it must be replaced with a new one and properly weatherproofed.

Operator Error

Error made by the customer can result in numerous service calls. This usually stems from an inadequate amount of time spent on education during the installation. This factor can be compounded by having additional electronic components such as VCRs and video games connected into the system.

Simple operational problems such as an unplugged converter or a television tuned to the wrong channel should have been identified by the customer service representative prior to the arrival of the service technician. If this has not been done, the technician should note this fact on the service record as well.

If the customer is unsure of the operation of the system, the service technician should demonstrate all items as outlined in the installation portion of this manual. After reviewing these items with the customer, the technician then should have the customer repeat each step to verify his understanding of the system and then complete the service record for customer signature.

Blown Power Supplies

If no signal is appears on any channel, a quick check with a volt ohm meter (VOM) can verify the status of the power supply. Although stray voltage resulting from power spikes and lightening storms can blow a power supply, the most common cause

RECEIVE-SITE INSTALLATION

occurs when a customer attempts to move a television set or hook up another line and inadvertently connects the power supply to a non-power passing device or unplugs it.

When this condition is encountered, install a new power supply with security sleeves around the power passing connectors, and document the repair in the service record. Also inform the customer what caused the problem and request that the lines are not disconnected again.

Spun Antennas

An antenna that has spun on its mast can result if the installation was inadequate. In this case, the mast must be lower to roof level and reinstalled properly. Merely rotating the base of the mast and then tightening it is not sufficient because the original installer may have forgotten to tighten the antenna itself or may have neglected to install the cotter keys in the mast sections.

Broken Masts

A broken mast is usually caused by improper guy wire placement or attachment. If the mast is bent but all wires are still attached, the guy wires are probably not spaced correctly around the base of the mast. If one of the guy wires has broken causing the mast to bend, the method of attachment or anchoring was not correct. In any case, a bent or broken mast is cause for complete replacement of the masting system.

Damaged Cable Drops

Low signal levels or poor pictures on some or all of the channels may indicate a damaged drop cable. Not all drop damage is evident from the outside. Some damage can be caused by water ingress into the cable. The best way for a technician to check for this condition is to measure the signal at the bottom of the mast and again at the input to the television set. If a signal differential of more than 6 dB per 100 feet is measured and if the fittings are apparently in good shape, this is an indication of a bad or damaged drop which should be replaced.

Drop damage can also be caused by pets or household maintenance and gardening activities. In many cases, it may a simple break that can be mended without replacement of the entire drop. However, as a rule, if the drop requires more than two new F-connectors or one length of new cable for repair, it should be replaced.

When installing joints and splices, a technician should place an air- shrink boot on each side of the F-81 barrel splice and allow it to set. If the drop requires replacement, the same type of cable and clips are required as in the initial installation. All cable repairs should be noted in the service record.

Blown Downconverters

Following a severe lightening storms, a service technician can expect to encounter damaged downconverters. These components may either be completely dead or may have an output level that is well below its installation specification. The status of the downconverter can be verified by attaching a signal level meter to the downconverter test point or test lead. This is the only point at which this test should be taken because other system problems could produce similar indications if tested at the wall plate or set-top box.

Water in the Antenna Dipole

Occasionally, water can accumulate in the dipole of the receive antenna. This often occurs after prolonged periods of severe storms. The most common indicator of this problem is adequate readings on the low channels with poor readings on the high channels as measured at the downconverter test point. If this problem is encountered, replace the entire dipole assembly because corrosion generally occurs in the unit once is has become wet.

Signal Blockage

If a system that had adequate levels at the time of installation starts to produce lower levels, this may be an indication of partial signal blockage. If the customer indicates that this problem has developed slowly over a period of time, then signal blockage should definitely be suspected.

A visual check with a pair of field glasses can uncover any obvious problems such as new tree growth or construction activity that may be blocking the signal. Fluctuations in signal levels may indicate a distant blockage problem.

RECEIVE-SITE INSTALLATION

In either case, the technician must move the antenna to a position where it can clearly receive the signal. First attempt to raise the antenna to clear the blockage. If this strategy is unsuccessful, use the fiberglass survey pole to determine if there is another location on the customers home where the signal can be received. If so, the then install a new antenna and mast system at this location. If the blockage cannot be overcome, the installer should contact the dispatch office and supervisor for further instruction.

G. SAFETY

The installation of wireless cable receive sites can create situations with the potential for injury to a technician. Safety should always be on the mind on the installer from leaving the shop until returning at the end of the day.

Most accidents occur when people act in a hurry. Concern for time becomes the most important item and shortcuts may be taken. Proper planning can help prevent accidents. Plan out the job thoroughly so that all required tools and materials are at hand when needed. Monitor the time spent at each job and notify the dispatch office of any excessive delays that may interfere with the completion of a subsequent job.

Safety is everyone's responsibility. Try to visualize the consequences of each risk taken. Imagine the bodily damage and the resulting pain that can result from taking an unnecessary risk. Then take the extra time required to perform your job in a safe manner. It will be time well spent.

Roof-Top Safety

Most wireless installations require that the antenna be roof mounted. This creates a potential for falls and serious injury as well as a possibility of electrocution from power lines and service drops.

Clearly, a technician must be constantly aware of safety at all times while on the roof-top. Always wear rubber soft sole shoes while working on a roof. These give the greatest traction and help to prevent slipping. When walking on roofs, stand upright and do not lean into the roof. Standing up straight increases the grip of shoes on a roof and helps to keep feet from slipping. Always be aware of the roof edge and of any power lines that are close by. It is all too easy to walk off of a roof top while preoccupied with work. Wet roofs can be extremely slick and dangerous. Never walk on a roof top that is slick with water. This is extremely important with steep pitched roofs.

Jobs on very steep pitched roofs should be handled only by a two man team. This makes it easier to handle tools, ladders and materials. In extreme cases, it may be necessary to use safety belts and lines.

Electrical Safety

The masts, guy wires and cables used in a wireless installation are highly conductive and can cause serious injury or death if they contact power lines. Always be aware of this potential hazard at all times and plan installation activity to minimize this risk.

All push up masts exceeding 30 feet in height require the help of a second technician to stabilize and prevent it from falling into a power line. Care must be taken with guy wires to prevent them from becoming loose during the installation and possibly hitting a power line. If the nearest power line is closer in feet than the height of the mast, call the supervisor for additional assistance.

Fiberglass ladders are the only kind that should be allowed on a job site. Aluminum ladders are highly conductive and can quickly electrocute an installer if they accidentally come in contact with a power line or service drop.

Service drops at the roofs edge are another potential safety hazard. If the service riser is being used for a ground point, the installer must take care not to touch or brush up against the drop. He must also avoid hitting the drop accidentally when climbing up or down any ladders near the drop.

Under no circumstance, should a technician attempt to install or service a system during a storm. In addition to the physical hazard of working on wet

RECEIVE-SITE INSTALLATION

and slippery roofs, there is a serious potential for electrocution from lightning strikes on the mast system.

Eye Safety

Eye damage is always painful and, in may cases, non-repairable. The best way to prevent eye injury is to wear safety glasses. It only takes a moment to put them on and the can prevent almost all types of accidents to the eyes.

Safety glasses should be worn at all times when working on a roof top. They will prevent damage from antenna elements, snapping guy wires as well as other sources. Glasses are also required when attaching cables to walls and drilling entry holes.

Night Operations

Under no circumstance should a technician attempt to install or service a system at night. Walking on roof-tops and raising masts in the dark are the quickest way to have a serious accident. If an installer is in the middle of a job and realizes that it will not be completed before dark, call the supervisor immediately. The only type of night work allowed is installation below roof level.

INTRODUCING

The PERFECT SOLUTION for small-scale, multi-unit wireless

All-Port Power Passing, Amplifying, 8-Way Splitter

3 dB gain on each port
Meets CATV specifications
Short-circuit isolation between ports
Surge protection on input and output
Passes dc power from each port to input
22 dB port-to-port and 50 dB input-to-output isolation
Encapsulated circuit enclosure for shock resistance & long life

CALL **303-449-4551**   or   FAX: **303-939-8720**

CHAPTER 15. SYSTEM PLANNING and OPERATION

A. THE BUSINESS PLAN

In order to accurately and completely evaluate the business opportunity of a proposed wireless system, it is crucial to research and write a business plan that outlines all aspects of the project. This document becomes the guide to developing the system and provides background information necessary to make sound business decisions.

A business plan is a blueprint for success. It details the opportunity at hand as well as the resources required to accomplish stated objectives. Creating a business plan is just as important as obtaining licenses or completing system designs or any other process associated with the overall plan. It is a document that is required by any potential investor.

A well conceived plan is a map to follow during business development. It also provides a method of control by establishing goals and expected costs. A well developed plan should address the following key points:

- Market size and subscriber potential
- System construction costs and financing
- Subscriber revenue and income

Market Study

The critical component of a wireless cable business plan is the market study, a reflection of total potential revenues. The key factors affecting total subscriber potential are numerous. Failure to properly evaluate them can and has often resulted in financial disaster.

This first step in market evaluation is the construction of a shadow zone map (see figure 15-1). This map is a computer projection of the line of site reception in the proposed area. It is based on tower height, receive site height and terrain topography.

This map highlights those areas that are blocked from receiving the wireless signal by intervening terrain. However, it does not identify blockage due to trees and other localized obstructions. If the proposed service area has a high concentration of tall trees, this will have to be factored into estimates of overall coverage area.

SYSTEM PLANNING AND OPERATION

After determining the coverage area, an estimate is made as to the total number of homes within that particular area. Local city governments, highway departments and planning commissions can provide detailed maps showing population concentrations. This data can then be compared to the shadow map and foliage assessment to develop an accurate estimate of the total number of homes in the area.

A competitive analysis is also crucial to determining ultimate customer potential. If the proposed service area is already served by cable television, the number of total potential customers would be greatly reduced. As a general rule of thumb, 10% of the television households in a cabled environment can be expected to subscribe to the wireless service. In contrast, in non-cabled areas, this penetration rate can be as high as 50%.

A detailed study is required if an existing cable service is present. In many areas, customers may be dissatisfied with the local CATV service and might be receptive to an alternative. The ability of wireless to provide comparable service at a reduced rate, combined with its inherently greater degree of reliability, can be quite attractive to many cable subscribers.

A package outlining channel line-up and pricing can be obtained by visiting the local cable office. A survey of the customer base then indicates by percentage the number of interested customers in the area. The local city council or city attorney's office may have a large collection of complaints registered against the cable company by angry citizens. All of this data can prove useful in preparing an educated estimate of the potential subscriber market.

A key to preparing the final subscriber estimate is to be conservative. Be aware that wireless cable does not require a huge customer base to be lucrative. A solid business plan, based on conservative estimates should be the bedrock upon which sound decisions are made and a strong business built.

Costs and Income

Once a good estimate of the subscriber potential has been completed, an accurate proforma that details expected revenue, costs and profits can be produced. This information then becomes the core of the business plan.

The business plan used as an example here in Table 15-1 details these items for a medium size system. The costs presented are relatively accurate and reflect current industry history. Note that they should not be used in the preparation of a final plan without contacting vendors directly to verify current prices.

A well developed financial analysis should contain a subscriber growth projection, income statement, cash flow analysis, and an analysis of construction and programming costs. Other factors such as inflation, churn, financing and related costs need to be included. Recruiting a qualified CPA or accountant to help in the preparation of this plan is a wise strategy.

Supporting Documentation

Upon completion of the financial projections, substantiating documentation should be presented in a standard business plan format including:

- Executive Summary
- Table of Contents
- Company Description
- Market Analysis
- Technology Description
- Operation Plan
- Review of Management and Ownership
- Organization and Personnel Structure
- Funds Required and Their Usage
- Financial Data
- Supporting Exhibits
- Press Articles

The complete business plan in presented at the end of this chapter. It utilizes this standard format in a simplified version to illustrate these points.

SYSTEM PLANNING AND OPERATION

TABLE 15-1. TEN YEAR FINANCIAL MODEL for a SAMPLE WIRELESS SYSTEM

ASSUMPTIONS FOR FINANCIAL PROJECTIONS:

1 SUBSCRIBER ANALYSIS

Total sub growth by year 5	10,000
Percentage of subs in year 1	30%
Percentage of subs in year 2	40%
Percentage of subs in year 3	15%
Percentage of subs in year 4	15%
Annual sub growth after year 4	15%

2 CHURN ANALYSIS

	YEARLY	MONTHLY
Addressable disconnects	2.50%	0.21%
Transfers	1.50%	0.13%
Moves	2.00%	0.17%
Non-pays	2.00%	0.17%
Other/dissatisfied	2.00%	0.17%
TOTAL CHURN	10%	0.83%
% OF EQUIPMENT LOST TO CHURN	15%	1.25%

3 INCOME ANALYSIS

SERVICE	RETAIL	MONTHLY COST	% OF SUBS
Basic only	$17.95	$3.50	23%
Basic plus one pay	$27.95	$9.00	80%
Basic plus two pays	$36.95	$14.50	40%
Basic plus three pays	$45.95	$20.00	20%
Additional outlet	$5.00	$0.00	50%
Pay-Per-View	$7.95	$4.50	13%
Primary outlet install	$49.95		
Secondary outlet install	$25.00		
Advertising revenue as % of gross	5%		
Copyright fees as % of basic service revenue		5%	
Annual price increase	5%	2.50%	

4 SALES AND MARKETING COSTS

% of commisioned sales	80%
Commision per sale	$25.00
Marketing cost per sub	$10.00

SYSTEM PLANNING AND OPERATION

ASSUMPTIONS FOR FINANCIAL PROJECTIONS:

5 PERSONNEL COSTS

	ANNUAL
General manager	$65,000
Chief tech	$40,000
2nd tech	$28,000
CSR	$18,000
Secretarial	$20,000
Payroll tax as salary %	7%
Benefits as salary %	15%
# of subs for each 2nd tech	5,000
# of subs for each CSR	3,000
Annual cost increase	5.00%

STAFFING ANALYSIS

	YEAR 1	YEAR 2	YEAR 3	YEAR 4	YEAR 5
General manager	1	1	1	1	1
Chief tech	1	1	1	1	1
2nd tech	1	1	1	1	2
CSR	2	3	3	3	4
Secretarial	1	2	2	2	2
ANNUAL PAYROLL	**$175,504**	**$234,290**	**$240,554**	**$245,940**	**$250,087**

	YEAR 6	YEAR 7	YEAR 8	YEAR 9	YEAR 10
General manager	1	1	1	1	1
Chief tech	1	1	1	1	1
2nd tech	2	2	2	2	2
CSR	4	4	4	4	4
Secretarial	2	2	2	2	2
ANNUAL PAYROLL	**$254,442**	**$259,014**	**$263,814**	**$268,855**	**$274,148**

ASSUMPTIONS FOR FINANCIAL PROJECTIONS:

6 BROADCAST COSTS

	YEARLY	MONTHLY
Transmitter site	$6,000.00	$500
Repair & maintenance – headend	3%	
Repair & maintenance – subscriber		3%
Vehicle operations – per vehicle	$8,000	$667
Utility costs – per channel	$240	$20
# of channels	31	
Total Utitlity cost	$7,440	$620
Channel lease cost		
MMDS	$4.80	$0.40
OFS	$1.80	$0.15
ITFS	$12.00	$1.00
Total channel lease costs	$18.60	$1.55

7 G&A COSTS

	YEARLY	MONTHLY
Billing cost per sub	$6	$0.50
Guides & postage cost per sub	$9	$0.75
Bad debt as % of revenue		3%
Collection fees–(reduces bad debt 50%	50%)	30%
Lost equipment as % of disconnects		15%
Insurance – as % of gross		5%
Management fee – as % of gross		0%
Attorney fees	$5,000	$417
Phone hardware	$5,000	$417
Network hardware	$5,000	$417
Leasehold improvements	$2,000	$167
Tools and test equipment	$10,000	$833
Office rent	$15,000	$1,250
Utitlities	$6,000	$500
Telephone	$20,000	$1,667
Warehouse	$15,000	$1,250
Dues and subscriptions	$2,000	$167
Travel and entertainment	$10,000	$833
Annual cost increase	2.00%	

SYSTEM PLANNING AND OPERATION

ASSUMPTIONS FOR FINANCIAL PROJECTIONS:

8 CAPITAL COSTS

A— INSTALLATION COSTS

SINGLE FAMILY HOME – Materials	OUTLET 1	OUTLET 2
Antenna, mount & cable	$65.00	$15.00
Down converter	$75.00	$0.00
Set top	$125.00	$125.00
TOTAL SFH MATERIALS	$265.00	$140.00

SINGLE FAMILY HOME – Labor	COST	% OF INSTALLS
10' mount	$60.00	30%
20' mount	$70.00	40%
30' mount	$80.00	20%
40' mount	$90.00	10%
Additional outlet	$30.00	
Disconnects	$25.00	

B— OTHER CAPITAL AND PRE–OPERATION EXPENDITURES

License acquisition costs	$0
Transmit station	$850,000
Furniture and fixtures	$5,000
Leasehold improvements	$5,000
Phone system	$30,000
Billing hardware	$25,000
Engineering	$10,000
Legal costs for organization	$5,000
Tools and test equipment	$25,000
TV's and VCR's	$2,000

9 EXIT PRICING DATA

Cashflow multiplier	10
System life – years	10
System sale price – per sub	$1,500

SYSTEM PLANNING AND OPERATION

PROJECTED CASHFLOW BUDGET

1 – REVENUE ANALYSIS

	YEAR 1	YEAR 2	YEAR 3	YEAR 4	YEAR 5
A – SUBSCRIBER PROJECTIONS					
Single Family Homes					
Beginning	0	2,802	6,146	6,686	7,150
Installed	3,000	4,000	1,500	1,500	1,073
Disconnected	198	656	960	1,036	715
ENDING	2,802	6,146	6,686	7,150	7,508
B – SYSTEM REVENUE					
Basic only	$65,360	$216,812	$317,064	$342,053	$354,227
Basic plus one pay	$353,991	$1,174,254	$1,717,223	$1,852,562	$1,918,495
Basic plus two pays	$233,989	$776,184	$1,135,088	$1,224,547	$1,268,129
Basic plus three pays	$145,491	$482,620	$705,782	$761,406	$788,505
Pay-Per-View	$16,362	$54,275	$79,372	$85,627	$88,675
Second outlet	$39,579	$131,290	$191,997	$207,129	$214,501
Annual subscriber rate increase revenue	$0	$141,772	$217,693	$235,409	$243,787
Primary outlet installation fee	$149,850	$199,800	$74,925	$74,925	$53,572
Secondary outlet installation fee	$75,000	$100,000	$37,500	$37,500	$26,813
Advertising revenue	$48,088	$161,381	$234,012	$252,261	$253,671
TOTAL REVENUE	$1,127,710	$3,438,387	$4,710,655	$5,073,419	$5,210,374

SYSTEM PLANNING AND OPERATION

PROJECTED CASHFLOW BUDGET

1 – REVENUE ANALYSIS

A – SUBSCRIBER PROJECTIONS

Single Family Homes	YEAR 6	YEAR 7	YEAR 8	YEAR 9	YEAR 10
Beginning	7,508	7,883	8,277	8,691	9,125
Installed	1,126	1,182	1,242	1,304	1,369
Disconnected	751	788	828	869	913
ENDING	7,883	8,277	8,691	9,125	9,582

B – SYSTEM REVENUE

	YEAR 6	YEAR 7	YEAR 8	YEAR 9	YEAR 10
Basic only	$371,938	$390,535	$410,062	$430,565	$452,093
Basic plus one pay	$2,014,420	$2,115,141	$2,220,898	$2,331,943	$2,448,540
Basic plus two pays	$1,331,535	$1,398,112	$1,468,018	$1,541,419	$1,618,490
Basic plus three pays	$827,930	$869,327	$912,793	$958,433	$1,006,354
Pay-Per-View	$93,108	$97,764	$102,652	$107,785	$113,174
Second outlet	$225,226	$236,487	$248,312	$260,727	$273,763
Annual subscriber rate increase revenue	$255,976	$268,775	$282,214	$296,325	$311,141
Primary outlet installation fee	$56,250	$59,063	$62,016	$65,117	$68,372
Secondary outlet installation fee	$28,153	$29,561	$31,039	$32,591	$34,220
Advertising revenue	$260,519	$273,253	$286,901	$301,245	$316,307
TOTAL REVENUE	$5,465,057	$5,738,018	$6,024,904	$6,326,148	$6,642,456

SYSTEM PLANNING AND OPERATION

PROJECTED CASHFLOW BUDGET

2 – EXPENSE ANALYSIS

	YEAR 1	YEAR 2	YEAR 3	YEAR 4	YEAR 5
A – PROGRAMMING COSTS					
Basic only	$12,744	$42,275	$61,823	$66,696	$69,069
Basic plus one pay	$113,986	$378,114	$552,952	$596,532	$617,762
Basic plus two pays	$91,822	$304,592	$445,434	$480,539	$497,642
Basic plus three pays	$63,326	$210,063	$307,196	$331,406	$343,201
Pay-Per-View	$9,261	$30,722	$44,927	$48,468	$50,193
Copyright fees	$637	$2,114	$3,091	$3,335	$3,453
Annual programming rate increase costs	$0	$24,197	$36,270	$39,153	$40,546
TOTAL PROGRAMMING COST	$291,778	$992,077	$1,451,693	$1,566,128	$1,621,868
B – DIRECT COSTS					
Transmitter site	$6,000	$6,000	$6,000	$6,000	$6,000
Total Utitlity cost	$7,440	$7,440	$7,440	$7,440	$7,440
Channel lease cost	$28,882	$86,582	$119,875	$129,140	$139,640
Guides & postage	$21,723	$68,774	$98,173	$105,845	$110,843
Annual direct cost rate increase	$0	$3,376	$4,722	$5,070	$5,386
TOTAL DIRECT COSTS	$64,045	$172,171	$236,210	$253,494	$269,309

	YEAR 1	YEAR 2	YEAR 3	YEAR 4	YEAR 5
TOTAL COST OF SALES	$355,823	$1,164,248	$1,687,904	$1,819,623	$1,891,177
GROSS MARGIN	$771,887	$2,274,138	$3,022,752	$3,253,796	$3,319,197

SYSTEM PLANNING AND OPERATION

PROJECTED CASHFLOW BUDGET

2 – EXPENSE ANALYSIS

A – PROGRAMMING COSTS

	YEAR 6	YEAR 7	YEAR 8	YEAR 9	YEAR 10
Basic only	$72,523	$76,149	$79,956	$83,954	$88,152
Basic plus one pay	$648,651	$681,083	$715,137	$750,894	$788,439
Basic plus two pays	$522,524	$548,650	$576,083	$604,887	$635,131
Basic plus three pays	$360,361	$378,379	$397,298	$417,163	$438,022
Pay-Per-View	$52,703	$55,338	$58,105	$61,010	$64,061
Copyright fees	$3,626	$3,807	$3,998	$4,198	$4,408
Annual programming rate increase costs	$42,573	$44,702	$46,937	$49,284	$51,748
TOTAL PROGRAMMING COST	**$1,702,961**	**$1,788,109**	**$1,877,515**	**$1,971,390**	**$2,069,960**

B – DIRECT COSTS

	YEAR 6	YEAR 7	YEAR 8	YEAR 9	YEAR 10
Transmitter site	$6,000	$6,000	$6,000	$6,000	$6,000
Total Utitlity cost	$7,440	$7,440	$7,440	$7,440	$7,440
Channel lease cost	$146,622	$153,953	$161,651	$169,733	$178,220
Guides & postage	$116,385	$122,205	$128,315	$134,731	$141,467
Annual direct cost rate increase	$5,642	$5,910	$6,192	$6,488	$6,798
TOTAL DIRECT COSTS	**$282,089**	**$295,508**	**$309,598**	**$324,392**	**$339,926**

	YEAR 6	YEAR 7	YEAR 8	YEAR 9	YEAR 10
TOTAL COST OF SALES	$1,985,050	$2,083,617	$2,187,112	$2,295,782	$2,409,886
GROSS MARGIN	$3,480,006	$3,654,401	$3,837,792	$4,030,366	$4,232,570

SYSTEM PLANNING AND OPERATION

PROJECTED CASHFLOW BUDGET

	YEAR 1	YEAR 2	YEAR 3	YEAR 4	YEAR 5
D – OPERATING EXPENSE					
Salaries and wages	$175,504	$234,290	$240,554	$245,940	$250,087
Employee benefits	$26,326	$35,143	$36,083	$36,891	$37,513
Payroll tax	$12,285	$16,400	$16,839	$17,216	$17,506
Marketing and advertising	$30,000	$40,000	$15,000	$15,000	$10,725
Sales commissions	$60,000	$80,000	$30,000	$30,000	$21,450
Bad debt	$11,578	$34,112	$45,341	$48,807	$49,788
Collection fees	$3,473	$10,234	$13,602	$14,642	$14,936
Disconnects	$4,947	$16,411	$24,000	$25,891	$17,875
Lost equipment	$7,866	$26,094	$38,159	$41,167	$28,421
Insurance	$38,594	$113,707	$151,138	$162,690	$165,960
Management fee	$0	$0	$0	$0	$0
Attorney fees	$5,000	$5,000	$5,000	$5,000	$5,000
Phone hardware	$5,000	$5,000	$5,000	$5,000	$5,000
Network hardware	$5,000	$5,000	$5,000	$5,000	$5,000
Leasehold improvements	$2,000	$2,000	$2,000	$2,000	$2,000
Tools and test equipment	$10,000	$10,000	$10,000	$10,000	$10,000
Repair & maintenance – headend	$1,930	$5,685	$7,557	$8,134	$8,298
Repair & maintenance – subscriber	$23,157	$68,224	$90,683	$97,614	$99,576
Vehicle operations	$20,483	$25,833	$26,697	$27,440	$27,440
Office rent	$15,000	$15,000	$15,000	$15,000	$15,000
Utitilities	$6,000	$6,000	$6,000	$6,000	$6,000
Telephone	$20,000	$20,000	$20,000	$20,000	$20,000
Warehouse	$15,000	$15,000	$15,000	$15,000	$15,000
Dues and subscriptions	$2,000	$2,000	$2,000	$2,000	$2,000
Travel and entertainment	$10,000	$10,000	$10,000	$10,000	$10,000
Annual cost increase	$0	$10,106	$10,807	$11,355	$10,805
TOTAL OPERATING EXPENSE	**$511,145**	**$811,240**	**$841,460**	**$877,787**	**$855,382**
NET INCOME	**$260,742**	**$1,462,899**	**$2,181,292**	**$2,376,009**	**$2,463,815**

SYSTEM PLANNING AND OPERATION

PROJECTED CASHFLOW BUDGET

	YEAR 6	YEAR 7	YEAR 8	YEAR 9	YEAR 10
D – OPERATING EXPENSE					
Salaries and wages	$254,442	$259,014	$263,814	$268,855	$274,148
Employee benefits	$38,166	$38,852	$39,572	$40,328	$41,122
Payroll tax	$17,811	$18,131	$18,467	$18,820	$19,190
Marketing and advertising	$11,261	$11,824	$12,416	$13,036	$13,688
Sales commissions	$22,523	$23,649	$24,831	$26,073	$27,376
Bad debt	$52,200	$54,816	$57,567	$60,455	$63,489
Collection fees	$15,660	$16,445	$17,270	$18,137	$19,047
Disconnects	$18,769	$19,707	$20,693	$21,727	$22,814
Lost equipment	$29,842	$31,335	$32,901	$34,546	$36,274
Insurance	$174,000	$182,720	$191,890	$201,518	$211,629
Management fee	$0	$0	$0	$0	$0
Attorney fees	$5,000	$5,000	$5,000	$5,000	$5,000
Phone hardware	$5,000	$5,000	$5,000	$5,000	$5,000
Network hardware	$5,000	$5,000	$5,000	$5,000	$5,000
Leasehold improvements	$2,000	$2,000	$2,000	$2,000	$2,000
Tools and test equipment	$10,000	$10,000	$10,000	$10,000	$10,000
Repair & maintenance – headend	$8,700	$9,136	$9,594	$10,076	$10,581
Repair & maintenance – subscriber	$104,400	$109,632	$115,134	$120,911	$126,977
Vehicle operations	$27,440	$27,440	$27,440	$27,440	$27,440
Office rent	$15,000	$15,000	$15,000	$15,000	$15,000
Utitilities	$6,000	$6,000	$6,000	$6,000	$6,000
Telephone	$20,000	$20,000	$20,000	$20,000	$20,000
Warehouse	$15,000	$15,000	$15,000	$15,000	$15,000
Dues and subscriptions	$2,000	$2,000	$2,000	$2,000	$2,000
Travel and entertainment	$10,000	$10,000	$10,000	$10,000	$10,000
Annual cost increase	$11,220	$11,667	$12,137	$12,631	$13,149
TOTAL OPERATING EXPENSE	$881,435	$909,368	$938,726	$969,554	$1,001,924
NET INCOME	$2,598,571	$2,745,033	$2,899,065	$3,060,812	$3,230,647

SYSTEM PLANNING AND OPERATION

PROJECTED CASHFLOW BUDGET

3 — CAPITAL EXPENDIDTURES

	YEAR 1	YEAR 2	YEAR 3	YEAR 4	YEAR 5
Initial capital expenditures	$957,000	$0	$0	$0	$0
Primary outlet – material	$795,000	$1,060,000	$397,500	$397,500	$284,214
Secondary outlet – material	$210,000	$280,000	$105,000	$105,000	$75,075
10' mounts	$54,000	$72,000	$27,000	$27,000	$19,305
20' mounts	$84,000	$112,000	$42,000	$42,000	$30,030
30' mounts	$48,000	$64,000	$24,000	$24,000	$17,160
40' mounts	$27,000	$36,000	$13,500	$13,500	$9,653
Additional outlets	$45,000	$60,000	$22,500	$22,500	$16,088
TOTAL CAPITAL EXPENDITURES	$2,220,000	$1,684,000	$631,500	$631,500	$451,524
NET OPERATING CASHFLOW	$260,742	$1,462,899	$2,181,292	$2,376,009	$2,463,815
YEARLY CASH REQUIREMENT	($1,959,258)	($221,101)	$1,549,792	$1,744,509	$2,012,291
CUMULATIVE CASH POSISTION	($1,959,258)	($2,180,359)	($630,566)	$1,113,942	$3,126,233

CASH OUT POSITION

	YEAR 1	YEAR 2	YEAR 3	YEAR 4	YEAR 5
NET OPERATING CASHFLOW MULTIPLIER	$2,607,425	$14,628,989	$21,812,922	$23,760,088	$24,638,149
PER SUBSCRIBER SALE	$4,203,160	$9,218,489	$10,028,509	$10,725,042	$11,261,294

SYSTEM PLANNING AND OPERATION

PROJECTED CASHFLOW BUDGET

3 – CAPITAL EXPENDIDTURES

	YEAR 6	YEAR 7	YEAR 8	YEAR 9	YEAR 10
Initial capital expenditures	$0	$0	$0	$0	$0
Primary outlet – material	$298,424	$313,345	$329,013	$345,463	$362,737
Secondary outlet – material	$78,829	$82,771	$86,909	$91,254	$95,817
10' mounts	$20,270	$21,284	$22,348	$23,465	$24,639
20' mounts	$31,532	$33,108	$34,764	$36,502	$38,327
30' mounts	$18,018	$18,919	$19,865	$20,858	$21,901
40' mounts	$10,135	$10,642	$11,174	$11,733	$12,319
Additional outlets	$16,892	$17,737	$18,623	$19,555	$20,532
TOTAL CAPITAL EXPENDITURES	$474,100	$497,805	$522,696	$548,831	$576,272

NET OPERATING CASHFLOW

	$2,598,571	$2,745,033	$2,899,065	$3,060,812	$3,230,647

YEARLY CASH REQUIREMENT

	$2,124,471	$2,247,227	$2,376,369	$2,511,981	$2,654,374

CUMULATIVE CASH POSISTION

	$5,250,704	$7,497,931	$9,874,300	$12,386,282	$15,040,656

CASH OUT POSITION

	YEAR 6	YEAR 7	YEAR 8	YEAR 9	YEAR 10
NET OPERATING CASHFLOW MULTIPLIER	$25,985,713	$27,450,326	$28,990,652	$30,608,119	$32,306,465
PER SUBSCRIBER SALE	$11,824,358	$12,415,576	$13,036,355	$13,688,173	$14,372,582

# SYSTEM PLANNING AND OPERATION

B. SYSTEM OPERATIONS PLANNING

The Operations Plan

Construction/Pre-Marketing Phase

The system construction and pre-launch activities should begin after all business planning is completed and investment funds and operating capital have been allocated.

The construction and/or preparation of the transmit station should be completed prior to equipment delivery. This insures that the installation, activation and proof of equipment performance are finished well in advance of launching the marketing effort. These activities should be completed under the supervision of the company's director of engineering or a consulting engineer and in concert with the appropriate equipment vendor or vendors.

During this initial period, field service technicians should be hired and trained to understand all aspects of the system. These employees should thereafter be used for various functions of system engineering and ongoing evaluations of signal availability and quality throughout the service area. In this initial phase vehicles, tools and test equipment should be acquired and configured for use in ongoing system maintenance.

Warehouse space, if possible contiguous to the central business offices, should be purchased or leased and equipped prior to the initial deliveries of subscriber reception equipment. At this time, Inventory control procedures should be in place and responsible personnel assigned prior to any materials deliveries. All necessary computer and telephone lines should also be installed and verified for proper operation prior to activation of the computer system. The company internal telephone system should also be installed in the central business office and all personnel should be trained in its operation.

The billing and addressing computer system should be installed early in this period in order to assure that both the software and hardware are performing accurately.

Customer service representatives (CSRs) should be hired and fully trained in all aspects of system operation during this time so they are fully prepared by system launch. These employees have the responsibility to build the system subscriber database, inform potential subscribers via telephone about available services, as well as receive and dispatch customer service orders. Their duties also include attending to customer billing inquiries and scheduling subscriber system installations.

Subscriber Installations

Both residential and multi-family wireless cable installations are similar to cable TV hook-ups with the major exception being the need to connect to an aerial versus an underground distribution system. Since wireless cable signals are distributed over-the-air, a short mast and a highly directional antenna replace the cable service drop. However installations in multi-family condominiums or apartments require more sophisticated attention to existing signal distribution systems.

A typical residential single-family unit installation should take approximately 1.5 to 2.5 hours to complete depending upon mast height requirements as well as the complexity of the particular installation. Multiple-outlet, VHF/UHF antenna or installations requiring a tall mast generally require additional time and charges to the subscriber. Residential installs should be completed by experienced, specialized in-house or contract labor. For the first 90 days of operation, and thereafter as determined by the director of engineering, all installations should be quality checked by company technicians for adherence to the following standards:

- Secure antenna and mast
- Proper system grounding
- Quality and tightness of cable connectors
- Overall neatness
- Reception quality
- Customer education about system use
- Customer satisfaction

Installations should be scheduled at each customer's convenience from 8 a.m. to 5 p.m. Monday through Saturday. These can be broken down into choices of A.M., P.M., all day, or first or last daily installation. It should certainly be company policy to meet every appointment so installations are completed on schedule. Nevertheless, some subscribers are not as punctual and may miss appointments, so procedures should be established for a timely rescheduling of installations. If a subscriber is not

SYSTEM PLANNING AND OPERATION

home when an installer arrives on schedule, attempts should be made to try at least one additional call back on that same day. Contract installers often have the advantage of not having constraints of an eight hour work day thus allowing more flexibility in making return calls and working late to complete installations.

In the event that all attempts to complete an installation on scheduled fail, CSRs should attempt to reschedule for the next available day. If this attempt also is not met with success, the work order must be returned to the sales department so the subscriber can be re-contacted to select another convenient installation date.

Upon completion of an installation, the subscriber must sign an installation work order. This confirms that all equipment was installed and received by the customer and that the system was left in good working order. If the system is not performing to the subscriber's satisfaction, the installer should be instructed to call the dispatch coordinator who in turn should immediately dispatch a service technician to the customer's residence.

The installation process should be designed to be as simple as possible. However, in order to satisfy each subscriber's desires, where practical and where time permits, additional work such as adding extra outlets and/or VCR installations may be necessary. This extra labor should result in an additional charge as determined by posted company rates.

Subscriber disconnects should generally be handled by company technical staff in order to oversee inventory control and reduce contract labor costs. Subscribers disconnecting from the system should be encouraged to return the installed equipment to company offices. Customers who return equipment should be given a receipt. Upon disconnect, arrangements should be made to retrieve the antenna and block downconverter, if required. When equipment is returned to the company, steps should be taken to refurbish and return these components to inventory. A completed disconnect work order then is passed on to the accounting department for customer account update; a copy is retained in subscriber files.

Customer Service

All technical service must be performed be qualified company technicians. These personnel should preferably have at least one year of experience in wireless cable operations and/or three years of CATV technical experience. If such personnel are not available in the local work force, a well-conceived and detailed training program should be established. Service technicians are responsible for resolving customer complaints concerning technical problems in a timely and efficient manner.

All service personnel should undergo detailed training about system operation, problem identification and resolution at one or more receive sites. Emphasis should be placed on customer relations since these people are direct representatives of the wireless company. All service personnel should wear uniforms and carry ID cards identifying them as company employees.

Scheduling of service calls is handled by customer service (see Figure 15-1). Optimally, most technical problems should be scheduled for same or next day service. Any customer who reports a complete interruption of service should, of course, receive the highest priority and be afforded same day service. Under no circumstances should any customer be left without service for any more than 24 hours.

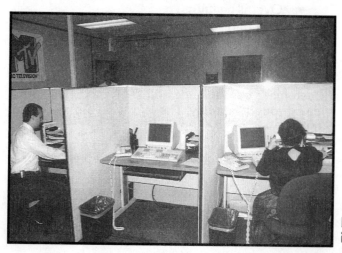

Figure 15-1. Customer Service Center at OmniVision in Corpus Christi, Texas.

Technical operations maintain a 24-hour, seven-day a week schedule. Subscribers then can have 24-hour telephone access to the service department. When a call is received during normal working hours, a work order should be generated and the problem should be noted. Then the information can be dispatched to the area service technician.

The time spent at a subscriber's residence must be minimized. As soon as a service technician has diagnosed the problem, faulty equipment must be either repaired or replaced. These components should then be returned to the company repair facility as soon as possible for testing and repair and subsequent return to inventory.

All vehicles, ladders, tools and test equipment necessary for efficient service by in-house technicians are generally provided by the company.

Industry experts estimate that one service technician is required for each 5,000 subscribers. This is a very flexible number that can increase or decrease as a service history for a given system is compiled. The type and amount of service required can depend upon a number of factors including the quality of the subscriber equipment and installations as well as factors such as local climate.

Billing

Subscriber account billings are processed by the billing and database management computer system located at the company's central office. The director of finance is responsible for all billing and collections activities. CSRs are responsible for account accuracy and general input to the database. Monthly statements are then processed by the computer and mailed to subscribers.

Billing for both subscription fees and service is generally scheduled near the end of each month. Pay-per-view charges are based on the number of events or movies viewed during the previous month. A collection period of 15 days from date of billing is considered to be a good operations policy. Subscribers accounts not paid within the billing period should be assessed a late fee and sent a delinquency letter requesting immediate payment. Should a subscriber fail to respond to this request within 10 days from date of mailing, service should be withheld by the addressing computer until such time that the customer pays. If the subscriber account remains delinquent 45 days from date of initial billing, service should be disconnected and the equipment retrieved. Subscribers who are disconnected for non-payment but who desire to be reinstated should be required to pay a reconnection fee, past due account charges and the first month's service fee prior to reconnection.

Organization and Personnel

Operations personnel necessary to perform the required functions within an wireless cable business can be categorized under (1) management, (2) administration, (3) technical/engineering and (4) sales and marketing. The job responsibilities for each of these categories are outlined below.

Management Personnel

Management personnel include the system general manager, marketing director, director of finance and director of engineering. They are compensated based on their position and their overall experience. Managers are salaried employees with, in certain cases, equity positions in the operating company via stock ownership or profit sharing.

System General Manager.

Responsible for overall system operations, formulation and performance of all budgets, system growth and profitability and planning and development of new projects relating to system operation. Hires key management staff to support the various operating activities within the company. These personnel report directly to the general manager. Reports to the Board of Directors or their representative.

Marketing Director

Responsible for formulation and performance of the marketing plan, including budget preparation and tracking, preparation and implementation of media and advertising campaigns and interface with programming suppliers. The marketing director hires, trains and supervises the sales and telemarketing forces and is responsible for their productivity. The marketing director reports to the system general manager.

SYSTEM PLANNING AND OPERATION

Director of Finance.

Responsible for tracking of capital and expense budgets, company financial records and preparation of balance sheets. Maintains company payroll and subscriber billing systems. Supervises the billing clerk(s) to ensure accurate input to the billing system. Also responsible for implementing procedures for customer billing and collections. Supervises the warehouse and operations and is ultimately responsible for inventory control and tracking. Reports to the system general manager.

Director of Engineering.

Responsible for operation and performance of the transmitter site and peripheral equipment, field service operations and subscriber installation activities. Supervises and trains technicians, installers and dispatchers. Formulates and controls capital and operations budgets within the Engineering Department. Reports to the system general manager.

Administration

Administrative personnel include an executive secretary, receptionist/typist, customer service representatives, computer operator/billing clerk and programming director. They are compensated based on their position and work experience. Compensation is hourly, except for the salaried programming director. Initial operations typically require one executive secretary, one receptionist, three customer service rep, one computer operator/billing clerk and one programming director.

Executive Secretary

Administrative assistant to the system general manager. Responsible for company correspondence, file maintenance and general secretarial duties. Reports to the system general manager.

Receptionist/Typist.

Greets the public in the lobby, handles questions and collects subscriber payments, runs the telephone switchboard and performs typing and general secretarial duties for the managerial staff. Reports to the system general manager.

Customer Service Representatives.

Interfaces by telephone with existing and prospective subscribers concerning order entry, service requests, billing questions and general subscriber accounts maintenance. Reports to the director of marketing.

Computer Operator/Billing Clerk.

Responsible for operation of the subscriber billing system. Performs daily printing of installation orders, service orders, marketing and administrative reports, preparation of subscriber bills and required computer maintenance. Reports to the director of finance.

Programming Director.

Responsible for all functions of the pay-per-view channel(s) including program acquisition, daily program compilation, events/movies scheduling, preparation of reports on channel(s) activity and maintenance of the equipment associated with the channel(s). Reports to the director of marketing. Only needed in larger scale systems.

Personnel Director

Selects and hires employees. Manages review programs and employee insurance. Only needed in larger scale systems.

Technical/Engineering

Technical staff personnel include dispatchers, field service technicians and warehouse supervisor. They are compensated on an hourly basis according to their position and general experience level. Initial operations in the first year of operations require two dispatchers, three field service technicians and one warehouse supervisor.

Dispatchers

Responsible for interfacing with field technicians and installers by two-way radio and telephone in order to expedite customer service requests and installations. Respond to customer complaint lines and schedule/route technicians on a daily basis. Report to the director of engineering.

Field Service Technicians.

Responsible for signal coverage evaluations, installation quality control and customer service in the home. Field techs report to the director of engineering.

Warehouse Supervisor

Responsible for receiving materials from vendors, issuing materials to installers and technicians and inventory control. Daily pre-programming of converters and minor troubleshooting of field returned inventory will also be a warehouse function. Reports to the director of finance.

Sales and Marketing

The sales and marketing staff may include the sales manager, outside sales counselors and telephone sales representatives.

Sales Manager.

Responsible for total sales department productivity requirements. Directly supervises the outside sales forces and telephone sales reps to ensure maximum sales productivity. Conduct initial and recurrent sales training and promotes the company services throughout the marketplace. Reports to the director of marketing.

Outside Sales Counselors.

Responsible for the direct sales of the programming services by means of door-to-door selling, remote sales locations and telemarketing. Additional responsibilities will involve market development to ensure cohesive selling efforts. Sales counselors are compensated on a commission basis and report to the sales manager.

Telephone Sales Representatives.

Responsible for receiving incoming sales calls from prospective subscribers, making sales presentations over the telephone, taking service requests (sales orders) and scheduling installations per customer service work schedule guidelines. Additional responsibilities include outbound telemarketing activities as required/assigned by the director of marketing. Report to the sales manager.

Construction Schedule

Permits and Licensing

Once a go-ahead has been granted to build and operate the proposed wireless cable broadcasting facility, all the necessary permits and licenses must be obtained. The complexity and time involved in this process have a direct impact on scheduling.

Capital Expenditure Schedule

When all permits have been obtained, equipment can be ordered. Generally 25% of the total invoice be paid upon ordering equipment. The wireless cable system should be operational from six to nine months from this time.

Programming

Availability

All programming that can be either re-used at no charge or purchased from other non-publicly available sources can be fed into the distribution system.

Packages and Rates

The programming menu should be structured with a basic package and additional "premium" channels available on a single purchase basis. Pay-per-view or special events programming can be offered as the need arises.

SYSTEM PLANNING AND OPERATION

CHAPTER 16. A LEGAL PERSPECTIVE on WIRELESS CABLE in the UNITED STATES

This chapter was kindly provided by Robert F. Corazzini and Jennifer Richter of Pepper & Corazzini.

PEPPER & CORAZZINI

Communications Law Specialists

Pepper & Corazzini's attorneys include recognized communications law experts, with the ability to provide vital assistance to clients confronting complex legal and business issues in the rapidly fluctuating communications marketplace. Practice specialities include Cable Television, MMDS, ITFS, Satellite, Radio, Television and Mobile Radio.

Pepper & Corazzini, 1776 K Street N.W., Washington, DC 20006
Telephone: (202) 296-0600   Telecopy: (202) 296-5572

Many of the legal complexities associated with multipoint distribution service, better known as wireless cable[1], are little understood by most users of the technology. Thus with good reason most wireless cable licensees hire attorneys to handle these legal matters. However, a basic understanding of some of the legal aspects of the technology can be useful in putting the applicable rules in a context that makes sense. The intent of this chapter is to provide the system operator with a greater understanding of the rules, their origin and their purpose, although no publication should be considered a substitute for retaining an attorney and an engineer who specialize in wireless cable regulation. This is especially true in a field such as wireless cable where the rules, regulations and policies are continually changing and only such professionals can provide a full analysis of the specific regulatory requirements.

The chapter is divided into three sections. The first section discusses the history of wireless cable and some of the more important rules pertaining to wireless cable and how they have evolved over the years. The second section details the current Federal Communications Commission (FCC) practices and procedures especially as relates to the application process for wireless cable licenses. The final section focuses on recent trends in the wireless industry.

A. AN HISTORICAL PERSPECTIVE

Wireless cable regulation has been driven, in large part, by three distinct goals. The primary objective was the enhancement and strengthening of wireless cable service as a viable multi-channel competitor to traditional cable service. The second goal, often-times conflicting with the first, has been the protection of the educational use of the television spectrum, also known as the Instructional Television Fixed Service (ITFS)[2]. The third interest, which theoretically at least, overrides the other two is the "public interest."

One of the duties of the FCC is to decide how to allocate the United States' portion of the electromagnetic spectrum ("the airwaves") among all its potential users. This task must be performed with the public interest as the paramount goal. That is, the spectrum must be allocated in such a way that communications services are made available to "all people". Because of the nature of the population distribution in our country, both urban and rural, this is a huge challenge, one that requires no spectrum space be left idle if it can be put to more productive use.

Since there is far more demand for the airwaves than there is space on the spectrum, the decision of how to allocate the space is sometimes difficult. The spectrum is finite, and competition between communications services for it can be fierce. Like so many other communications battles, this is a battle which begins on a theoretical plane and over time evolves into one of sheer strength. Ultimately, the stronger, more viable industry, the one more likely to make full use of the spectrum space allotted, is the one given entree to the greater number of frequencies.

One such battle has been played out, over the years, between wireless cable and educational television. Wireless cable may in some sense be considered to be gaining in the struggle having acquired much of its spectrum space at the expense of educational television. But that is not to say that the FCC's overall commitment to the development of educational television has been diminished in any way. The FCC's commitment is probably as strong as ever, as evidenced by the protections built in to the allowance of wireless to use more ITFS spectrum. The educational television spectrum usage simply has not grown at the same pace as wireless cable, making full utilization of the space originally allocated unlikely. Wireless cable, on the other hand, is premised upon a multitude of channels and has developed to a point where it is in dire need of frequencies. The FCC has attempted to accommodate this need by reallocating the spectrum, and then reallocating it again.

Allocation of the Electromagnetic Spectrum

In 1962, the frequency band from 2150-2160 MHz was allocated for the development of omnidirectional, non-broadcast systems. The maximum authorized bandwidth available for use was 3.5 MHz. Neither of these facts would be of particular note except for the fact that the allocation and bandwidth limitations rendered this portion of the spectrum useless. No FCC licensees had operated on this band on a regular basis for the eight years it was available.

Then in the late sixties, the FCC received several applications proposing to use this spectrum space for the distribution of television programming. The FCC wanted to oblige, but there was one problem. The 1962 order, which limited bandwidth use to 3.5 MHz, prevented use of the frequencies in this way because a standard television signal requires 6 MHz of bandwidth. Therefore in July of 1970, in the interest of making full use of the spectrum, the FCC removed the bandwidth limitation and opened the door to a new "common carrier" service which through the years evolved into the wireless cable service of today[3].

The nature of the new service was "point-to-multipoint." That is, television programming could be transmitted from one central location or "point" to several households or "points" in a given market area. But the service was not regulated as such or even termed "multipoint." Rather, largely because the scope of the service was not yet fully understood, it was regulated as a point-to-(single)point microwave radio service. In two years time, this sort of regulation proved insufficient for the multipoint service, and in 1972 the FCC gave the service a name of its own – "Multipoint Distribution Service." Rules governing the service were adopted in 1974, and the two channels designated for use were Channel 1 which operates at 2150-2156 MHz, and Channel 2 which operates at 2156-2160 MHz. Therefore, in the 50 largest markets the cities were provided with two wireless cable channels each. However, in rural areas where the 2160-2162 MHz band was, and is, used for rural telephone service, operating on channel 2 is impossible. Channel 2A, therefore, is substituted in those areas. However, since Channel 2A operates on just four MHz from 2156-2160 MHz it cannot be used to transmit television signals.

Education Television

In 1963, the FCC created educational television[4]. Its purpose was, and is, to "transmit instructional material [via the airwaves] for the formal education of students and to perform directly related services."[5] Educational television is viewed by many as the sort of service that can provide improved educational opportunities for large numbers of students delivering classroom instruction from the origination source to schools scattered throughout an area or district. Educational television was assigned to the same spectrum space allocated for Operational Fixed Stations (OFS). The two information services were to share the thirty-one 6 MHz channels available between 2500 MHz and 2690 MHz on the spectrum. Not knowing what to expect in terms of educational television industry growth, the FCC froze applications for new OFS stations for 3 years to "observe the amount of use of these channels by educators and ... determine what course of action should be taken to encourage the fullest development of the ... band ...."[6] After eight years the evidence favored educational television use over OFS use, and the FCC allocated 28 of the 31 channels to educational television[7].

Reallocation

In 1980 the FCC began to consider further reallocation of the 2500-2690 MHz band this time from educational television to wireless cable[8]. Three justifications were put forth in favor of reallocation: (1) the demand for wireless cable service was exceeding the supply of available frequencies; (2) the 2500-2690 MHz band was under-utilized; and (3) the removal of existing restrictions on the use of OFS channels would increase demand for these channels[9].

To a large degree, all of these justifications rang true. But it was the third justification that carried the most weight. At that time, there were only two wireless cable channels available in each market, and a single licensee could generally use just one. The FCC precluded a licensee from obtaining a second channel in the same metropolitan area until it had operated one channel for at least one year, and could show a public demand for additional service not likely to be met by a competing carrier[10]. This rule proved an impediment to wireless cable growth. Virtually all industry members agreed that if wireless cable was to survive, multi-channel operation was an absolute necessity[11]. Without it, wireless could never compete with traditional cable systems or other services offering multi-channel capability.

The FCC eventually was to find reallocation of the spectrum from educational television to wireless cable to be in the public interest for a number of reasons. First, it appeared there was a large unmet demand in rural areas for multiple channels of premium television that was unlikely to be met by cable television or other available technologies[12]. Wireless cable in a multi-channel format could meet this demand. This led to the second public interest to be served – providing competition to "monopolistic" cable companies:

> "Since there are no alternative distribution systems authorized to provide multi-channel broadband service, cable has been able to behave as a monopoly industry, building at a schedule suited to its own pace with little incentive to upgrade antiquated systems... [A]n expanded MDS would provide a competitive spur to cable, thereby moderating its monopoly characteristics and speeding its growth."[13]

The major argument against reallocation was that multi-channel wireless cable "was not needed." Dissenters argued that other technologies, like low-power television (LPTV), satellite television (STV) and direct broadcast services (DBS), would be more than sufficient alternatives to meet service demands in both urban and rural areas[14]. But the FCC did not find this argument sufficiently compelling.

The Reallocation Plan

In 1983, the FCC adopted a reallocation plan that it believed struck a reasonable balance between the need to make spectrum available for educational television use and, at the same time, make spectrum available for multichannel wireless cable. In all, just eight channels were reallocated for wireless use, leaving twenty channels for the needs of educational television. The eight channels were comprised of four "E group" channels and four "F group" channels. At that time a wireless cable operator could hold a license for no more than seven channels in a single market: either of the four E group channels or the F group channels, either or both of the wireless cable single channels – Channels 1 & 2, and one OFS (H) channel.

The reallocation plan "grandfathered" existing ITFS licensees, permittees, and applicants. These users or potential users could renew their licenses for E and F group channels, thus not suffering displacement of their services as a result of allocation of their frequencies to wireless cable. Educational television licensees were precluded, however, from assigning their licenses, and no new educational television applications for E or F group channels would be allowed.

The FCC provided that each service area would have two separately owned wireless systems. It was thought that in some large cities, two wireless systems could be viable, and the FCC believed that two systems was an attractive option because of the potential benefits that competition between two systems would bring. This turned out to be one of the most seriously flawed projections that the FCC would ever make in its effort to encourage competition to wired cable. Fortunately, however, the Commission recognized its mistake and corrected the situation in 1990.

Finally, in 1991 the Commission modified its spectrum allocation again[15]. Eight of the 20 channels allocated for educational television use were made available for wireless cable use upon application, conditioned upon a number of significant restrictions designed to ensure that wireless cable's direct use of ITFS channels would only occur where the ITFS frequencies continued to be un-used or under-utilized by the educational institutions. An application for vacant ITFS channels would only be granted, if after the grant a minimum eight channel reserve for educational television remained[16]. To be included in this 8 channel reserve, no licensee or applicant for the educational television frequency could exist within 50 miles of the wireless cable transmission point.

To be eligible for "available" ITFS channels, a wireless cable operator must hold a conditional license, license or lease, or have on file an unopposed application, for at least four wireless cable channels. The wireless cable licensee must show that there are no wireless channels available that could be used in lieu of the ITFS frequencies applied for. Likewise, educational television licensees are prohibited from applying for the same channels as wireless cable entities if additional ITFS channels are available. Finally, the ITFS channels licensed to wireless cable entities are, under certain conditions, preemptible by ITFS entities.

In addition to the eight ITFS channels made available, the three OFS H-Channels which the FCC had previously permitted to used as a part of a wireless cable system, were reassigned directly the MDS spectrum.

Today, the frequencies used for wireless cable are those in the Multipoint Distribution Service, and the Instructional Television Fixed Service. Use of these frequencies is governed by two different sets of rules: Part 21 applies to the wireless cable service channels and Part 74 applies to the ITFS channels.

The channels currently available for wireless cable include eight wireless channels (Channels E1-E4 and F1-F4), two single- channel wireless cable channels (Channels 1 & 2 or 1 & 2A), three previous OFS channels (Channels H1-H3) and 8 of the 20 ITFS channels (from Channels A1-A4, B1-B4, C1-C4, D1-D4, and G1-G4). In addition as discussed later, wireless cable lease of unused capacity on the 12 remaining ITFS channels is also permitted. Thus, wireless cable can amass usage of a maximum total of 21 full time and 12 part time channels. Through a process known as "channel mapping" the subscriber receives full time channels in the channel package.

How the Rules Evolved

Several of the rules instituted in the MMDS Allocation Order have changed and evolved over the years, while others were newly promulgated since then. The following is a discussion of the most significant of those rules.

Lottery

The question of how the FCC should handle applications for wireless cable technology was raised in the 1983 allocation order. Prior to then, when two applications were received for the same channel, a comparative hearing was held to determine which applicant would be awarded the license. Concern was raised that this procedure would not work for newly available wireless cable channels, because a large number of applications would be filed for each of the newly allocated channels. Using comparative hearings each time would "involve the applicants in costly and lengthy comparative hearing procedures and thereby unnecessarily delay availability of the service to the public."[17]

The FCC therefore concluded that the comparative hearing procedure was not the best suited method to resolve mutually exclusive wireless applications. The lottery method was proposed for the selection of both multi-channel and single channel wireless permittees. On November 21, 1984, the Commission approved lotteries as the method to select multi-channel MDS operators[18] but declined to authorize lotteries for single channel wireless applicants. The lottery rules are entitled "Grants by Random Selection" and are found at 21.33 of the rules. Two important policies with respect to the lottery were adopted: (1) minority preferences would be given; and (2) cumulative chances would be given to applicants that entered into settlement agreements with each other[19]

Leasing

Another significant rule resulting from the 1983 allocation order pertained to "leasing." The rules pertaining to leasing are found at 74.931(e) of the Commission's rules, and are under the heading "Purpose and Permissible Service." If the number of channels allocated for wireless did not prove sufficient for the programming needs of a wireless cable licensee, it could deal with the licensee of the local ITFS channels, who had at its option the right to lease its excess channel capacity. The Commission thought leasing space made sense for a variety of reasons.

The first is that educational use of frequencies occurs primarily during daytime hours. Entertainment television occurs predominantly in the evening. Sharing space between educational television and wireless cable for these dual purposes seemed practical, feasible and in no way detrimental to the service offered by either entity. Second, because federal funding for educational television had decreased, it was desirable for these licensees to generate revenue by leasing "excess capacity".[20] "The income derived from [leasing] could enable stations to be on the air for a greater portion of the day and to increase programming availability. In addition, new revenues might prove sufficient to bring currently vacant ITFS channels on the air."[21] The Commission failed to define "excess capacity" but it was apparently willing to authorize a 75/25 split. ITFS licensees were allowed to lease up to 75% of their channel capacity for just about any purpose, including wireless cable. In point of fact, the leasing rules have proven to be an economic boon to ITFS licensees providing them not only with revenue but allowing the construction of ITFS channels which would otherwise not have been built, with state of the art studio, production and transmission facilities, all at no cost to the ITFS licensee.

Prior to 1990, the Commission's rules required that educational television licensees preserve forty hours per week on each channel for their own use. The licensee had to actually use twenty hours of air time per week before it could lease its facilities to wireless cable users. This requirement was considered by many as prohibitive. It was pointed out to the FCC that requiring twenty hours of usage per channel per week actually prevented the development of new systems which could not utilize that much time to start with. Also, the licensees needed the revenues from leasing to help develop their systems. In view of this, the FCC resolved to relax its rules, permitting new entities to use just twelve hours per week of air time for the first two years, and twenty hours per week thereafter[22]. A maximum lease term of ten years was retained. In 1991, time and day restrictions on leasing were lifted. Now the twenty hours per week the educational television licensee must program can be scheduled any time[23]. ITFS licensees are free to make their own decisions about the wireless cable licensees they lease space to and on what terms. They are under no legal compulsion to deal with any wireless cable licensee as they are not considered common carriers.

Common Carrier Status

Initially all MDS licenses were issued as common carrier licenses. However, in 1987 the FCC granted the licensees and applicants the right to choose to operate either as a common carrier or a non-common carrier. This right of election took on an added significance when in 1991, the FCC released new guidelines for fines and forfeitures. The rules had some interesting implications for the importance of choosing either non-common carrier status, or common carrier status. The fines for violations by common carriers were made significantly higher than for non-common carrier licensees, about ten times higher. Further, common carriers must comply with rules that don't apply to non-common carriers. The option of choosing non-common carrier status, therefore, is much more attractive, and in fact, the vast majority of wireless cable operators and applicants now opt for non-common carrier status.

Non-Interference

The 1983 allocation order established critical "non-interference." rules to protect ITFS users. With the exception of E and F group frequencies subject to grandfathering, all E and F channels were reallocated for wireless cable use nationwide. The use was limited to a "strict noninterference basis." This meant that frequencies would not be authorized for wireless until non-interference with adjacent channel and co-channel ITFS users with transmitters within 50 miles was proved unless the ITFS operator accepted any interference caused (or of course, if there was no ITFS facility within 50 miles). Potential wireless cable licensees had to prove non-interference after their construction permit was granted. If they could not make this proof, then they were required to negotiate with affected educational television entities. Because many ITFS licensees were reluctant to deal with the wireless cable licensees, this requirement proved to be a barrier to the construction of many wireless cable facilities.

Therefore, in 1990 and again in 1991, the FCC amended the non-interference rules in its Wireless Cable Service Report and Order, simplifying them considerably. Under the new rules, a wireless cable licensee is no longer required to file a letter of non-interference from all co-channel and adjacent channel licensees[24]. Instead, the wireless cable applicant must serve a copy of the engineering analysis on any affected educational television entity by certified mail, return receipt requested, certifying to the FCC that such service has been initiated. The ITFS licensee is then afforded an opportunity to file any objection.

Each wireless cable license is afforded a "protected service area" which in very general terms is a fifteen mile radius from the transmitter for a licensee proposing to use an omni-directional transmitting antenna. No later proposed wireless cable facility is allowed to cause interference within the first station's protected service area. In filing a new wireless cable application, the applicant must also analyze each previously filed or existing wireless cable facility within 50 miles and prove that it will not cause any interference to the protected service area of those facilities. In a Notice of Proposed Rule Making released May 8, 1992[25] discussed in the next section, the FCC has proposed additional measures of protection to ITFS registered receive sites and has proposed strict separation standards between all MDS proposals and existing or proposed ITFS stations.

Resale of MDS Licenses

In addition, the FCC imposed two significantly important restrictions on the sale of an MDS license. If a lottery was held to choose a wireless cable permittee and the successful applicant was the beneficiary of a preference or if a comparative hearing was used, then the licensed facility is not freely transferable. That is, it can be sold only for reimbursement of expenses. Instead, the station must be constructed and operated for one year before it can be sold for a profit. In no event may a construction permit (conditional license) be sold for a profit.

Recent Changes in the Rules

In the 1990 *Wireless Cable Service Report and Order*[26], the Commission sought once again to remove barriers to the development of wireless cable in an effort to help wireless reach its fullest possible potential.

One problem sought to be remedied was the potential warehousing of wireless cable channels by franchised cable systems to prevent competition by wireless. The FCC decided to prohibit franchised cable operators from acquiring wireless cable licenses or leasing wireless cable channels where the protected service area overlaps the cable system's franchise area, unless there is a competing franchised cable system also in the franchise area. An exception to the cross ownership rule applies to rural areas that would remain unserved by wireless cable.

Even under the exemption, the application process a franchised cable entity must go through to acquire a wireless license differs from the usual process in one important way. Franchised cable applications for wireless cable are put on public notice for 30 days. If no non-cable party files a competing application, then the cable application can be granted. If a wireless cable applicant does respond, however, then its application supplants the cable operator's application and is considered regardless of the cable application[27].

Another problem that was addressed was the fact that prior to 1990, the rules limited output power to 10 watts for wireless cable transmitters. In many instances, however, 10 watts proved insufficient for wireless cable operators to reach their subscribers. This was especially true when storms or other natural and manmade obstacles were present that attenuated the signal. Under the new rules, an operator can increase output power for wireless cable to 100 watts although an application still must be filed. In addition, applicants for higher power must execute the standard interference analysis and notification procedures, specifically serving stations that could experience interference due to the increased power load[28].

A provision for signal "boosters" was also included in the rules as amended in 1990. The special importance of boosters to wireless cable lies in the basic nature of the technology. Since wireless cable is a line of sight technology, if that line is obstructed in some way, the ability to deliver television programming is impeded. Boosters are useful because they fill in gaps in a service area caused by the shadowing effects of tall buildings, bridges, ground depressions or other obstructions. The FCC was persuaded by the usefulness of this technology and approved it for use. The power output, or EIRP, from a booster was limited to a low power (18 dbW), deemed sufficient to fill in shaded areas within a protected service area and not extending beyond it[29]. Because making use of a booster is considered a major modification, applicants must submit an interference analysis and make the necessary service certifications[30] prior to implementation. In 1991 the FCC modified these rules, allowing pre-authorization construction and operation of low power signal boosters, and decreasing the acceptable power level to 9dbW[31]. However, a booster may not be used to extend a signal beyond the station's protected service area.

Filing Procedures

The filing procedures were also modified in 1990. Previously, the FCC would announce the filing of a wireless cable application in its "public notice" publication. Any one wanting to file for that same market would then have 60 days in which to prepare and file a competing application thus subjecting the two applications to a lottery to choose the winner. The FCC chose to move from a 60 day filing period to a one-day filing window to hold down the number of applications submitted and to eliminate the plagiarizing of applications with the simple use of a copy machine. The rules were changed in response to a concern that the filing procedures were an incentive for "insincere applicants to file competing applications to induce the legitimate wireless cable operator into a financial settlement."[32]

The new rules are referred to as "first come, first served." Applications accepted for filing that are for the same service area[33] are mutually exclusive only if received by the Commission on the same calendar day. Since there could be no prior public notice of the application, the FCC assumed that they would be dealing with only one application. The Commission stated: "We expect that such a change will eliminate the opportunity for application mills to merely copy applications that have previously been filed and re-submit them with the name changed."[34] The effort has been only partly successful since many application mills now file multiple applications on the same day thus assuring the need for a lottery.

B. THE APPLICATION PROCESS FOR AN MDS OPERATION

Eligibility to Apply

Any individual, partnership, joint venture or corporation that is a citizen of the United States may be licensed as a provider of Multipoint Distribution Service. Foreign citizens and corporations are precluded from holding station licenses, as are convicted felons.

The Market

As a general rule for new applications, the market chosen must be at least 50 miles away from all locations for existing or previously proposed stations. The location must also be at least 15 miles away from the boundary of a metropolitan statistical area for which a license has been granted or there are MMDS applications pending.

Each applicant for wireless facilities may apply only once for the same group of channels in each service area. Most individual MMDS applicants file on a single date for both the E and F channel groups, the three H channels and channels 1 and 2/2A in order to secure a minimum of thirteen channels. Of course, often all 13 channels are not available in a given market.

Pending Applications

Because of the previously discussed first come-first served rule and 50 mile restriction, it is important to make sure there are no pending applications for the location and channels of the proposed operation. At present the best source for this information is the MDS computer printout listing of pending or granted applications in the FCC's Public Reference Room, 1919 M Street, N.W., Room 6220, Washington, D.C. There is also a commercial computer service which may be used for determining channel availability. Unfortunately however, in both cases the data is usually between 2 and 6 weeks old and thus, applications may have been filed within the last few weeks for the desired market which have not yet appeared on the listings. As a result, it is not now possible to file an application with certainty that it is the first application for a market. The FCC is presently studying the feasibility of a software system that would provide information current within 48-64 hours. The availability of such a data base would obviously help avoid useless expenditures of time and money for both prospective applicants and the FCC as well. Furthermore, the FCC announced in its May 8, 1992 rule making proposal that it intended to create a consolidated data base of all existing and proposed MDS, ITFS and H-channel stations.

The Application

An application for a conditional license is filed on FCC Form 494, "Application for a New or Modified Microwave Radio Station License." Whether the licensee elects Common Carrier Status or Non-Common Carrier Status, its application must be submitted with an FCC Form 430, "Licensee Qualification Report." The filing fee is $155.

Applications that are not accepted for filing will be returned to the applicant with a brief explanation of the application's deficiencies. Common deficiencies are listed at rule 21.20(b), and include insufficient filing fee and the existence of a previously filed application within the restricted area.

Interference Analysis

Each applicant must demonstrate that operation of the applicant's transmitter will not cause harmful interference to co-channel or adjacent-channel ITFS or MMDS stations with transmitters within 50 miles of the proposed MMDS transmitter site. All applications must contain a showing of how interference will be avoided, and the steps the applicant has taken to comply. Alternatively, the MMDS applicant can submit a statement from the ITFS or MMDS licensee that it does not object to operation of the proposed MMDS station and will accept whatever interference may occur. The interference analysis must be delivered to the co-channel and adjacent-channel parties affected. These parties, of course, have the right to file a protest against the application with the FCC if they believe that the proposal will cause harmful interference to their station. However as noted previously, on May 8, 1992 the FCC released a proposal that would significantly modify the interference criteria through use of a strict mileage separation standard with added protection standards for registered receive sites.

Applications are considered mutually exclusive if they apply for the same channels in the same market, or if grant of one application effectively precludes grant of another application because of harmful electrical interference.

Settlement Groups

Competing applicants can form an alliance, or settlement group, after applications have been filed and before the lottery is held. Once joined, settlement group members have as many chances to win the lottery as its members would have had individually. If there are no "holdouts," that is, applicants who refuse to join the group, the settlement group is granted a conditional license once it forwards its single application to the FCC and it is accepted. Again however, the FCC is presently considering adopting rules to prohibit settlements as a means of deterring application mills.

Selection Process

If mutually exclusive applications are for a multi-channel MDS, or a single H-channel MDS, the Commission will use a lottery or random selection process to award a conditional license. In a lottery, minority preferences and/or diversity preferences may be claimed by the applicants. The factors used to determine if an applicant qualifies for either type of preference are explained in section V of FCC Form 346.

Minority preferences are given to Blacks, Hispanics, American Indians, Alaska Natives, Asians and Pacific Islanders. No other groups are recognized for the purposes of the lottery. Diversification preferences are given to applicants who own no other media interests (full preference) or who own no more than 3 other media interests (partial preference).

If the application is for a single-channel MDS, the Commission currently uses a comparative hearing to award a conditional license. This is a considerably more expensive procedure for both the applicants and the FCC and will undoubtedly be eliminated in favor of a lottery selection process under the May 8, 1992 rule making. In fact, that rule making proceeding could lead to a number of significant changes in the entire MDS lottery procedure.

If there is only one eligible application filed on a single day for any particular market, the Commission will notify the wireless cable applicant by public notice that their application has been preliminarily accepted for a market area. In addition, the FCC rules state that no application will be granted until thirty days following the public notice identifying the potential licensee. In this thirty day period, the public can voice its opposition to the application which could conceivably lead to a denial of the application. Generally however, disputes are limited to either contending applicants or licensees who fear harmful interference.

If the MMDS applicant avoids or overcomes any challenges, then it is awarded a conditional license. The new licensee now has twelve months in which to construct the station to an operational status. If the licensee fails to do this, the license is automatically forfeited. The licensee can file FCC Form 701 prior to the construction deadline for additional time to construct the station. However, the FCC does not grant such construction time extensions liberally and will grant extensions only if a compelling showing is made.

After the system is constructed the applicant must file FCC Form 494A, "Certification of Completion of Construction Under Part 21." This will serve to certify completion of station construction and authorize operation of the station. The filing fee is $455 per channel. Failure to file this Form in a timely fashion (within 5 days of the construction deadline) will result in an automatic forfeiture of the license even though the station has in fact been constructed.

Station Operation

After the station has been constructed, the FCC will issue a full (or unconditional) license and the licensee is required to commence service within one year after the completion of construction. Once the station is operating, each licensee is required to file an annual report on March 1, detailing certain statistical and programming information contained in rule 21.911. In addition, FCC Form 430 must be filed on March 31 of each year, showing all changes in information required by the form from the previous year or a statement that there have been no changes. If, during its period of operation, the licensee wishes to assign or transfer the station license, it can do so by filing FCC Form 702 (for assignment of license) or FCC Form 704 (for transfer of the licensee company). The filing fee is $55.00. MMDS licenses granted after May 1, 1991 are valid until May 1, 2001. Thirty to sixty days prior to this date, FCC Form 405 "Application for Renewal of Station License" must be filed. The filing fee is $155.00.

It is extremely important that the reader realize that the wireless cable rules have been and will continue to be subject to numerous changes, as have been detailed in this chapter. As a result, many of the procedures in place today will change in the future as the FCC continues to wrestle with this dynamic medium. In fact, the rules proposed in the May 8, 1992 rule making, if adopted in their most restrictive form, would radically alter the present processing procedure affecting both pending applications thus requiring the filing of amendments, as well as new applications which may be filed after the FCC lifts its application freeze imposed beginning April 9, 1992. It is for this reason that no attempt should be made to prepare or file an MDS application without the assistance and counsel of a communications lawyer and a consulting engineer.

C. RECENT TRENDS

As mentioned, on April 9, 1992, the FCC imposed a temporary freeze on all new wireless cable applications. The freeze is projected to last through at least August of 1992. In addition, in the rule making proposal adopted that day and released May 8, 1992, drastic changes to the wireless cable licensing procedures were proposed. The FCC hopes that these two measures will reduce the backlog of current applications and reduce the time necessary to process an application.

Among the proposed procedural changes are modification of interference standards, elimination of certain non-technical questions from the current application form, elimination of partial settlement agreements among lottery participants, prohibition of settlement groups altogether, and establishment of an application processing priority that places consideration of lottery applications last. Obviously, if all of the proposals are adopted, many of the existing procedures will be changed or eliminated.

The change that could potentially have a significant impact on the industry is the prohibition of settlement groups. Settlement groups, formed by so-called "application mills," have been a sore spot in the industry for some time now, and have gained a great deal of media attention in recent months. Before discussing "application mills" in more depth, it is important to note that not all wireless application preparation services have been accused of serving as "boiler rooms" for wireless applications. The number of true "application mills" are great enough in number, however, that it is a multi-million dollar industry.

Since 1984, when the FCC approved lotteries as the method for choosing between mutually exclusive applicants and awarding licenses, "application mills" began cropping up, soliciting investors to apply for wireless cable licenses. The practice interjected a virtual "explosion of applications" into the industry. At times, more than 5,000 applications for extremely small markets were received solely as the result of these application preparation companies.

The sheer volume of applications may result in administrative overload, but the filing of numerous applications is not, in and of itself illegal. However, it is illegal to perpetrate a fraud on applicants by such means as misrepresenting their chances of success or the likely value of the license or the facilities.

Typically, a preparation company charges $5,000 to prepare a would-be licensee's wireless application. The company often purports to use the money for "research" into the potential for wireless in a certain market or for the technical preparation of the application. Because many applicants are placed in the same market, the research and technical preparation need only be done once. But the company does not prorate the cost of the research and preparation among the number of potential applicants. Instead, the full $5,000 is charged each applicant, covering the cost of the research and preparation a multitude of times over. The North American Securities Administrators Association (NASAA) view of the application mills was described in Broadcasting Magazine:

> "Application mills, NASAA said, inflate the prospects of an investor in a wireless cable television lottery, gloss over the complicated mechanics of the FCC lottery process, understate the risks, exaggerate the potential value of a license, overstate the availability of necessary financing and make it seem that fat profits are all but certain and will start rolling in almost immediately."[38]

Largely because of concerns like these, the Wireless Cable Association asked the FCC to prohibit settlement groups altogether. The FCC has asked Congress for the authority to switch from the lottery process to an auction process. Many in the industry welcome such a change, fearing that without it the "application mills" will continue to scare away lenders and equity investors, and give the industry a black eye in general.

Many methods have been tried in an attempt to stem the attractiveness and profitability of the "application mills". In 1990, the FCC eliminated the sixty day filing window, limiting applicants to the first qualified candidate in the market and any other similarly qualified candidate filing within 24 hours. It was hoped that the rule change would clean up the process, but to no avail. In fact, this year alone the FTC has filed complaints against two "application mills", sued three others, and obtained preliminary injunctions against still others. But the abuse continues.

Although the FCC has tried to "blunt" the progress of "application mills" by various rules many have blamed the FCC for creating the "application mill" problem. In setting up the lottery system and providing for the viability of settlement groups, critics charge that the FCC has created the perfect environment for such fraud to take place. They say that the FCC should not be in the lottery business if it cannot police it and keep it reasonably free from fraud and abuse. But the FCC counters these arguments by stating it has done all it can to institute measures designed to thwart "application mills" and abuse of the settlement group rule. No amount of rules will stop the professional con-artist from finding a way around them.

Whether, in the large scope of things, the auction method will better the process is questionable. It may abolish the fraudulent practices of "application mills", but awarding spectrum licenses to the highest bidder, will only ensure that the very rich attain a voice in the marketplace. The very rich will not always represent the best qualified applicants, nor will they always provide a diversity of programming choices. These are concerns the FCC must address. Whatever method is chosen, the FCC must not lose sight of the fact that the most important objective is maximizing service to the public.

Robert F. Corazzini
is a partner in Pepper & Corazzini, a Washington, DC communications law firm. He served as Director Southern Satellite Systems, Inc., the satellite carrier for Superstation WTBS, TEMPO Enterprises, Inc. and WTPO-AM Inc., from 1979 until sale of companies to TCI in 1988. Named Chairman of the Legal, Litigation and Legislation Subcommittee of the Satellite Broadcasting and Communications Association of America for 1988. Named to Who's Who in the East 1980, Captain United States Air Force Discg. 1965, J.D. Georgetown University Law School, 1961, Law Journal 1969-1961, B.S. Cum Laude Mt. St. Mary's College 1958. Winner Hogan Award for highest average in the pre-law course.

Jennifer Richter
is an attorney and an associate with the law firm of Pepper & Corazzini in Washington, DC She graduated cum laude from the University of Wisconsin in Milwaukee with a B.A. in Journalism in 1988. She received both her Juris Doctor and her M.A. in Mass Communications from Drake University in 1991. Prior to joining Pepper & Corazzini, she served as federal judicial clerk to Commissioner Julie Carnes of the United States Sentencing Commission.

LEGAL PERSPECTIVES on WIRELESS CABLE

Footnotes to Chapter

1. For purposes of this chapter, "wireless cable" will refer collectively to single-channel and multiple channel MDS.
2. For purposes of this chapter, ITFS will also be referred to as "educational television."
3. Amendment of Part 12.703(g), 47 FCC 2d 957 [19 RR 2d 1847] (1970).
4. Educational Television, 39 FCC 846 [25 RR 1785] (1963), recon. denied, 39 FCC 873 [2 RR 2d 1615] (1964).
5. Id. at 853.
6. Second Report and Order, FCC 91-302, Gen Docket No. 90-54 (October 25, 1991) at 851.
7. Instructional Television, 30 FCC 2d 197 [22 RR 2d 1635] (1971).
8. Notice of Inquiry and Proposed Rule Making and Order, in Gen Docket #80-112, 45 Fed. Reg. 29323 (1980).
9. MMDS Allocation Order, in Gen. Docket No. 80-112, 94 FCC 2d 1203 [54 RR 2d at 113] (1983).
10. 47 C.F.R. §21.901(d).
11. MMDS Allocation Order at 119.
12. Id. at 120.
13. Id. at 123.
14. Id. at 123.
15. See, Wireless Cable Service Reconsideration, in Gen. Docket No. 90-54, 56 Fed. Reg. 57596 [69 RR 2d 1477] (1991); Wireless Cable Service Second Report and Order, in Gen. Docket No. 90-54, 56 Fed. Reg. 57808 [69 RR 2d 1499] (1991).
16. Market is defined as a 50 miles radius from the wireless cable operator's transmitter site. A wireless entity must conduct a 50 miles search from its transmit site to determine the use and availability of ITFS frequencies.
17. MMDS Allocation Order, 54 RR 2d at 144.
18. Second Report and Order, in Gen. Docket No. 80-112, 50 Fed. Reg. 5983 [57 RR 2d 943] (11984).
19. Id. at 948-956.
20. MMDS Allocation Order at 138.
21. Id. at 139.
22. Wireless Cable Service Report and Order in Gen. Docket No. 90-54, 55 Fed. Reg. 46006 [68 RR 2d 429] (1990) at 19.
23. Wireless Cable Service (Reconsideration) at 1489.
24. Wireless Cable Service Report and Order at 10.
25. PR Docket No. 92-80 RM 7909.
26. Wireless Cable Service Report and Order in Gen. Docket No. 90-54, 55 Fed. Reg. 46006 [68 RR 2d 429] (1990).
27. Wireless Cable Service Report and Order at 21.
28. Wireless Cable Service Report and Order at 25.
29. Id. at 35-36.
30. Pursuant to 47 C.F.R. § 21.902.
31. Wireless Cable Service (Reconsideration) at 1483.
32. Wireless Cable Service Report and Order at 39.
33. Service area is defined in 47 C.F.R. § 21.31(a).
34. Wireless Cable Service Report and Order at 39.
35. Flint, J., "Wireless Cable Lotteries Attacked by FTC: Trade commission says FCC's method of awarding licenses attracting con artists, "Broadcasting, p. 42 (April 20, 1992).

CHAPTER 17. The WIRELESS DEAL in the U.S.

What does one do when the license to broadcast one or more channels of wireless television has been obtained or when one wants to be either an active or passive investor in the wireless cable industry? This chapter explores this question. The first section, contributed by Ted Tarver of the "Wireless Cable Connection," explores the wireless deal. The role of the Wireless Cable Association International (WCA) is outlined in the second section. It was contributed by the WCA.

A. THE WIRELESS DEAL

What do you want and what will you take? What can you stand, what would you make? These are question that must be answered when considering entering into a deal for the lease and/or purchase or rights to wireless cable frequencies.

Wireless frequencies fall into two basic categories: commercial (MMDS) and educational (ITFS) channels, each having different requirements and expectations.

The first thing to be considered is that without the Federal Communications Commission's "Authorization to Transmit," a wireless cable system cannot be developed. Call them whatever you want frequencies, channels, spectrum, etc., they are still authorized by the FCC and without them a system cannot be built.

The second factor is that virtually all commercial frequencies have been licensed or applied for in the U.S. This means that in most cases the frequencies will have to be negotiated for rather than filed for.

There are just about as many types of deals as there are markets, although most are similar in structure. In this section we address various aspects of different agreements pertaining to the development of a wireless cable system, including the licensee's, operator's and educational entities perspective.

Some licensees want to develop the system themselves and are really not interested in relinquishing their rights to the channels. Others have applied speculatively for the frequencies in hopes of finding someone who really wanted to build a wireless system, in order to lease or sell the rights.

Some licensees realize that this is a growing industry and are willing to expand their horizons as the system grows, as long as the operator strives to develop the system. Others believe that they have won the New York lottery rather than an FCC lottery, and want an arm, a leg and a first born to lease the channels. The latter people are best to be avoided while another market is explored for development.

Some basic considerations deal with whether or not the licenses are going to be purchased outright or leased on a long term basis. However, most lease agreements include some form of an option buy-out at some negotiated future date and price.

THE WIRELESS "DEAL"

The most important factors to consider when trying to determine the fair value of the frequencies in an outright purchase are:

- The number of households that can potentially be served
- The type of terrain in the intended market
- The status and availability of other frequencies for which licenses have been applied or granted
- The current status of the license or grant
- The asking price

The number of households in the market that can receive the service is the single most important factor in determining the value of a license. A prospective operator or licensee should carefully analyze the real market potential by performing a series of propagation studies from several potential transmit sites. The operator should do so to make sure he is getting all the potential subscribers for whom he is paying. The licensee should do so to make sure that every serviceable household is counted in determining the purchase price.

The type of terrain is important because of the shadowing effect of the transmission signals and and because of the amount and cost of the consumer equipment required at the subscriber's home. Flat terrain generally requires shorter, less costly masts for the receive antenna, while mountainous, hilly terrain requires taller, more expensive masting for the same.

For example, a five-foot "Jiffy" mount costs about five dollars, while the additional cost to install a receive antenna thirty feet above the roof is approximately fifty dollars more when including the required tri-pod, push-up mast, guy wire and additional labor. Multiply this additional cost by the number of subscribers and the importance of terrain evaluation becomes quite apparent.

The availability of additional channels and the ease of acquiring those channels is also a very important consideration. Should all the educational channels be licensed and in use, the greatest number of frequencies available to the wireless system would be the thirteen commercial channels.

The current status of the frequencies is also important because the FCC application process may pass through one or more of several phases that include: tendered for filing, accepted for filing, tentatively selected, conditionally licensed, certified licensed, dismissed, returned, reinstated, dismissed, forfeited, etc. One can easily understand that the current status is crucial knowledge.

For example, a conditional license typically requires the licensee to construct the transmitters for the entire group of channels for which the license has been granted, within one year from the date of the grant. Extensions are not normally available. Naturally, the purchase of a conditional license that was granted eleven and a half months previously is not usually a wise move.

In the final outcome, value and therefore price is in the eye of the beholder. There are two questions that must be asked in every deal: What is the asking price and what is it really worth? Answers to each must be determined by both the buyer and seller.

Most of these issues outlined above must be examined and satisfactory answers must be achieved by lessor and lessee, or licensee and operator.

The Deal from both Buyer's and Operator's Perspective

The optimal scenario from the buyer's perspective would have been to have personally filed for the licenses. Should this option not be available, the only alternative is a lease or purchase agreement.

Most licensees require several assurances before committing to lease the rights to their frequencies such as a proof of financial wherewithal to construct and operate the system for at least the first year. In addition, a signing fee that can range from a few thousand dollars for smaller markets to tens of thousands of dollar for larger markets is normally required. A letter of credit, escrow account or similar collateral may also be required to ensure timely payment of monthly subscriber royalty payments.

Typical agreements also require that the operator pay for the construction of the transmitting facilities and guarantee the timely completion thereof. A penalty equaling the option buyout multiple (discussed below) in the third or fifth year, using the projected number of subscribers or a predetermined amount, is often assessed against the operator if construction is not complete and the licensee has to forfeit the license due to the operator's failure to construct in a timely manner.

Typical royalty payments range from twenty to fifty cents as calculated from the cost per subscriber, per month, per group. A good rule of thumb for the operator to follow in determining royalty payments is to use a percentage of gross revenues. Royalty payments for all channels, including all ITFS and MMDS frequencies, should never exceed ten percent of gross revenues.

Typical royalty fees for MMDS channels are based per group and run from 1.5 to 2% of gross for E or F group channels and 1 to 1.5% of gross for all three H-group channels, at a maximum. This breaks down to one-half of 1% per channel for commercial channels. MDS-1 and MDS-2 or 2A fees would be slightly lower due to the additional cost associated with the dual-band receive equipment required.

Some licensees prefer a flat rate per subscriber, per channel, per month, which translates to approximately 5 to 12.5 cents per subscriber, per channel, per month or 20 to 50 cents per group.

By whatever method the royalty fee is calculated for the commercial channels, a monthly minimum is usually required to ensure that the operator actively builds the system. This minimums typically starts at five hundred to one thousand dollars per month in years one and two and increases in years three through five. A provision may be included in the agreement whereby the licensee can option the license back, should the subscriber royalty fee not exceed the minimum after year five.

When an option buyout provision is included in the agreement, it is normally a multiple of cash flow. The normal range for this multiple is five to ten times annual cash flow. The following example outlines these considerations.

Assume the operator is leasing one commercial group of four channels at ten cent per channel, per sub, per month, for a total of 40 cents for the entire group. When he exercises the option, ten thousand subscribers are served. Therefore, the operator pays $4,000 per month for the group in royalty fees. The annual total for the lease of the group would then be $48,000. Assuming that the negotiated multiple for the buyout is eight times cash flow, the option price to purchase the group would be $384,000.

The licensee usually allows the option to be exercised between the third and fifth year of operation. It can be readily seen that the sooner the option is exercised, the more cost effective it is to the operator.

Some licensees would agree to an outright purchase of the license. In this case, a purchase price is negotiated then the funds are escrowed. The money would then be disbursed to the licensee over time, based on certain events that trigger payment. Benchmarks for payment are usually at the time of signing, when the conditional license is granted, when the construction of the transmission system is certified and when the assignment paperwork is filed with the FCC. The final payment is usually made when the FCC approves the transfer of ownership.

Some additional costs for the operator may include any engineering and legal expenses for modifications to the transmit facility as well as the FCC licensing fee. The licensee typically requires that the operator name the licensee as an additional insured party on his insurance policy.

Educational entities have their own needs and motivations for leasing their excess air-time. The educational licenses typically require a signing fee and operator supply of all equipment necessary to fulfill the institution's obligations relating to transmitting and receiving the educational programming. In other words, the operator most likely has to supply all of the transmit and receive equipment for the school's needs as well as his own.

Educational institutions also receive a per-subscriber royalty fee for leasing their excess air time. The royalty fee usually paid to the educator is typically much less than that paid to a commercial licensee. The range of royalty fees for ITFS channels is 8 to 20 cents per group, depending on market size. In the case where the royalty fee is based on a percentage of gross revenue, the typical fee is 1% of gross.

In some instances, a school may have additional needs for equipment such as televisions or VCRs. Perhaps buildings may need to be wired to accommodate an MATV system, cameras and editing. These additional components or services may not absolutely be required of the operator but may serve as the enticement necessary to land the deal.

Educational institutions are required to transmit a minimum number of hours of educational programming that is to be incorporated into their curriculum. If the institution does not already have access to such programming, the operator may be required to obtain a source or to even supplement the programming via a grant or some other form of financial aid to the school.

THE WIRELESS "DEAL"

In Conclusion

Hopefully what appeared complicated at the onset of this section has been simplified. Remember, "typical" is the keyword, not the gospel. If nothing else, this information above should give the novice a head start in negotiating his first wireless deal or fortune.

Ted Tarver is president and CEO of Wireless Cable Connection, Inc., a Houston-based firm, founded in 1985, which specializes in the design, construction, implementation and operation of wireless cable systems. He began his cable experiences in 1979 as a cable TV installer and later became manager of the Dallas/Ft. Worth Metropolitan wireless cable system. He also serves as Technical Editor for *Private Cable plus Wireless Cable Magazine*.

B. THE WIRELESS CABLE ASSOCIATION INTERNATIONAL, INC.

The wireless cable industry has entered the 90s as one of the new and significant multichannel video providers of programming throughout the world. Established in 1988, the Association recently changed its name to Wireless Cable Association (WCA) International, Inc. to help expand its efforts and benefits to the rapidly growing international market. Today, over 800,000 subscribers in 40 countries worldwide are enjoying wireless cable services as some of the most reliable, cost-efficient, multichannel home video service available.

Headquartered in Washington, D.C., the Wireless Cable Association International is made up of various components of the industry: program suppliers, system operators, equipment and service suppliers, MMDS lottery awardees, ITFS licensees/operators, academics, engineers, attorneys and other interested parties.

The WCA International sponsors an Annual International Convention and Exposition featuring the latest technology and trends in the business, as well as periodic seminars. Members of the WCA International register for these events at reduced fees. A general membership meeting is held in conjunction with the annual convention.

The WCA International publishes a newsletter for members six times a year with informative articles and updates on the state of the industry. Also included with the membership is a complimentary subscription to *Private Cable plus Wireless Cable Magazine*. Membership lists, attendees' lists and the like are available only to WCA International members.

Most importantly, the WCA International has a staff working full time to serve the needs of its members. Outside counsel who specialize in FCC, Congressional, local or international matters represent the WCA International before various government agencies such as lobbying and contributing information to standardize MMDS/Wireless cable frequencies in the 2.5 - 2.7GHz band as presented at WARC '92. The WCA International Board of Directors, which meets bimonthly, includes a cross-section of the wireless cable industry, and directors are elected for two-year terms. There are also committees within the WCA International open exclusively for member participation.

Membership in the WCA International insures your best representation in the marketplace for the wireless cable industry. Call 1-202-452-7823 to join.

*The Association Devoted Exclusively
to the MMDS/Wireless Cable Industry Worldwide*

- Membership insures your best representation in the market place
- Sponsor of an Annual International Convention and Exposition
- Sponsor of seminars and educational sessions
- Serving the industry for over 5 years
- Lobbied to standardize MMDS frequencies in the 2.5 - 2.7GHz band at WARC '92
- Quarterly newsletters
- Source of updated industry information

With over 40 countries currently using MMDS/Wireless Cable......

WE CAN HELP SUPPORT YOUR EFFORTS AROUND THE WORLD!

Call or write today and join WCA International, Inc.

Wireless Cable Association International, Inc.
2000 L Street
Suite 702
Washington, D.C. 20036
Phone 202-452-7823
FAX 202-452-0041

THE WIRELESS "DEAL"

SECTION III

PRIVATE CABLE SECURITY SYSTEMS

PROGRAM SECURITY

CHAPTER 18. PRIVATE CABLE SECURITY

Once the construction of an SMATV distribution plant or wireless broadcast facility has been completed, a signal security system must be installed with two goals in mind:

- To control the access of customers to programming
- To provide a method of increasing sales and reducing installation labor costs.

The choice of security system must ultimately be based on the menu of services to be offered both at system turn on and in the future as well as on economic constraints. A properly engineered system can furnish an operator with an effective means of controlling billing for basic and premium services and, if desired, pay-per-view movies and a host of other salable products.

Theft of service has been a serious concern of cable TV operators for years. An estimated 2% to 25% of pay-TV programming is lost to theft. This problem is an even greater concern for smaller-scale, private cable operators who usually have a less stable financial base upon which to build than that available to the larger, franchised CATV operators.

There are numerous choices of scrambling systems available today that range from very low to high levels of security. Depending on the types of services offered as well as the underlying contractual details, an operator may chose to install a system ranging from one having no security at all to a fully addressable design with instant pay-per-view (IPPV) and full interactive capability.

The need for security must be analyzed for each particular operating environment. In some cases, the possibility of theft is lower than others. For example, one would expect that pirating would be lower in a smaller-scale, high-cost retirement community than in a center-city, university housing complex. An SMATV operator is in an enviable position compared to a franchised operator who must cable all areas of his territory. The private cable operator can turn down any job that does not meet the necessary criteria for success.

It is interesting to note that most CATV operators do not opt for "hard" high-security scrambling systems on the basis of a well-considered marketing strategy. They have concluded that if their customers can just barely see the picture and hear the sound of a channel to which they do not yet subscribe, this partial information may entice them to complete the purchase.

In some cases, program security is not required. For example, if an SMATV operator has negotiated a contract to deliver bulk services to a specific apartment or institution, signals can simply be transmitted to all units "in the clear." In most such cases the operator would "cherry pick" a small group of basic programs and perhaps a premium service to distribute. This arrangement is popular in hospitals, motels and hotels where management is paying a flat monthly fee and where there is no need to control access to programming. This is one of the best arrangements for an operator because it reduces equipment costs, simplifies collection of funds and assures a steady cash flow.

PROGRAM SECURITY

Unfortunately, most system operators require a method to control subscriber access to programming. This is mandatory for a wireless cable operator because it is quite simple to intercept unscrambled microwave transmissions with just a few low cost components.

In general, security systems should optimally be compatible with the installation of multiple outlets at one subscriber location, VCRs, stereo FM broadcasts and cable-ready television sets. However, not all systems can be used with all user technologies. Clearly a system that is more flexible is advantageous. This allows an operator to authorize or deauthorize customer or re-configure the tiering structure easily and quickly.

A. Wireless Cable Security Systems

In the past, the original single channel MDS operators in the United States relied on the fact that their microwave transmissions could not be received on equipment that was readily available to the general public. However, a surge in availability of "mail order" antennas and downconverters led to a proliferation of pirate antennas in systems throughout this nation in the late 70s and early 80s. This caused serious financial losses for many operators and the need for more secure systems became quite apparent. Today, most commercial programmers would simply not license a wireless cable operator without a programmer-approved signal security system in place.

The preferred wireless cable security system today employs an addressable set-top converter. While these are essentially the same as set-tops used in traditional CATV systems, they have been specifically adapted for use in wireless cable. The modifications can include internal power supplies for providing voltage to the downconverter, channel mapping systems for utilization of the ITFS frequencies and a host of pay-per-view features including store-and-forward modules that allow for instant pay-per-view (IPPV) capability.

Set-top boxes work in conjunction with the encryption system installed at the wireless cable headend (see Figures 18-1 and 18-2). This system encodes all of the broadcast channels into a secure format that cannot be viewed without the use of the set-top decoders. This encryption usually takes place either in the baseband signal before the modulator or following the modulator IF stage but before the microwave transmitter. A dedicated encoder is used for each channel to be transmitted. As a result, all channels are broadcasted in a secured mode and unauthorized reception is prevented.

Control data for each individual subscribers box is inserted into the vertical blanking interval (VBI) of the secured channel and then sent to the subscribers set-top via the microwave transmission path. This data is individually addressed to each subscriber and each box is refreshed with this data on a regular basis. This information is decoded in the set-top box and provides the commands that allow the box to tune to specified programming services that the customer has ordered as well as special pay-per-view events (PPV).

Many addressable system today also include provisions for pay-per-view control through the remote addressing controls. In some instances the customer must call up and order these events from a customer service rep (CSR) at the wireless cable company. A more advanced version of this process utilizes an automatic number identification (ANS) system that allows the customer to use the touch tone pad of the telephone to order without directly speaking to anyone.

However, both of these processes can be time consuming and may result in lower PPV rates if the customer loses the desire to order an event as it becomes too much of a bother. To make it easy for a customer to satisfy the instant desire to purchase a film or special event, impulse PPV (IPPV) has been developed. With IPPV a customer can order by merely pushing a button on the remote control. This transaction is processed automatically for instant viewing of the desired event.

PROGRAM SECURITY

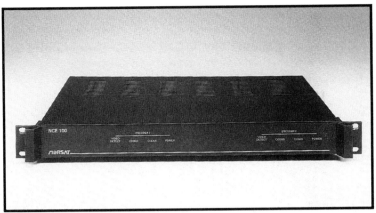

Figure 18-1. Norsat N*Code System. *The Norsat encryption system can be used in VHF, UHF or MMDS bands on NTSC or PAL broadcast formats. It consists of the NCE 100 headend encoder and the STV 50 subscription television decoder. The encoder accepts signals from satellite receivers, VCR or any other baseband audio/video sources as well as a data signal from the headend controller and outputs an audio and video signal to the channel modulator. The addressable decoder has separate inputs for wireless and off-air signals and may be configured to receive data from any one of the headend channel encoders. (Courtesy of Norsat International)*

To insure that the customer is billed for these purchases, wireless cable systems are no employing a technology known as store-and-forward whereby the customer's purchases are stored in the set-top's internal memory. At the end of the day, usually in the very early morning hours, the set-top box uses an internal modem to dial the billing computer at the wireless cable company and forward the billing information. This requires that the set-top be connected to a phone jack at the time of installation.

Another important function of the set-top box is its ability to "channel map" ITFS frequencies onto unused carriers. Because of FCC regulations regarding minimum levels of educational broadcasting time on ITFS carriers, a wireless operator may not always be able to transmit a specific program on the same ITFS channel but may be required to shift between three or four channel in a certain group over the course of a day to achieve an uninterrupted program broadcast. Channel mapping allows the operator to switch a specific channel between a primary carrier and one or more secondary channels through the use of an automatic, computer driven system. The channel map system is connected to the converter control computer and switches the output of the receivers and processors to various transmitters as required. The computer system also generates a control signal that is transmitted in the VBI to the set-top box. This signal instructs the box to which channel to switch in order to be able to automatically follow the channel map without interruption of the programming service. The box changes to the

Figure 18-2. Zenith Z-TAC Headend Computer and Set-Top Converter. *The Zenith Z-TAC baseband addressable system is based on SSAVI encryption. It can accept inputs from wireless as well as off-air broadcasts. (Courtesy of Zenith Cable Products)*

new tuning frequency while the actual channel indicator remains on the primary channel designation to avoid confusing the customer.

339

PROGRAM SECURITY

The same secured transmission path that controls a wireless broadcast also presents one of its biggest drawbacks, the need for an individual box for each television set in a home. Because all signal are encrypted, simply splitting the downconverter output does not provide independent channel selection capability for the additional sets. Instead, a separate set-top must be installed on each additional television. Given that the cost of set-tops range from $95 to over $135 each, installing additional boxes can add significantly to system cost. In general, at least half of wireless subscribers request service to an additional TV set in their homes. Some operators try to overcome this limitation by simply splitting the output of the set-top and feeding both sets from one unit. While this results in cost savings, both televisions must be tuned to the same channel, not an acceptable solution for most customers.

Most set-tops available today scramble the signal at either the baseband or RF level. At either level, a combination of techniques is used to prevent the signal from being viewed by the unauthorized subscriber, and to protect the system itself from tampering by electronic hackers and cable thieves.

Very simple scrambling systems use a technique the moves the sync pulse to a different position in the TV signal. The TV set cannot detect the sync pulse in this new position, and as a result, cannot create a watchable picture. A simple set-top decoder replaces the sync signal in its original position, and the picture is tunable by a standard TV set. This technique, one of the first scrambling systems used in CATV, is relatively easy to defeat.

Modern set-top systems expand on this concept with a combination of techniques. The sync pulse is dynamically manipulated so it is relocated to various positions on a constant or random basis. An active data stream required for the set-top for descrambling. Active video inversion is also used in conjunction to invert the video signals at random intervals as well. Other processes mask or distort the audio channel. Dynamic modification of both the phase and amplitude of the RF signal is also combined with the above techniques to achieve a higher multi-mode security level.

B. SMATV Security Systems

Set-top technology similar to that used in wireless cable is also available for use in SMATV systems. However, the cost of the encryption equipment, usually in excess of $1500 per channel makes this solution impractical for all but the largest projects. A drawback similar to that occurring in a wireless system exists because a set-top is required for each TV set to be connected. This need is an additional liability in the SMATV environment since most apartment dwellers frequently move and, in the process, set-tops tend to get lost or stolen.

Passive Traps

The most effective security systems used in SMATV projects are those that utilize an off-premise device to control signal access. In its simplest form, such a system can consists of a series of distribution taps that have signal traps installed to control premium services.

Traps are low cost filters that are inserted into a subscriber line to block the passage of a specific frequency, such as a pay service or an entire group of channels such as the mid-band frequency range. When secured in a locked enclosure, traps provide the lowest cost security available to an operator.

One common problem with traps is the need to install one or more on every drop leading to each customer. When a large number of drops terminate in one distribution cabinet, the resulting number of traps can be considerable and can quickly turn the cabinet into an installer's nightmare. A simpler solution is to arrange a series of splitters on a board with various traps connected to the input of the splitters (see Figure 18-3). This creates a "punch board" to which the subscribers drop can be connected and allows for rapid service changes while minimizing the number of traps in any one box.

PROGRAM SECURITY

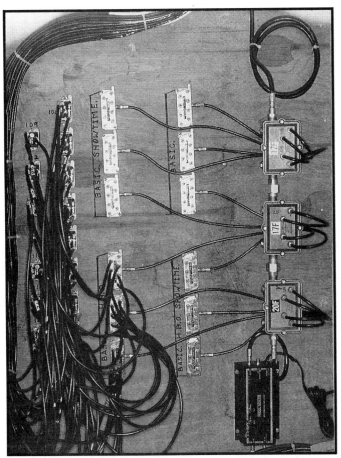

Figure 18-3. "Tap and Trap" Distribution Control System. *Three subscriber service levels are controlled from this junction box. The jumpers can be moved to any of four customer tiers, Basic, Basic/Showtime, Basic/HBO and Basic/Showtime/HBO. The unit at the bottom is Pico Macom MCM-55 line amplifier.*

Active traps are more expensive that passive traps and are usually only used to secure premium services. The number of active jamming carriers in any system is usually limited to three or four at the most because these encoders tend to produce additional unwanted intermodulation products that increase geometrically as the number of carriers increases.

Although trap systems can provide a high level of security at a low cost, they are labor intensive and tend to limit sales. Each time a service change is required, a technician must physically visit the site and install or remove the device. In addition, traps do not allow for the provision of pay-per-view services. In order for an operator to fully realize the full profit potential of a system, addressable security is required.

Off-Premise Addressable Security

An off-premise, addressable security system allows an operator to fully control access to all programming via a remote computer that communicates directly with the subscriber control device. The two most common systems used today are remote switching and interdiction. Using both of these technologies together allows for the greatest flexibility in designing a subscriber control system.

Remote switching systems use a pin diode switch to turn the signal on and off. These switches are also used to route the signal through a channel or band trap to control an additional tier of service. Given this technology, an operator can easily activate or deactivate subscribers and also provide access to one or two premium services.

For more elaborate control functions, interdiction technology is used whereby jamming carriers are inserted onto individual channels. These carriers are created in the control device by frequency specific oscillators. They are switched onto those channels to which a customer has not subscribed. While this jamming carrier renders the picture unwatchable, it does not interfere with any adjacent channels.

Active Traps

Another type of trap commonly used is the active trap. In this system, a headend encoder creates a jamming signal that is inserted into the modulated carrier of a particular channel. This signal is slightly offset from its center frequency and results in an unwatchable picture. The active trap removes this jamming signal before it reaches the television receiver.

PROGRAM SECURITY

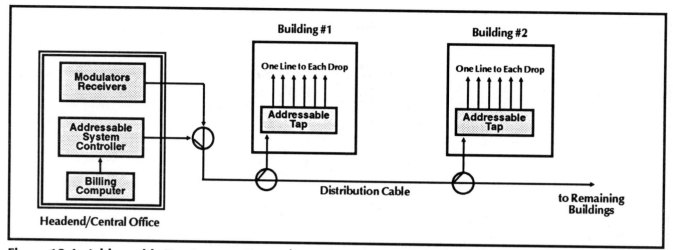

Figure 18-4. Addressable-Tap System. *The schematic illustrates the operation of an addressable tap system. The billing computer interfaces between the headend receivers and modulators to control access throughout the system. An example of such a system, the Electroline EAS system, is outlined in Figures 18-5 through 18-8)*

When these two technologies are combined the highest level of system control is achieved (see Figures 18-4, 18-5, 18-6, 18-7 and 18-8). By using the pin-diode switches, overall access to the programming can be authorized or denied. An expanded tier of service can be programmed onto, for example, the mid-band channels and this service can be controlled by switching through a mid-band trap. Individual premium services can be protected via interdiction carriers set to specific frequencies. With the combination of these two technologies, up to five or six different levels of service can be controlled and delivered to each individual subscriber.

These system are remotely controlled from the central office via a computer and modem (see Figure 18-4). The computer interfaces with the billing software and instructs the control system which customers to turn on and what level of service they are to receive. The modem transmits this information to the headend where it is modulated onto a carrier that is transmitted downline in the SMATV RF distribution plant. Each addressable station has a demodulator the receives these signals and routes the commands to the individual switches. Each subscriber has a specific address and, in most cases, can be controlled from the main office within a matter of seconds.

Such an addressable security system provides an operator with the ultimate SMATV subscriber control tool (see Figures 18-5. The subscriber control units are always locked up in a security cabinet and thus are beyond the reach of the customer. If and when a customer is late in paying a bill, the operator can disconnect the service at a touch of a button without having to dispatch a technician to the site. As soon as the customer settles his or her account, the operator can reactivate the service just as easily while collecting a $10 to $25 reconnect charge. The use of this system helps to maximize cash flow while minimizing costs and bad debt.

Another feature of the addressable system is its ability to provide instant installation of new subscribers if all of the units in a complex had been wired to the system from the start. When a new tenant moves in, the property manager can provide a small kit that contains jumper cables, a matching transformer and a simple instruction sheet on how to connect a TV to the wall outlet. After the television set has been connected, all they need to do is to call the main office and sign-up for the requested services. After a credit verification has been completed, the customer service rep can authorize the service via the computer. A subscriber can be watching TV within a matter of minutes without having to wait all day for an installer to arrive. This minimizes the manpower requirements for the operator and thus serves to cut direct labor and overhead costs.

Addressability also creates tremendous sales opportunities that cannot be achieved with standard trap technology. Pay-per-view (PPV) can add significantly to the bottom line with services such as first run movies, sporting events and concerts. The PPV industry is promoting more events every year and the revenue shares from subscriber sales can increase gross income by five to ten percent. However, PPV can only accomplished in a system equipped with addressable technology.

PROGRAM SECURITY

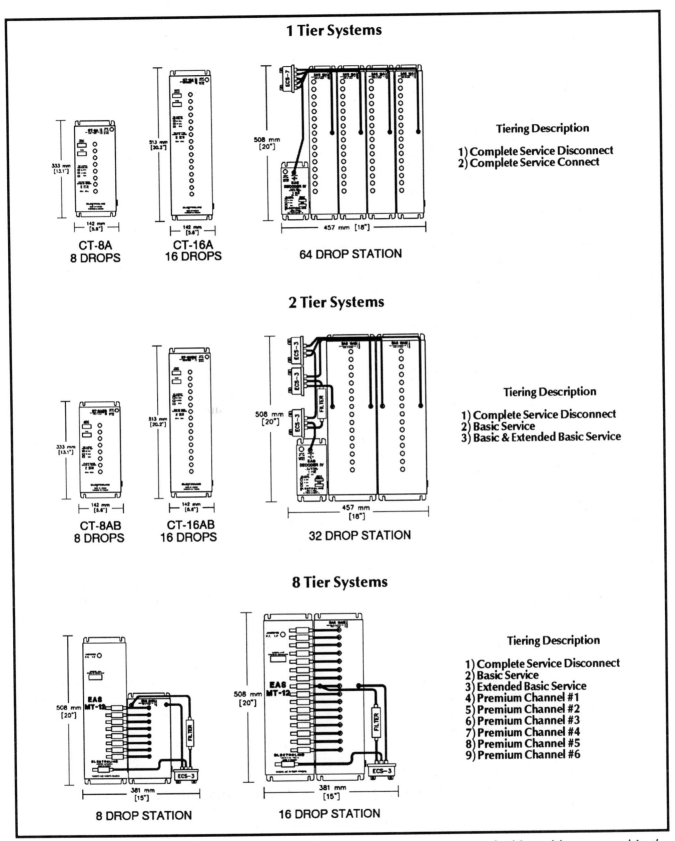

Figure 18-5. Electroline EAS Addressable Taps. *The various types of addressable taps used in the Electroline addressable system are illustrated here. These vary from one to eight tier systems and can manage signals in the 5 to 862 MHz range. Some of these components are shown in the following figures. (Courtesy of Electroline Equipment, Inc.)*

PROGRAM SECURITY

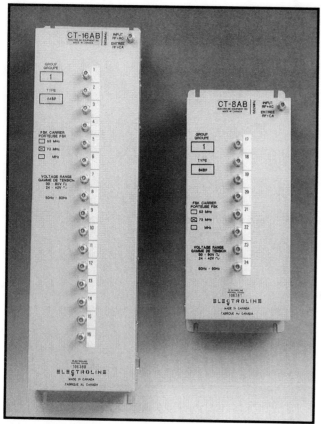

Figure 18-6. Addressable Taps for the Electroline EAS System. *This photo shows two addressable taps an 8-port and a 16-port model. Each of these is available in an on/off and a 2 tier/on-off variety. (Courtesy of Electroline Equipment, Inc.)*

An operator can also expect to increase premium movie subscribers with use of an addressable system. Premium to basic service ratios in the CATV industry have dropped over the past couple of years and programmers are working hard to increase their market share. Today many programmers broadcast weekend-long promotions of their services during periods of the year. With an addressable system, all subscribers can be instantly authorized to receive the special promotion. This can result in additional exposure and subscription to these services that an operator could not achieve with a trap system.

The availability of essentially instant installations in addressable systems provides further sales leverage. Sales personnel generally experience higher closing rates with new customers if they can arrange to turn on the service as soon as the customer signs the order and pays a deposit. In addition, by eliminating the need for the customer to spend a day waiting at home for an installation technician, the operator can experience higher subscriber penetration and fewer customer cancellations after the sale.

The cost to add off-premise addressability to an SMATV project can range from $70 to $100 extra per unit at the time of installation. Although this number can represent a 50% increase in per unit construction costs, the additional revenue that the system can generate along with the cost savings in manpower resources can more than offset the initial investment. According to industry averages, over 30% of all multi-unit residents move on an annual

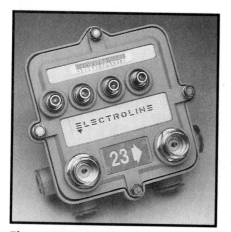  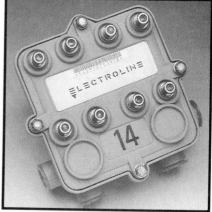

Figure 18-7. "Tap and Trap" Distribution Control System. *Three addressable taps are shown here. The "super tap" on the left has an input and output port at the bottom front, 4 on/off ports and 4 ports that control 2 tiers. The middle unit is available in an 8 drop - on/off, a 4 drop - on/off and a 4 drop - 2 tier plus on/off variety. The unit on the right has 8 on/off ports. (Courtesy of Electroline Equipment, Inc.)*

PROGRAM SECURITY

basis. Thus an SMATV operator could expect a complete turnover in subscribers once every three years. When using a trap system, one trip would be required to install each new customer. Given an average contract life of seven years per complex, this turnover would require at least four trips per unit for service changes. With service calls today running a minimum of $35 to $50 per truck, the installation of an addressable system can ultimately save an operator at least twice its cost.

In addition to tenant turnover there is the issue of disconnects. Again, the operator would be required to roll a truck to disconnect the service and to reinstall if the customer settles the account. With an addressable system, the operator can perform these functions from the main office and still collect a service fee from the subscriber.

The other financial consideration of an addressable system is the additional revenue generated from an increase in the number of basic subs, additional premium service sales and pay-per-view specials. Add to this is the increased value of the system to a potential buyer at the time of sale and the case for addressable technology becomes quite clear. In summary, given the reduced labor costs, greater service options, increased sales and higher system valuations installation of an addressable control system leads to an SMATV system with a greater value.

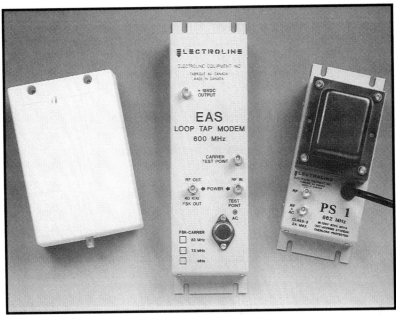

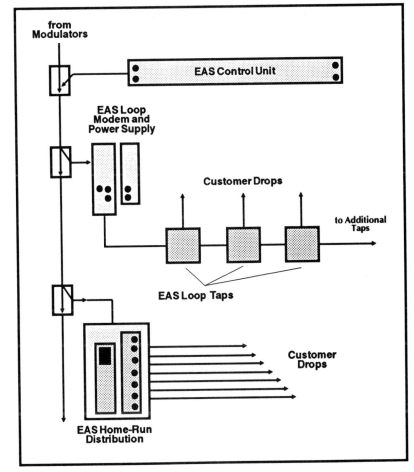

Figure 18-8. Addressable Taps for Loop-Run System. *The photo above shows the components used in a loop-run system. The tap (left) is functionally identical to a series run passive tap, in on/off and on/off plus pay configurations. The EAS modem and power supply, the middle and right units, respectively, powers and maintains up to 72 addressable loop taps. The bottom illustrations outlines how the EAS control unit can manage either loop-run or home-run*

PROGRAM SECURITY

APPENDIX A. ADDITIONAL EXAMPLES of SMATV DISTRIBUTION SYSTEMS

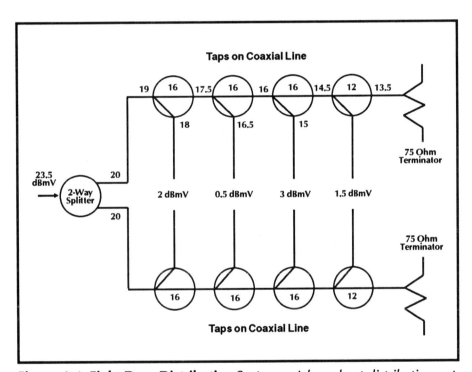

Figure A-1. Eight-Drop Distribution System. *A broadcast distribution network can be compared to a water sprinkler system in which each outlet should have as much water pressure as any other. Each leg in a distribution network should have enough power to properly drive the audio and video equipment. In this example, the 23.5 dBmv signal would reach the 2-way splitter and feed 20 dBmv into each branch. A 1 dB cable loss is incurred on the way to the first tap resulting in a 19 dBmv input. Then a 16 dB tap which has an insertion loss of 1 dB lowers the signal level to 18 dBmv. Finally, 18 dBmv less the 16 dB tap value or 2 dBmv reaches the first television receiver. 17.5 dBmv arrives at the second tap after a 0.5 dB cable loss. Then 16.5 dBmv less 16 dB tap value equal to 0.5 dBmv enters the second TV. 16 dBmv is relayed to the input of the third tap. Inserting another 16 dB tap here would cause levels to fall below the designed 0 to 3 dBmv required television input, so therefore a 12 dB tap is used. Subsequently 15 dBmv less 12 dB or 3 dBmv is fed to the television receiver. 14.5 dBmv less a 1 dB tap insertion loss less the 12 dB tap value equal to 1.5 dBmv arrives at the final television set. A 75 ohm terminator is then used on the output.*

APPENDIX A

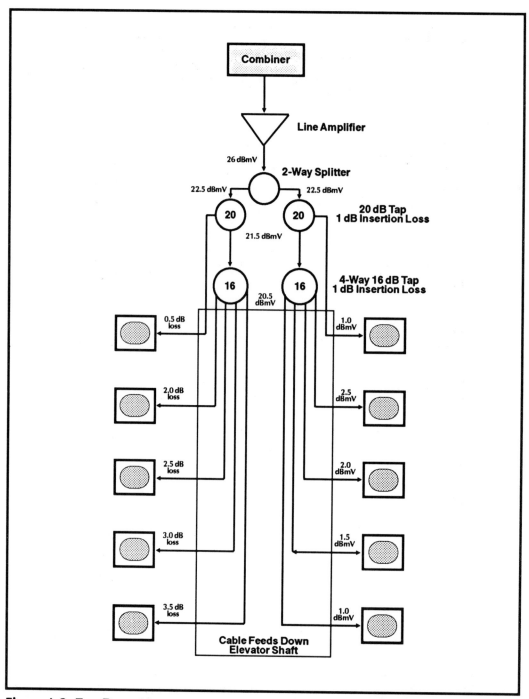

Figure A-2. Ten-Drop, Homerun System. *The combined outputs from the modulators are fed into a distribution amplifier having an output of 26 dBmV. The 2-way splitter attenuates the signal by 3.5 dB so that 22.5 dBmV is present on the output ports. These signals are fed into 20 dB taps each having 1 dB of insertion loss resulting in a 21.5 dBmV input to a 4-way, 16 decibel tap, also having a 1 dB insertion loss. Cable losses to television receivers on the building's first floor are 0.5 dB in each drop so that a 1.0 dBmV signal level arrives at the set. Second, third, fourth and fifth floor televisions receive 2.5, 2, 1.5 and 1 dBmV after 2, 2.5, 3 and 3.5 dB cable losses, respectively. All these signal levels are within the 0 to 3 dBmV input requirements for a TV set.*

APPENDIX A

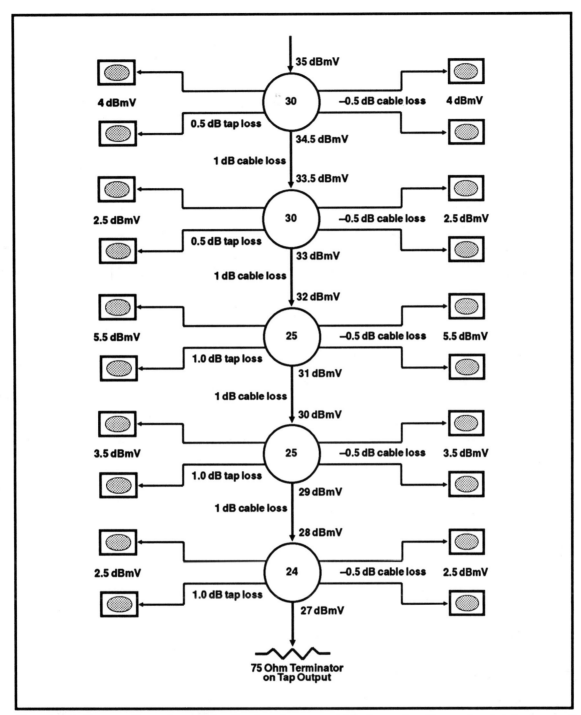

Figure A-3. Twenty Drop-Loop Distribution System. *A 35 dBmv signal is received at the input of the first 4-way tap. The tap insertion loss is 0.5 dB giving an output of 34.5 dBmv. Each cable run from the tap to the TV set also has 0.5 dB loss. Therefore, 34.5 dBmv less 0.5 dB less the 30 dB tap value results in 4 dBmv reaching each television set. A 2-way splitter having a 3.5 dB insertion loss installed in each unit would yield a final signal level is 0.5 dBmv. There is 1 dB of cable attenuation loss between each tap. 33.5 dBmv enters the second 30 dB tap which has a 0.5 dB insertion loss. Therefore the tap output is 3 dBmv less the 0.5 dB cable loss so that 2.5 dBmv signal is available to these televisions. The input to the 25 dB tap is 32 dBmv and following an insertion loss of 1 dB the output is 6 dBmv less 0.5 dB for cable loss or 5.5 dBmv. 30 dBmv arrives at the second 25 dB tap which has a 1 dB insertion loss. Therefore, 4 dBmv less 0.5 dB cable losses or 3.5 dBmv enters these sets. A 28 dBmv signal arrives at the 24 dB tap. Following a 1 dB insertion loss and 0.5 dB cable loss the output is 2.5 dBmv. A 75 ohm terminator must be installed at the output of this final tap to avoid signal reflections and unwanted interference.*

APPENDIX A

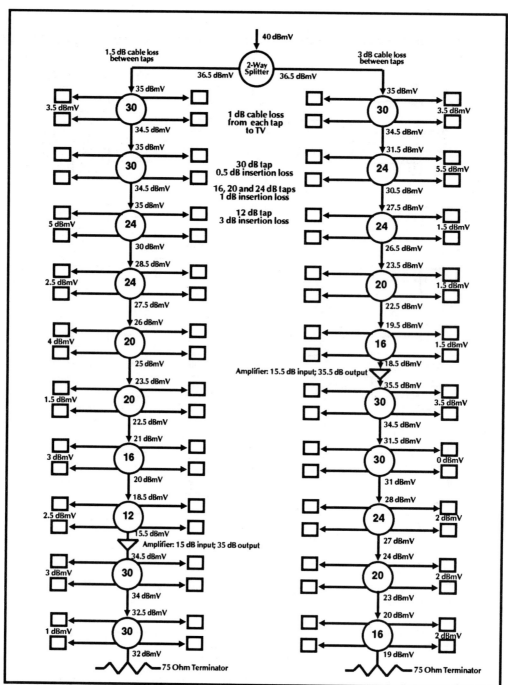

Figure A-4. Eighty-Drop Apartment Complex. *This example expands upon the tap and cable loss configuration presented in Figure 7-54. Cable losses between the taps on the right-hand leg are 3 dB and from each tap to each television receiver are 1 dB. Insertion losses for the 30, 24, 20, 16 and 12 dB distribution taps are 0.5, 1.0, 1.0, 1.0 and 3.0 dB, respectively. The output power on the right hand leg following the fifth tap is, by subtraction, therefore 18.5 dBmv. An additional 3 dB is lost to cable attenuation. The net result is a 15.5 dBmv input to the line amplifier, whose gain must be chosen to be 20 dB in order to generate a 35.5 dBmv output for the remaining drops. Cable and insertion losses must be subtracted between the remaining five 30, 30, 20, 20 and 16 taps. A 75 ohm terminator is installed at the final output.*

The left hand leg has only half the cable attenuation of 1.5 dB between taps because runs are shorter than those on the right hand leg. After the eighth 12 dB tap on the left hand side the signal level is 15.0 dBmv. A short jumper from the output of the tap to an amplifier results in another 0.5 dB loss. The amplifier output is 35 dBmv. Subsequently, signal levels decrease through each successive tap and cable run until the 75 ohm terminator installed on the output of the last 30 dB tap. If additional apartment buildings were to be added to this system, the terminator could be removed and a further series of taps with drop cables running into television sets could be added until the signal level dropped to 15 dBmv. At this point, an additional line amplifier would be installed on this feeder line. For this example, it is clear that designing even the most complex system is simply a matter of addition or subtraction to account for all signal losses in cable runs and taps. Actual measurements are completed by a field strength meter, an essential tool when installing headend and distribution systems..

APPENDIX A

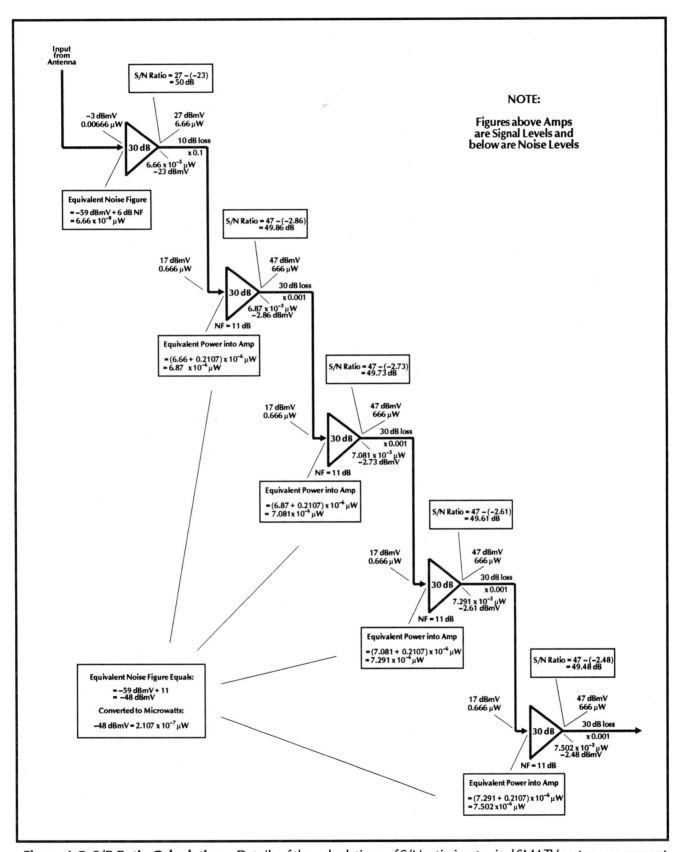

Figure A-5. S/R Ratio Calculations. *Details of the calculations of S/N ratio in a typical SMATV system are presented here. Signal levels are to the above right of the amplifiers symbols; noise levels to the bottom left.*

APPENDIX A

APPENDIX B. EQUATIONS and TECHNICAL DETAILS

A. Satellite and Cable TV Equations

The performance of a satellite TV system can be theoretically calculated by using some general equations. Some of these are presented here.

Wavelength to Frequency Conversion

Wavelength

The wavelength of an electromagnetic signal is given by

$$\lambda = v_{light} / f$$

where λ is the wavelength of the signal, v_{light}, the velocity of light, equals 186,000 miles/second or $300 \times 10^{+6}$ meters/sec, and f is the signal frequency.

Ohm's Law

The basic law of passive dc circuits, Ohm's law, relates the voltage and current by

$$E = I R$$

alternatively $I = E/R$

or $R = E/I$

where E is the voltage in volts, I is the current in amperes and R is the resistance of the circuit.

Ohm's law for active ac circuits is expressed as

$$E = I Z$$
$$Z = E/I$$

where E is the voltage in volts, I is the current in amperes and the Z is the impedance in ohms.

Power

The power flowing through any circuit is given by

$$P = V^2/R = E I$$

where power is expressed in watts, voltage V, in volts, the current I, in amperes and resistance, R, in ohms.

Inductive and Capacitive Reactance

Inductive and capacitive reactance in a circuit are given by

$$X_L = 6.28 \, f L$$
$$1/X_C = 6.28 \, f C$$

where X_L is the inductive reactance, X_C is the capacitive reactance, f is the signal frequency, L is the inductance in henries, and C is the capacitance in farads.

The current lags voltage by 90 degrees in an inductive circuit. As the signal frequency increases, inductive reactance increases and capacitive reactance decreases. The current leads voltage by 90 degrees in a capacitive circuit.

APPENDIX B

Resonant Frequency

The resonant frequency of an inductive-capacitive circuit is given by

$$f_r = 1 / (6.28 \, LC)$$

where f_r is the resonant frequency, L is the inductance in henries and C is the capacitance in farads.

Link Equations

The link equations are used to calculate the ratio of carrier-to-noise (C/N ratio) power reaching the input of a satellite receiver. The link equation is:

$$\text{C/N ratio} = \text{EIRP} - \text{Path Loss} + G/T_{sys} - 10\log B + 228.6$$

EIRP is the effective isotropic radiated power directed by a downlink antenna to a location below. It is expressed in dBw, decibels relative to one watt.

Path loss measures how much signal is lost on the journey from the communication satellite to the receiving antenna. Losses are mainly due to the "spreading out" of the signal on its long journey. The amount of spreading out is determined by the distance the signal travels from a satellite to a receiving antenna. Path loss in dB is given by:

$$\text{Path Loss} = 20 \log 4\pi S f$$

where S is the slant range and f is the signal frequency in Hertz. Slant range in kilometers is given by:

$$S = [R^2 + (R+h)^2 - 2R(R+h)\cos\Phi \cos\Delta]^{1/2}$$

where R equals 6,367 km, the radius of the earth, h equals 35,803 km, the distance of a satellite above the center of the earth, Φ is the site latitude and Δ is the absolute difference between the site and satellite longitude. Substituting R and h in this equation gives:

$$S = 1000 [58.32 - 53.69 \cos\Phi \cos\Delta]^{1/2}$$

If this value of S is substituted into the above equation for path loss we find:

$$\text{Path Loss} = 185.05 + 10 \log[1 - 0.295 \cos\Phi \cos\Delta] + 20\log f$$

where f is expressed in GHz. At 12 GHz the path loss equals 205.11 dB at an earth station located on the equator directly below a satellite. This equation also shows that signals from a Ku-band satellite 10° in longitude away from an earth station at 40° latitude would suffer a 205.54 dB free space path loss.

Atmospheric absorption causes additional path losses. Absorption increases with slant range because the satellite signal must pass through a greater thickness of atmosphere. As described in Chapter III, variations between rainy or overcast days and clear days are substantial for Ku-band broadcasts. An absorption loss of 0.5 dB is assumed in most calculations for reception on average clear days.

G/T_{sys}, the ratio of antenna gain to system noise temperature, is the figure of merit of an antenna/feed/LNB system. It is expressed in decibels as:

$$G - 10 \log T_{sys}$$

The system noise temperature primarily depends upon both the antenna and LNB noise temperatures. However, components further downstream towards the receiver also contribute small amounts of noise. This term is given by:

$$T_{sys} = T_{ant/feed} + T_{LNB}/G_{feed} + \frac{T_{rec/coax}}{G_{LNB} + G_{feed}}$$

where G refers to gain. The gain of a typical feed is in the vicinity of 0.99 while LNB gain is usually about 50 dB, equal to a factor of 100,000. This equation clearly shows why the noise contributed to T_{sys} by the satellite receiver and coaxial cable are negligible. The LNB amplifies both the signal and noise so much that any later contributions are of minor importance.

The second of last term in the link equations above adjusts for the system bandwidth and the final is a constant, called Boltzman's constant.

Antenna Gain

Antenna gain relative to an "isotropic" antenna, one that radiates equally in all directions, is given by:

$$G = E(\pi D/\lambda)^2$$

where E is the antenna efficiency, D is the diameter and λ is the wavelength of the incoming radiation. The wavelength in centimeters can be simply calculated by dividing 30 by the frequency expressed in GHz. The wavelength of 12 GHz microwaves is 2.50 centimeters or just under one inch.

For example, a 2 meter antenna with an efficiency of 55% operating at 12 GHz has a gain given by:

APPENDIX B

$$G = 0.55 \, (3.14 \times 200 \text{ cm} / 2.5 \text{ cm})^2$$
$$= 34,706$$

Expressed in decibels this gain or signal concentration of a factor of 34706 relative to an isotropic antenna is given by:

$$G = 10 \log 34706 = 45.4 \text{ dBi}$$

Loss of Gain with Surface Irregularities

The decrease of gain relative to a perfect antenna having no surface irregularities is given by:

$$\text{Loss of Gain} = e^{-8.80(\text{RMS})/\lambda}$$

where RMS is the root mean square deviation from a perfect geometrical shape and λ is the wavelength of the incoming signal. The RMS is a measure of the "tightness" of the surface or its average tolerance.

For example, a Ku-band antenna operating at 12 GHz, where wavelength equals 2.5 cm, having a RMS tolerance of 0.15 cm has a decrease in gain relative to a perfect antenna given by:

$$\begin{aligned}\text{Loss of Gain} &= e^{-8.80 \times 0.15 / 2.50} \\ &= e^{-0.53} \\ &= 0.59\end{aligned}$$

or a 41% decrease in gain. This equates to a 2.3 dB loss of gain given by:

$$\begin{aligned}\text{Decibel Loss in Gain} &= 10 \log 0.59 \\ &= -2.3 \text{ dB}\end{aligned}$$

Antenna Beamwidth

An approximate but very useful formula for 3 dB antenna beamwidth is:

$$\text{Beamwidth} = 70 \lambda / D$$

where λ is the wavelength of the microwave radiation and D is antenna diameter. For example, a 2 m antenna has a 3 dB beamwidth given by:

$$\text{Beamwidth} = 70 \times 2.5 / 200 = 0.88°$$

Similarly, a 1 m antenna would have a calculated beamwidth given by:

$$\text{Beamwidth} = 70 \times 2.5 / 100 = 1.75 \text{ degrees}$$

Noise Temperature and Figure

The noise any system generates is proportional to its ambient temperature and the bandwidth of the signal it processes. The larger either of these two quantities, the greater the contributed noise.

$$\text{Noise} = kTB$$

where k is Boltzman's constant, T is the ambient temperature and B is the system bandwidth.

A quantity called noise factor is defined by the ratio of the noise at the output on an electronic component to the noise at its input. This quantity measures, in essence, the amount of noise internally generated in any device. In a perfect device whose electronic circuits added no extra noise to a signal, the noise factor would be one.

$$\text{Noise Factor} = \frac{(\text{Ideal Noise} + \text{Internal Noise})}{\text{Ideal Noise}}$$

$$= (kBT_{\text{Ideal}} + kBT_{\text{Eq}}) / kBT_{\text{Ideal}}$$

$$= (T_{\text{Ideal}} + T_{\text{Eq}}) / T_{\text{Ideal}}$$

$$= 1 + T_{\text{Eq}} / T_{\text{Ideal}}$$

$$= 1 + T_{\text{Eq}} / 290$$

T_{Eq}, termed the equivalent noise temperature. The reference noise temperature, T_{Ideal}, is usually taken to be 290°K, equal to an average room temperature of about 63 °F.

Noise figure is the decibel equivalent of noise factor (see Table B-1) as is given by:

$$\text{Noise Figure} = 10 \log \text{Noise Factor}$$

For example, if the noise figure is 1.9 dB, the equivalent noise temperature is:

$$1.9 = 10 \log (1 + T_{\text{Eq}} / 290)$$

turning this equation inside out:

$$1 + T_{\text{Eq}} / 290 = 10^{0.19} = 1.55$$

$$T_{\text{Eq}} / 290 = 0.55$$

$$\therefore T_{\text{Eq}} = 159°K$$

APPENDIX B

TABLE B-1. NOISE TEMPERATURE and NOISE FIGURE

Noise Figure (dB)	Noise Temperature (°K)	Noise Figure (dB)	Noise Temperature (°K)
2.0	170	0.9	67
1.9	159	0.8	59
1.8	149	0.7	51
1.7	139	0.6	43
1.6	129	0.5	35
1.5	120	0.4	28
1.4	110	0.3	21
1.3	101	0.2	14
1.2	92	0.1	7
1.1	84	00	
1.0	75		

The Effect of Bandwidth on System Noise Power

The noise power in any communication system is given by:

$$\text{System noise power} = kT_{sys}B$$

where T_{sys} is the system noise temperature in degrees Kelvin mainly determined by antenna and LNB noise, k is Boltzman's constant equal to 1.381×10^{-23} and B is the communication bandwidth. The change in noise power between two systems can be computed as follows:

$$\text{Change in noise power} = (kT_1B_1/kT_2B_2)$$

$$= T_1B_1/T_2B_2$$

Therefore, if the noise temperature remains constant the change in noise power is simply the ratio of bandwidths. If the bandwidth were cut from 36 to 18 MHz as would be the case in half transponder formats, the noise power would be reduced by 50 percent or 3 decibels. The resulting doubling of the signal-to-noise ratio sometimes makes the difference between a watchable picture and a sparklie, faint ghost of a picture. But reducing the bandwidth will also result in a "softening" of the video as well as smearing and streaking of pictures having fast changes.

Declination Angle

The declination angle for a polar mount can easily be found from the tables and figures in the early chapters. It can also be calculated from:

$$\text{Declination} = \tan^{-1} \frac{3964 \sin L}{22300 + 3964(1 - \cos L)}$$

where L is the site latitude. The two numbers in this equation are the radius of the earth and the distance from the surface of the earth to the arc of satellites. For example, the declination angle at 40° latitude is:

$$\text{Declination} = \tan^{-1} \frac{3964 \sin 40}{22300 + 3964(1 - \cos 40)}$$

$$= \tan^{-1} 0.11$$
$$= 6.26 \text{ degrees}$$

Azimuth and Elevation Angles

Antenna pointing angles can be calculated in degrees from true north from the following equations:

$$\text{Azimuth Angle} = \cos^{-1}[-\tan\Phi/\tan Y]$$

$$\text{Elevation Angle} = \tan^{-1}[(\cos Y - 0.15116)/\sin Y]$$

$$Y = \cos^{-1}[\cos\Phi \cos\Delta]$$

where Δ is the absolute value of the difference between satellite and TVRO site longitudes and Φ is the site latitude.

Voltage Standing Wave Ratio

The voltage standing wave ratio, VSWR, is a measure of the amount of input signal reflected back and lost. A perfect device would have no reflective losses and have a VSWR of 1:1. Table B-2 shows how reflected signal power and transmission losses vary with VSWR.

TABLE B-2. VSWR and REFLECTED SIGNAL LOSS

VSWR	% Loss	dB Loss
1.1:1	0.2	0.01
1.2:1	0.9	0.03
1.3:1	1.6	0.07
1.5:1	4.0	0.18
2.0:1	11.0	0.50

APPENDIX B

Antenna Parabolic Geometry

The basic equation for a parabolic reflector is:

$$y = x^2/4f$$

where f is the focal distance. Another useful formula gives the focal distance f in terms of the antenna diameter and depth:

$$f = \text{diameter}^2 / 16 \times \text{depth}$$

Coaxial Cable Attenuation

The characteristic impedance of coaxial cables can be found from the follow equation:

$$z = (L/C)^{0.5}$$

where z is the characteristic impedance in ohms, L is the cable inductance in Henry's per unit length, and C is the capacitance is Farads per unit length.

Alternatively characteristic impedance can be calculated from a second equation:

$$z = \log(D/d) \times 138/(E)^{1.2}$$

where D is the inside diameter of the outer conductor, d is the outer diameter of the inner cable and E is the dielectric coefficient.

If cable attenuation at one frequency is known, then the loss characteristics at a second frequency can easily be calculated by:

$$C = (F_a/F_b)^{0.5}$$

where C is the change in attenuation. For example, attenuation is doubled when signal frequency increases from 50 to 200 MHz.

B. Wireless Cable System Parameters

Performance

The performance of a wireless cable system is determined by the signal-to-noise (S/N) ratio at the receive site. This ratio is determined by broadcasted power levels, by losses incurred on its voyage between the transmit and receive antennas, by the receive antenna gain, by the downconverter noise figure and by other losses such as those experienced in coaxial lines and connectors. S/N ratio at the downconverter output is given by:

> Transmitter Line Loss
> less
> Transmitter Line Loss
> plus
> Transmitter Antenna Gain
> less
> Path Loss
> plus
> Receive Antenna Gain
> less
> Receive Line Loss
> less
> Downconverter Noise Figure
> less
> Noise Floor

These quantities are all expressed in decibels. For example, transmitter power of 10, 50 and 100 watts equals 40, 47 and 50 dBm, respectively.

Each of these quantities are examined and a sample value is chosen to outline an example of this method to calculate S/N ratio. Note that while a S/N of 45 dB results in snow-free, class "A" pictures, this value is often unattainable due to intervening foliage, excessive distance from the transmitter or intervening topographical features. While the objective in designing a receive site should be to attain a 45 dB S/N, installations with S/N ratios of as low as 30 dB can be marginally successful.

The power leaving the transmit antenna is determined by transmit power, signal loss is incurred in the cable from the transmitter output to the antenna input terminal and transmit antenna gain. In the example below, transmit power is 47 dBm, transmit antenna loss is 3 dB and transmit antenna gain is 13 dBi.

The path loss is the amount of signal lost in the voyage between the transmit and receive antennas. This increases both as frequency and path length increases. Figure B-1 shows this loss for 2.156 and 2.69 GHz. A path loss of 131 dB is assumed in the following example.

APPENDIX B

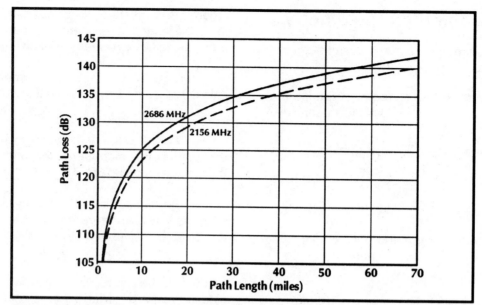

Figure B-1. Path Loss versus Distance between Transmit and Receive Antennas. *As path length and frequency increases, the path loss in decibels increases.*

At the receive site the signal is amplified in power by the receive antenna, relayed on a length of cable between the antenna dipole and the downconverter and processed by the downconverter. Receive antenna gains are typically 15, 18 and 21 dBi. The loss enroute to the downconverter is typically quite small. The downconverter noise figure is typically in the range of 2 dB. A receive antenna gain of 18 dBi, a receiver line loss of 0.1 dB and a downconverter noise figure of 1.8 dB is assumed in the example here.

The "noise floor" is the amount of inherent random noise in any electronic system. The noise floor is relatively constant at approximately –108 dBm. It does vary somewhat with ambient temperature. The signal-to-noise ratio is a comparison between the signal power at the output of the downconverter to the noise contributed by both the downconverter and the noise floor.

The S/N ratio at the downconverter output can now be easily calculated using the values listed in the text above:

S/N = 47 dBm – 3 dB + 13dB – 130 dB + 18 dB – 0.1 dB – 1.8 dB – (–108 dB)
= 51.1 dB

This S/N power is more than sufficient to deliver excellent quality pictures at the receive site

Conversion of FSM Voltage to Field Intensity Readings

When measuring egress levels with a FSM and dipole antenna the length of each of the 1/4 wavelength arms with a 0.95 correction factor is given by:

Dipole Length (inches) = 234 / Frequency (MHz)

The field intensity measured by this dipole antenna can then be calculated from:

Field Intensity = 0.0207 x Voltage x Frequency

where voltage readings are in microvolts and frequency is measured in MHz.

APPENDIX C. PROGRAMMERS

All News Channel
401 North Michigan Avenue
Suite 1600
Chicago, IL 60611
Telephone: 312-645-1122

American Movie Classics
150 Crossways Park West
Woodbury, NY 11797
Telephone: 516-364-2222

Arts & Entertainment (A&E)
2450 Broadway, Suite 500
Santa Monica, CA 90404
Telephone: 213-453-4492
or
555 Fifth Avenue
New York, NY 10017
Telephone: 212-661-4500

Black Entertainment Network (BET)
1212 New York Avenue
Suite 430
Washington, DC 20005
Telephone: 202-408-5480

Bravo Cable Network
150 Crossways Park West
Woodbury, NY 11797
Telephone: 516-364-2222

CNBC
3000 West Alameda Avenue
Burbank, CA 91523
Telephone: 818-840-3333

Cable News Network (CNN)
One CNN Center
100 International Blvd.
P.O. Box 105366
Atlanta, GA 30348-5366
Telephone: 404-827-1500

Comedy Central
1633 Broadway
New York, NY 10019
Telephone: 212-708-1257

Country Music TV
P.O. Box 10210
Stamford, CT 06904
Telephone: 203-965-6419

C-SPAN
400 North Capital Street, N.W.
Suite 650
Washington, DC 20001
Telephone: 202-737-3220

Court TV
600 Third Avenue, 3rd floor
New York, NY 10016
Telephone: 212-973-3345

Discovery (The Learning Channel)
7700 Wisconsin Drive
Bethesda, MD 20814-3522
Telephone: 301-986-0444 ext. 5456

Disney
3800 West Alameda Avenue
4th floor
Burbank, CA 91505
Telephone: 818-569-7650

E! Entertainment
5670 Wilshire Blvd.
Los Angeles, CA 90036-3709
Telephone: 213-954-2400

Eastern Microwave (WWOR, WSBK)
112 Northern Concourse
P.O. Box 4872
Syracuse, NY 13221
Telephone: 800-448-3322 ext. 236

EPG, Jr.
3801 South Sheridan Road
Tulsa, OK 74145
Telephone: 800-331-4806

ESPN
605 Third Avenue
New York, NY 10158
Telephone: 212-916-9200

ETN
P.O. Box 398
Branson, MO 65616
Telephone: 417-335-8600

Eternal Word TV Network
5817 Old Leeds Road
Birmingham, AL 35210
Telephone: 205-956-9537

Family Channel
5835 Highway 18, Suite A
Jackson, MS 39209
Telephone: 601-922-0181

Galavision
2121 Avenue of the Stars
Suite 2300
Los Angeles, CA 90067
Telephone: 310-286-0122

Home Box Office (HBO)
1100 Avenue of the Americas
New York, NY 10036
Telephone: 212-512-1000

International Channel
12401 West Olympic Blvd.
Los Angeles, CA 90064
Telephone: 213-826-2429

Learning Channel (see Discovery)
7700 Wisconsin Drive
Bethesda, MD 20814-3522
Telephone: 301-986-0444

Lifetime
15301 Dallas Parkway
Suite 1020, LB-28
Dallas, TX 75248
Telephone: 214-458-4600

Monitor Channel
Two Greenwich Plaza
Suite 100
Greenwich, CT 06830
Telephone: 203-622-4930

APPENDIX C

MTV, WH-1, Nickelodeon
See Showtime/TMC

Nashville Network (TNN)
P.O. Box 10210
Stamford, CT 06904
Telephone: 203-965-6419

Netlink
7951 East Maplewood Avenue
Suite 200
Englewood, CO 80111
Telephone: 800-832-4321

Nostalgia
125 East John Carpenter Freeway
Suite 670
Irving, TX 75062
Telephone: 214-506-7300

Outlaw Channel
Route 1, Briarcliffe #13
Spicewood, TX 78669
Telephone: 512-264-2922

Playboy
8560 Sunset Blvd., 7th floor
Los Angeles, CA 90069
Telephone: 213-659-4080

PrimeTime 24
342 Madison Avenue, Suite 1520
New York, NY 10019
Telephone: 212-599-4440

Showtime/The Movie Channel
1633 Broadway
New York, NY 10173
Telephone: 212-708-1257
or
15301 Dallas Parkway
Suite 1000, LB 29
Dallas, TX 75248
Telephone: 214-788-5000

Silent Network
1777 Northeast Loop 410, Suite 1401
San Antonio, TX 78217
Telephone: 512-824-7446

SPICE - Graff Pay-Per-View
532 Broadway, 6th floor
New York, NY 10010
Telephone: 212-941-1434

Travel Channel
1370 Avenue of the Americas
27th floor
New York, NY 10019
Telephone: 212-603-4500

Traveler's Program Guide
3801 South Sheridan Road
Tulsa, OK 74145
Telephone: 800-331-4806

Trinity Broadcasting
2442 Michelle
Tustin, CA 92680
Telephone: 714-832-2950

United Video
3801 South Sheridan
Tulsa, OK 74145
Telephone: 800-331-4803

USA Network
2049 Century Park East
Suite 2550
Los Angeles, CA 90067
Telephone: 213-201-2314

Weather Channel
2800 Cumberland Parkway
Atlanta, GA 30339
Telephone: 404-434-6800

WTBS (Southern Satellite)
7951 East Maplewood
Englewood, CO 80111
Telephone: 303-771-3800

APPENDIX D. SATELLITE TV GUIDES and PUBLICATIONS

Satellite TV Guides

Magazin, NB Fortag
Kjelsasvn. 51
0488 Oslo 4
Norway

On Sat, Triple-D Publishing
P.O. Box 2384
Shelby, NC 28151
USA
TEL: (800) 234-0023
    (704) 482-9673

OnSat, Triple-D Publishing Canada
9780 Bramalea Road, North
Suite 406
Brampton, Ontario L6S 2P1
Canada
TEL: (416) 458-9429

Satellite Entertainment Guide
Vogel and Son Publishing Company
P.O. Box 8266TU
Edmonton, Alberta T6H 4P1, Canada
TEL: (403) 425-1169

Satellite Orbit
CommTek Publishing
8330 Boone Blvd., Suite 600
Vienna, VA 22182, USA
TEL: (800) 234-4220
    (703) 827-0511

Satellite TV PreVue
Terra Publishing, Inc.
P.O. Bo x 460
Salamanca, NY 14779-0460
TEL: (800) 992-3499

Satellite TV Week
Fortuna Communications
P.O. Box 308
928 Main Street
Fortuna, CA 95540, USA
TEL: (800) 345-8876
    (707) 725-1185

SuperGuide, Triple-D Publishing
P.O. Box 2384
Shelby, NC 28151
TEL: (800) 234-0139

Cable and Satellite Europe
21st Century Publishing, 533 Kings Road
London, SW10 0BR, UK
TEL: 071-351-3612

Trade Publications

Cable and Satellite Europe
533 Kings Road
London, SW10 0BR
United Kingdom

CQ TV Magazine
British Amateur TV Club
Grenehurst, Pinewood Road
High Wycomb, Bucks HP12 4DD, UK

Electronics & Wireless World
Reed Business Publications Ltd.
Stewart House, Perrymount Road
Hayward Heath, W.Sussex RH16 3DH
United Kingdom

Elektor Electronics
Worldwide Subscription Services
Rose Hill, Ticehurst
East Sussex TN5 7AJ, UK

Feedback – CAI
Fulton House Business Centre
Rulton Road, Wembley Park
Middlesex HA9 0TF, UK
Telephone: 081-902-8998
FAX: 081-903-8719

McCormac's Hack Watch News
22 Viewmount
Waterford, Ireland
Telephone: 353-51-72640

Private Cable Magazine, NSPN
1909 Avenue G
Rosenburg, TX 77471, USA
Telephone: (800) 622-5990
FAX: (713) 342-7016

Satellite Business News
1050 17th Street, N.W., Suite 1212
Washington, DC 20036
Telephone: (202) 785-0505
FAX: (202) 785-9291

Satellite Communications
Cardiff Publications
6430 S. Yosemite Street
Englewood, CO 80111, USA
Telephone: (303) 694-1522

Satellite Retailer, Triple D Publishing
P.O. Box 2384
Shelby NC 28151, USA
Telephone: (704) 482-9673

Signal Magazine
Fernwood Publishing
P.O. Box 238, Station D
Scarborough, Ontario M1R 5B7
Canada
Telephone: (416) 759-6639

Television
Royal TV Society
Tavistock House East, Tavistock Square
London WC1H 9HR, UK
Telephone: 071-485-0011

The Transponder
Terra Publishing, Inc.
P.O. Box 460
Salamanca, NY 14779-0460
U.S.A.
(800) 992-3488

What Satellite
WV Publications Ltd.
57/59 Rochester Place
London NW1 9JU
England
Telephone: 071-485-0011

APPENDIX D

APPENDIX E. OFF-AIR and CABLE TV CHANNELS

NORTH AMERICA

Audio-Video Separation – 4.5 MHz

Channel	Video Frequency (MHz)	Wavelength (Inches)	Channel	Video Frequency (MHz)	Wavelength (Inches)
SUB CHANNELS			**SUPERBAND CHANNELS**		
T-7	7.00	1,686.7	J	217.25	54.5
T-8	13.00	908.2	K	223.25	52.9
T-9	19.00	621.4	L	229.25	51.5
T-10	25.00	472.3	M	235.25	50.2
T-11	31.00	380.9	N	241.25	49.0
T-12	37.00	319.1	O	247.25	47.8
T-13	43.00	274.6	P	253.25	46.6
			Q	259.25	45.6
VHF-LOW CHANNELS			R	265.25	44.5
2	55.25	213.7	S	271.25	43.5
3	61.25	192.8	T	277.25	42.6
4	67.25	175.6	U	283.25	41.7
5	77.25	152.9	V	289.25	40.8
6	83.25	141.8	W	295.25	40.0
FM BAND					
FM-1	89.25	132.3	**HYPERBAND CHANNELS**		
FM-2	95.25	124.0	AA	301.25	39.2
FM-3	101.25	116.6	BB	307.25	38.4
			CC	313.25	37.7
VHF MIDBAND			DD	319.25	37.0
A-1	109.25	108.1	EE	325.25	36.3
A-2	115.25	102.5	FF	331.25	35.7
A	121.25	97.4	GG	337.25	35.0
B	127.25	92.8	HH	343.25	34.4
C	133.25	88.6	II	349.25	33.8
D	139.25	84.8	KK	361.25	32.7
E	145.25	81.3	LL	367.25	32.2
F	151.25	78.0	MM	373.25	31.6
G	157.25	75.1	NN	379.25	31.1
H	163.25	72.3	OO	385.25	30.7
I	169.25	69.8	PP	391.25	30.2
			QQ	397.25	29.7
VHF-HIGH CHANNELS			RR	403.25	29.3
7	121.25	97.4	SS	409.25	28.9
8	127.25	92.8	TT	415.25	28.4
9	133.25	88.6	UU	421.25	28.0
10	139.25	84.8	VV	427.25	27.6
11	145.25	81.3	WW	433.25	27.3
12	151.25	78.1	XX	439.25	26.9
13	157.25	75.1	YY	445.25	26.6
			ZZ	451.25	26.2
			AAA	457.25	25.9
			BBB	463.25	25.5
			CCC	469.25	25.2
			DDD	475.25	24.9
			EEE	481.25	24.6
			FFF	487.25	24.3

APPENDIX E

UHF CHANNELS – NORTH AMERICA

Channel	Video Frequency (MHz)	Wavelength (Inches)	Channel	Video Frequency (MHz)	Wavelength (Inches)
14	471.25	25.1	50	687.25	17.2
15	477.25	24.7	51	693.25	17.0
16	483.25	24.4	52	699.25	16.9
17	489.25	24.1	53	705.25	16.7
18	495.25	23.8	54	711.25	16.6
19	501.25	23.6	55	717.25	16.5
20	507.25	23.3	56	723.25	16.3
21	513.25	23.0	57	729.25	16.2
22	519.25	22.7	58	735.25	16.1
23	525.25	22.5	59	741.25	15.9
24	531.25	22.2	60	747.25	15.8
25	537.25	22.0	61	753.25	15.7
26	543.25	21.7	62	759.25	15.6
27	549.25	21.5	63	765.25	15.4
28	555.25	21.3	64	771.25	15.3
29	561.25	21.0	65	777.25	15.2
30	567.25	20.8	66	783.25	15.1
31	573.25	20.6	67	789.25	15.0
32	579.25	20.4	68	795.25	14.9
33	585.25	20.2	69	801.25	14.7
34	591.25	20.0	70	807.25	14.6
35	597.25	19.8	71	813.25	14.5
36	603.25	19.6	72	819.25	14.4
37	609.25	19.4	73	825.25	14.3
38	615.25	19.2	74	831.25	14.2
39	621.25	19.0	75	837.25	14.1
40	627.25	18.8	76	843.25	14.0
41	633.25	18.7	77	849.25	13.9
42	639.25	18.5	78	855.25	13.8
43	645.25	18.3	79	861.25	13.7
44	651.25	18.1	80	867.25	13.6
45	657.25	18.0	81	873.25	13.5
46	663.25	17.8	82	879.25	13.4
47	669.25	17.6	83	885.25	13.3
48	675.25	17.5			
49	681.25	17.3			

EUROPE and OTHER 625-LINE COUNTRIES

System	Audio-video Carrier Separation (MHz)	Countries
G,H	5.5	Most of Western Europe
I	6.0	UK, Eire, South Africa
D,K,L	6.5	France, Eastern Europe, USSR

UHF CHANNELS

Channel	Frequency (MHz)
21	471.25
22	479.25
23	487.25
24	495.25
25	503.25
26	511.25
27	519.25
28	527.25
29	535.25
30	543.25
31	551.25
32	559.25
33	567.25
34	575.25
35	583.25
36	591.25
37	599.25
38	607.25
39	615.25
40	623.25
41	631.25
42	639.25
43	647.25
44	655.25
45	663.25
46	671.25
47	679.25
48	687.25
49	695.25
50	703.25
51	711.25
52	719.25
53	727.25
54	735.25
55	743.25
56	751.25
57	759.25
58	767.25
59	775.25
60	783.25
61	791.25
62	799.25
63	807.25
64	815.25
65	823.25
66	831.25
67	839.25
68	847.25

JAPAN

Audio-video Carrier Separation – 4.5 MHz

Channel	Frequency (MHz)
13	471.25
14	477.25
15	483.25
16	489.25
17	495.25
18	501.25
19	507.25
20	513.25
21	519.25
22	525.25
23	531.25
24	537.25
25	543.25
26	549.25
27	555.25
28	561.25
29	567.25
30	573.25
31	579.25
32	585.25
33	591.25
34	597.25
35	603.25
36	609.25
37	615.25
38	621.25
39	627.25
40	633.25
41	639.25
42	645.25
43	651.25
44	657.25
45	663.25
46	669.25
47	675.25
48	681.25
49	687.25
50	693.25
51	699.25
52	705.25
53	711.25
54	717.25
55	723.25
56	729.25
57	735.25
58	741.25
59	747.25
60	753.25
61	759.25
62	765.25
63	765.25

APPENDIX F. GLOSSARY

A/B Switch
A switch that selects one of two inputs (A or B) for routing to a common output while providing adequate isolating between the two signals.

AFC (Automatic Frequency Control)
A circuit which locks an electronic component onto a chosen frequency.

AGC (Automatic Gain Control)
A circuit that uses feedback to maintain the output of an electronic component at a constant level with a variable input.

Absolute Zero
The coldest possible temperature at which all molecular motion ceases. It is expressed in degrees Kelvin as measured from absolute zero. Zero degrees Kelvin equals minus 273.16 °C or minus 459.69 °F.

Adjacent Channel
An adjacent channel is immediately next to another channel in frequency. For example, NTSC channels 5 and 6 as well as 8 and 9 are adjacent. However, channels 4 and 5 or channels 6 and 7 are separated by signals used by non-TV media.

Agile Receiver
A satellite receiver which can be tuned to any desired channel.

Alignment
The process of fine tuning a dish or an electronic circuit to maximize its sensitivity and signal receiving capability.

Ambient Temperature
The existing dry bulb temperature.

Amplifier
A device used to increase the power of a signal.

Analog
A system in which signals vary continuously in contrast to a digital system in which signals vary in discrete steps

Analog-to-Digital Converter
A circuit that converts analog signals to an equivalent digital form. The varying analog signal is sampled at a series of points in time. The voltage at each of these points is then represented by a series of numbers, the digital value of the sample. The higher this sampling frequency, the finer are the gradations and the more accurately is the signal represented.

Antenna
A device that collects (and transmits) and focuses electromagnetic energy, i.e., contributes an energy gain. Satellite dishes, broadband antenna and cut-to-channel antennas are some types of antennas encountered in private cable systems. In the case of satellite antennas, gain is proportional to the surface area of the microwave dish.

Antenna Efficiency
The percentage of incoming satellite signal actually captured by an antenna.

Antenna Illumination
Describes how a feedhorn "sees" the surface of a dish as well as the surrounding terrain.

Aperture
The collection area of a parabolic antenna.

Aspect Ratio
The ratio of television screen width to height. The standard aspect ratio is 4 to 3.

Attenuation
The decrease in signal power that occurs in a device or when a signal travels to reach a destination point (path loss).

Attenuator
A passive device which reduces the power of a signal. Attenuators are rated according to the amount of signal attenuation.

Audio Subcarrier
The carrier wave that transmits audio information within a video broadcast signal. Satellite transmissions can relay more than a single audio subcarrier in the frequency range between 5 and 8.5 MHz.

Automatic Brightness Control
A television circuit used to automatically adjust picture tube brightness in response to changes in background or ambient light.

Automatic Fine Tuning
A circuit that automatically maintains the correct tuner oscillator frequency and compensates for drift and for moderate amounts of inaccurate tuning. Similar to AFC.

APPENDIX F

Automatic Brightness Control
A television circuit used to automatically adjust picture tube brightness in response to changes in background or ambient light.

Automatic Frequency Control (AFC)
A circuit that locks onto a chosen frequency and will not drift away from that frequency.

Automatic Fine Tuning
A circuit that automatically maintains the correct tuner oscillator frequency and compensates for drift and for moderate amounts of inaccurate tuning. See automatic frequency control (AFC)

Automatic Gain Control (AGC)
A circuit that locks the gain onto a fixed value and thus compensates for varying input signal levels keeping the output constant.

Azimuth-Elevation (Az-El) Mount
An antenna mount which tracks satellites by moving in two directions: the azimuth in the horizontal plane; and elevation up from the horizon.

Azimuth
A compass bearing expressed in degrees of rotation clockwise from true north. It is one of the two coordinates (azimuth and elevation) used to align a satellite antenna.

Back Match
The matching of the resistive values of the input and output of electronic devices to reduce signal reflection and ghosting. Also known as impedance matching.

Back Porch
That portion of the horizontal blanking pulse that follows the trailing edge of the horizontal sync pulse.

Balun
A transformer used to match 75 ohm coaxial cable to a 300 ohm input on older televisions. This acronym is derived from combining the two words, BALanced and UNbalanced.

Band
A range of frequencies.

Band Separator
A device that splits a group of specified frequencies into two or more bands. Common types include UHF/VHF, High/Low-band and FM separators. This device is essentially a set of filters.

Bandpass Filter
A circuit or device that allows only a specified range of frequencies to pass from input to output.

Bandstop
See bandpass filter

Bandwidth
The frequency range allocated to any communication circuit.

Baseband
The raw audio and video signals prior to modulation and broadcasting. Most satellite headend equipment utilizes baseband inputs. More exactly, the composite unclamped, non-de-emphasized and unfiltered receiver output. This signal contains the complete set of FM modulated audio and data subcarriers.

Beamwidth
A measure used to describe the width of vision of an antenna. Beamwidth is measured as degrees between the 3 dB half power points.

Bird
Jargon or nickname for communication satellites.

Blanking Pulse Level
The reference level for video signals. The blanking pulses must be aligned at the input to the picture tube.

Blanking Signal
Pulses used to extinguish the scan illumination during horizontal and vertical retrace periods.

Block Downconversion
The process of lowering the entire band of frequencies in one step to some intermediate range to be processed inside a satellite receiver. Multiple block downconversion receivers are capable of independently selecting channels because each can process the entire block of signals.

BNC Connector
A weatherproof twist lock coax connector standard on commercial video equipment and used on some brands of satellite receivers.

Boresight
The direction along the principle axis of either a transmitting or a receiving antenna.

Broadband
A device that processes a signal(s) spanning a relatively broad range of input frequencies.

Buttonhook Feed
A rod shaped like a question mark supporting the feedhorn and LNA. A buttonhook feed for use with commercial grade antennas is often a hollow waveguide that directs signals from a feedhorn to an LNA behind the antenna.

CATV
An abbreviation for Community Antenna Television - another name for cable TV.

CCD
Charge coupled device. In this device charge is stored on a capacitor which are etched onto a chip. A number of samples can be simultaneously stored. Used in MAC transmissions for temporarily storing video signals.

C-Band
The range of frequencies that includes 3.7 to 4.2 GHz for downlink and 5.7 to 6.2 GHz for uplink.

Cable-Ready Television
A television receiver that can receive unscrambled cable television channels without the use of a converter.

Cardioid Pattern
A heart-shaped pattern that is typical of some directional antennas.

Carrier
A pure-frequency signal that is modulated to carry information. In the process of modulation it is spread out over a wider band. The carrier frequency is the unmodulated frequency on any television channel.

APPENDIX F

Carrier-to-Noise Ratio (C/N)

The ratio of the received carrier power to the noise power in a given bandwidth, expressed in decibels. The C/N is an indicator of how well an earth receiving station will perform in a particular location, and is calculated from satellite power levels, antenna gain and the combined antenna and LNA noise temperature.

Cassegrain Feed System

An antenna feed design that includes a primary reflector, the dish, and a secondary reflector which redirects microwaves via a waveguide to a low noise amplifier.

Channel

A segment of bandwidth used for one complete communication link.

Chrominance

The hue and saturation of a color. The chrominance signal is modulated onto a 4.43 MHz carrier in the PAL television system and a 3.58 MHz carrier in the NTSC television system.

Chrominance Signal

The color component of the composite baseband video signal assembled from the I and Q portions (NTSC) or U and V (PAL). Phase angle of the signal represents hue and amplitude represents color saturation.

Circular Polarity

Electromagnetic waves whose electric field uniformly rotates along the signal path. Broadcasts used by Intelsat and other international satellites use circular, not horizontally or vertically polarized waves as are common in North American and European transmissions. Circularly polarized waves are used for satellite telephony because Faraday rotation does not alter their behavior.

Clamp Circuit

A circuit that removes the dispersion waveform from the downlink signal.

Clarke Belt

The circular orbital belt at 22,247 miles above the equator, named after the writer Arthur C. Clarke, in which satellites travel at the same speed as the earth's rotation. Also called the geostationary orbit.

Color Bars

A test pattern of specifically colored vertical bars used as a reference to test the performance of a color television and transmission paths.

Coaxial Cable

A cable for transmitting high frequency electrical signals with low loss. It is composed of an internal conducting wire surrounded by an insulating dielectric which is further protected by a metal shield. The impedance of coax is a product of the radius of the central conductor, the radius of the shield and the dielectric constant of the insulation. In an SMATV system, coax impedance is 75 ohms.

Color Sync Burst

A "burst" of 8 to 11 cycles in the 4.43361875 MHz (PAL) or 3.579545 MHz(NTSC) color subcarrier frequency. This waveform is located on the back porch of each horizontal blanking pulse during color transmissions. It serves to synchronize the color subcarrier's oscillator with that of the transmitter in order to recreate the raw color signals.

Combining Network

An active or passive network that serves to combine several signals into one output while maintaining a high degree of isolation between each input.

Composite Baseband Signal

The complete audio and video signal without a carrier wave. Satellite signals have audio baseband information ranging in frequency from 55 to about 10,000 Hertz. NTSC video baseband is from zero to 4.2 MHz. PAL video baseband ranges from 0 to 5.5 MHz.

Composite Video Signal

The complete video signal consisting of the chrominance and luminance information as well as all sync and blanking pulses.

Companding

A form of noise reduction using compression at the transmitting end and expansion at the receiver. A compressor is an amplifier that increases its gain for lower power signals. The effect is to boost these components into a form having a smaller dynamic range. A compressed signal has a higher average level, and therefore, less apparent loudness than an uncompressed signal, even though the peaks are no higher in level. An expander reverses the effect of the compressor to restore the original signal.

Cone

An abbreviation for the European continent.

Converter

A device used to transfer signals from a channel of one frequency to another.

Contrast

The ratio between the dark and light areas of a television picture.

Conus

An abbreviation for the continental United States.

Cross Modulation

A form of interference caused by the modulation of one carrier affecting that of another signal. It can be caused by overloading an amplifier as well as by signal imbalances at the headend.

Cross Polarization

Term to describe signals of the opposite polarity to another being transmitted and received. Cross-polarization discrimination refers to the ability of a feed to detect one polarity and reject the signals having the opposite sense of polarity.

Crosstalk

Interference between adjacent channels often caused by cross modulation. Leakage can occur between two wires, PCB tracks or parallel cables.

dc Power Block

A device which stops the flow of dc power but permits passage of higher frequency ac signals.

Decibel (dB)

The logarithmic ratio of power levels used to indicate gains or losses of signals. Decibels relative to one watt, milliwatt and millivolt are abbreviated as dBw, dBm and dBmV, respectively. Zero dBmV is used as the standard reference for all SMATV calculations.

Declination Offset Angle

The adjustment angle of a polar mount between the polar axis and the plane of a satellite antenna used to aim at the geosynchronous arc. Declination increases from zero with latitude away from the equator.

Decoder

A circuit that restores a signal to its original form after it has been scrambled.

APPENDIX F

De-emphasis
A reduction of the higher frequency portions of an FM signal used to neutralize the effects of pre-emphasis. When combined with the correct level of pre-emphasis, it reduces overall noise levels and therefore increases the signal-to-noise ratio.

Dehydrator
A device used to compress and dry the air that is contained inside an air dielectric transmission line.

Demodulator
A device which extracts the baseband signal from the transmitted carrier wave.

Detent Tuning
Tuning into a satellite channel by selecting a preset resistance.

Digital
Describes a system or device in which information is transferred by electrical "on-off," "high-low," or "1/0" pulses instead of continuously varying signals or states as in an analog message.

Digital-to-Analog Converter
A circuit that converts digital signals into their equivalent analog form.

Diplexer
A device used to combine the video and audio components of a wireless signal. This component is internal to the transmitter at lower powers ranging from 1 to 50 watts but is external at higher powers.

Dipole Antenna
A two-element receiving or transmitting antenna that is typically center-fed and half a wavelength in length.

Direct Broadcast Satellite (DBS)
A term commonly used to describe Ku-band broadcasts via satellite directly to individual end-users. The DBS band ranges from 11.7 to 12.2 GHz.

Direct Pickup
Signal ingress from local services into a distribution system.

Directional Coupler
See Tap

Dish
Jargon for a parabolic microwave antenna.

Distribution System
A communication system consisting of coax but occasionally of line-of-sight microwave links that carries signals from the headend to end-users.

Domsat
Abbreviation for domestic communication satellite.

Downconverter
A circuit that lowers the high frequency signal to a lower, intermediate range. There are three distinct types of downconversion used in satellite receivers: single downconversion; dual downconversion; and block downconversion.

Downlink Antenna
The antenna on-board a satellite which relays signals back to earth.

Drifting
An instability in a preset voltage, frequency or other electronic circuit parameter.

Drop Cable
The cable used to route a signal from outside distribution equipment to a subscriber television receiver and other equipment.

Dual-Band Feedhorn
A feedhorn which can simultaneously receive two different bands, typically the C and Ku-bands.

Earth Station
A complete satellite receiving or transmitting station including the antenna, electronics and all associated equipment necessary to receive or transmit satellite signals. Also known as a ground station.

Effective Isotropic Radiated Power (EIRP)
A measure of the signal strength that a satellite transmits towards the earth below. The EIRP is highest at the center of the beam and decreases at angles away from the boresight.

Elevation Angle
The vertical angle measured from the horizon up to a targeted satellite.

Encoder
A device for scrambling a signal.

Energy Dispersal
The modulation of an uplink carrier with a triangular waveform. This technique disperses the carrier energy over a wider bandwidth than otherwise would be the case in order to limit the maximum energy compared to that transmitted by an unclamped carrier. By spreading the spectrum, there is less chance of interfering with other users of the same frequencies. This triangular waveform is removed by a clamp circuit in a satellite receiver.

Equalizing Pulses
A series of six pulses occurring before and after the serrated vertical sync pulse to ensure proper interlacing. The equalizing pulses are inserted at twice the horizontal scanning frequency.

F-Connector
A standard RF connector used to link coax cables with electronic devices.

FCC
The Federal Communications Commission, the regulatory board which sets standards for communications within the United States.

f/D Ratio
The ratio of an antenna's focal length to diameter. It describes antenna "depth."

Feedhorn
A device that collects microwave signals reflected from the surface of an antenna. It is mounted at the focus of all prime focus parabolic antennas.

Field
One half of a complete TV picture or frame, composed of 262.5 scanning lines. There are 60 fields per second for black/white TV and 59.94 fields per second for color TV in NTSC transmission. In the PAL broadcast system there are 50 fields per second.

Filter
A device used to reject all but a specified range of frequencies. A bandpass filter allows only those signals within a given band to be communicated. A rejection filter, the mirror image of a bandpass filter, eliminates those signals within a specified band but passes all other frequencies.

Focal Length
The distance from the reflective surface of a parabola to the point at which incoming satellite signals are focused, the focal point.

Footprint
The geographic area towards which a satellite downlink antenna directs its signal. The measure of strength of this footprint is the EIRP.

Forward Error Correction
FEC is a technique for improving the accuracy of data transmission. Excess bits are included in the out-going data stream so that error correction algorithms can be applied upon reception.

Frame
One complete TV picture, composed of two fields and a total of 525 and 625 scanning lines in NTSC and PAL systems, respectively.

Frequency
The number of vibrations per second of an electrical or electromagnetic signal expressed in cycles per second or Hertz.

Front Porch
The portion of the horizontal blanking pulse that precedes the horizontal sync pulse.

Gain
The amount of amplification of input to output power often expressed as a multiplicative factor or in decibels.

Gain-to-Noise Temperature Ratio (G/T)
The figure of merit of an antenna and LNA. The higher the G/T, the better the reception capabilities of an earth station.

Geostationary Orbit
See Clarke Belt.

Ghost
A weak image either offset to the left or right of the primary image in a received television picture, a result of multipath conditions.

GigaHertz (GHz)
1000 MHz or one billion cycles per second.

Global Beam
A footprint pattern used by communication satellites targeting nearly 40% of the earth's surface below. Many Intelsat satellites use global beams.

Ground Noise
Unwanted microwave signals generated from the warm ground and detected by a dish.

Guard Band
An unused band of frequencies between two active channels that can serve to prevent mutual interference.

Hall Effect Sensor
A semiconductor device in which an output voltage is generated in response to the intensity of a magnetic field applied to a wire. In an actuator, the varying magnetic field is produced by the rotation of a permanent magnet past a thin wire. The pulses generated serve to count the number of rotations of the motor.

Hardline
A low-loss coaxial cable that has a continuous hard metal shield instead of a conductive braid around the outer perimeter. This type of cable was used in the pioneer days of satellite television.

Headend
The portion of an SMATV system where all desired signals are received and processed for subsequent distribution.

Heliax
A thick low-loss cable used at high frequencies; also known as hardline.

Hertz
An abbreviation for the frequency measurement of one cycle per second. Named after Heinrich Hertz, the German scientist who first described the properties of radio waves.

High Definition Television (HDTV)
An innovative television format having approximately twice the number of scan lines in order to improve picture resolution and viewing quality.

High Pass Filter
An electronic circuit designed to pass all frequencies above a specified frequency while attenuating signals with frequencies below the rated value.

High Power Amplifier (HPA)
An amplifier used to amplify the uplink signal.

Home Run
A method to wire a distribution system using individual drop cables to each subscriber.

Horizontal Blanking Pulse
The pulse that occurs between each horizontal scan line and extinguishes the beam illumination during the retrace period.

Horizontal Sync Pulse
A 5.08 microsecond (4.7 microsecond in the PAL system) rectangular pulse riding on top of each horizontal blanking pulse. It synchronizes the horizontal scanning at the television set with that of the television camera.

Hum Bars
A form of interference seen as horizontal bars or black regions passing across the field of a television screen caused by the mains 50 or 60 cycle power.

I Signal
One of the two color NTSC video signals which modulate the color subcarrier. It represents those colors ranging from reddish orange to cyan.

IF Rejection
Rejection of signals that fall in the same band as the desired signal.

ITFS
An abbreviation for Instructional Television Fixed Services, a block of frequencies allocated for broadcast of wireless signals by educational institutions in the United States.

APPENDIX F

Image Frequency
An undesired signal obtained in frequency conversion using a mixing or heterodyning process.

Impulse Pay-Per-View
Impulse pay-per-view (IPPV) is a feature of a decoder that allows an authorized subscriber to purchase a one-time scrambled program at will. IPPV shows are selected by a button on the decoder or its remote control unit.

Inclinometer
An instrument used to measure the angle of elevation to a satellite from the surface of the earth.

Input Impedance
The effective impedance measured across the input terminals of an electronic device.

Interference
An undesired signal intercepted by a TVRO that causes video and/or audio distortion.

Interlaced Scanning
A scanning technique to minimize picture flicker while conserving channel bandwidth. Even and odd numbered lines are scanned in separate fields both of which when combined paint one frame or complete picture.

Intermediate Frequency (IF)
A middle range frequency generated after downconversion in any electronic circuitry including a satellite receiver. The majority of all signal amplification, processing and filtering in a receiver occur in the IF range.

Intermodulation
A form of interference that is generated from creation of two or more beats between carrier. Beat frequencies are given by fbeat = nf1 + mf2 where n and m are integers.

Insertion Loss
The amount of signal energy lost when a device is inserted into a communication line. Also known as "feed through" loss.

INTELSAT
The International Telecommunication Satellite Consortium, a body of 154 countries working towards a common goal of improved worldwide satellite communications.

Isolator
A device that allows signals to pass unobstructed in one direction but which attenuates their strength in the reverse direction.

Isolation Loss
The amount of signal energy lost between two ports of a device. An example is the loss between the feed through port and the tap/drop of a top-off device.

Jumper
A short piece of transmission line used to interface components in a broadcast system.

Kelvin Degrees (°K)
The temperature above absolute zero, the temperature at which all molecular motion stops, graduated in units the same size as degrees Celsius (°C). Absolute zero equals -273 °C or -459 °F.

Kilohertz (kHz)
One thousand cycles per second.

Ku-Band
The microwave frequency band between approximately 11 and 13 GHz used in satellite broadcasting.

Latitude
The measurement of a position on the surface of the earth north or south of the equator measured in degrees of angle.

Leakage
Egress interference, namely an undesired leakage of radiation from a cable or other closed circuit system.

Line Amplifier
An amplifier in a transmission line that boosts the strength of a signal.

Line Splitter
An active or passive device that divides a signal into two or more signals containing all the original information. A passive splitter feeds an attenuated version of the input signal to the output ports. An active splitter amplifies the input signal to overcome the splitter loss.

Local Oscillator
A device used to supply a stable single frequency to an upconverter or a downconverter. The local oscillator signal is mixed with the carrier wave to change its frequency.

Longitude
The distance east or west of the prime meridian, zero degrees, as measured in degrees.

Low Band
Television channels 2 through 6.

Low Noise Amplifier (LNA)
A device that receives and amplifies the weak satellite signal reflected by an antenna via a feedhorn. C-band LNAs typically have their noise characteristics quoted as noise temperatures rated in degrees Kelvin. K-band LNA noise characteristics are usually expressed as a noise figure in decibels.

Low Noise Block Downconverter (LNB)
A low noise microwave amplifier and converter which downconverts a block or range of frequencies at once to an intermediate frequency range, typically 950 to 1450 MHz or 950 to 1750 MHz.

Low Noise Converter (LNC)
An LNA and a conventional downconverter housed in one weatherproof box. This device converts one channel at a time. Channel selection is controlled by the satellite receiver. The typical IF for LNCs is 70 MHz.

Low Pass Filter
An electronic circuit designed to pass all frequencies below a specified frequency while attenuating signals with frequencies above the rated value.

MDS
An abbreviation for multipoint distribution service, a band of frequencies in the United States allocated for wireless television broadcast service.

MMDS
An abbreviation for multichannel multipoint distribution service, a band of frequencies in the United States allocated for wireless television broadcast service.

Magnetic Variation
The difference between true north and the north indication of a compass.

APPENDIX F

Master Antenna TV (MATV)
Broadcast receiving stations that use one or more high-quality centrally located UHF and/or VHF antennas which relay their signals to many televisions in a local apartment/condo or group-housing complex.

Match
The condition that exists when 100 percent of available power is transmitted from one device to another without any losses due to reflections.

Matching Transformer
A device used to match impedance between devices. A matching transformer is used, for example, when connecting a 75 ohm coax to a television 300 ohm input terminal.

MegaHertz (MHz)
One millions cycles per second.

Microprocessor
The central processing unit of a computer or control system, either on a single integrated (IC) circuit chip or on several ICs.

Microwave
The frequency range from approximately 1 to 30 GHz and above.

Midband
Cable television channels A through I that lie between VHS channels 6 and 7.

Mixer
A device used to combine signals together.

Modulation
A process in which a message is added or encoded onto a carrier wave. Among other methods, this can be accomplished by frequency or amplitude modulation, known as AM or FM, respectively.

Monochrome
A black and white television picture.

Mount
The structure that supports an earth station antenna. Polar and az-el mounts are the most common variety.

Multiple Analog Component (MAC) Transmissions
An innovative television transmission method which separates the data, chrominance and luminance components and compresses them for sequential relay over one television scan line. There are a number of systems in used and under development including A-MAC, B- MAC, C-MAC, D-MAC, D2-MAC, E-MAC and F-MAC.

Multiplexing
The simultaneous transmission of two or more signals over a single communication channel. The interleaving of the luminance and chrominance signals is one form of multiplexing, known as frequency multiplexing. MAC transmissions make use of time division multiplexing.

N-Connector
A low-loss coaxial cable connector used at the elevated C-band microwave frequencies.

NTSC
The National Television Standards Committee which created the standard for North American TV broadcasts.

NTSC Color Bar Pattern
The standard test pattern of six adjacent color bars including the three primary colors plus their three complementary shades.

Negative Picture Phase
Positioning the composite video signal so that the maximum level of the sync pulses is at 100% amplitude. The brightest picture signals are in the opposite negative direction.

Negative Picture Transmission
Transmission system used in North America and other countries in which a decrease in illumination of the original scene causes an increase in percentage of modulation of the picture carrier. When demodulated, signals with a higher modulation percentage have more positive voltages.

Noise
An unwanted signal which interferes with reception of the desired information. Noise is often expressed in degrees Kelvin or in decibels.

Noise Figure
The ratio of the actual noise power generated at the input of an amplifier to that which would be generated in an ideal resistor. The lower the noise figure, the better the performance.

Noise Temperature
A measure of the amount of thermal noise present in a system or a device. The lower the noise temperature, the better the performance.

OFS
An abbreviation for Operational Fixed Services, a block of three channels designated with the letter H, originally intended for use by banks and other businesses for transmission of high speed data.

Odd Field
The half frame of a television scan which is composed of the odd numbered lines.

Offset Feed
A feed which is offset from the center of a reflector for use in satellite receiving systems. This configuration does not block the antenna aperture.

Omnidirectional Antenna
An antenna that radiates power with equal strengths in all directions.

Orthomode Coupler
A waveguide, generally a three-port device, that allows simultaneous reception of vertically and horizontally polarized signals. The input port is typically a circular waveguide. The two output ports are rectangular waveguides.

PAL
Phase Alternate Line. The European color TV format which evolved from the American NTSC standard.

Pad
A concrete base upon which a supporting pole and antenna can be mounted.

Passband
The band of frequencies that a circuit passes with little attenuation.

APPENDIX F

Path Loss
The attenuation that a signal undergoes in traveling over a path between two points. Path loss varies inversely as the square of the distance traveled.

Parabola
The geometric shape that has the property of reflecting all signals parallel to its axis to one point, the focal point.

Pay-Per-View
Pay-per-view is a method of purchasing programming on a per-program basis.

Persistence of Vision
The physiological phenomena whereby a human eye retains perception of an image for a short time after the image is no longer visible.

Phase
A measure of the relative position of a signal relative to a reference expressed in degrees.

Phase Distortion
A distortion of the phase component of a signal. This occurs when the phase shift of an amplifier is not proportional to frequency over the design bandwidth.

Picture Detail
The number of picture elements resolved on a television picture screen. More "crisp" pictures result as the number of picture elements is increased.

Polar Mount
An antenna mount that permits all satellites in the geosynchronous arc to be scanned with movement of only one axis.

Polarization
A characteristic of the electromagnetic wave. Four senses of polarization are used in satellite transmissions: horizontal; vertical; right-hand circular; and left-hand circular.

Positive Picture Phase
Positioning of the composite video signal so that the maximum point of the sync pulses is at zero voltage. The brightest illumination is caused by the most positive voltages.

Preamplifier
The first amplification stage. In an SMATV system, it is the amplifier mounted adjacent to an antenna to increase a weak signal prior to its processing at the headend.

Pre-emphasis
Increases in the higher frequency components of an FM signal before transmission. Used in conjunction with the proper amount of de-emphasis at the receiver, it results in combating the higher noise detected in FM transmissions.

Primary Colors
Red, green and blue.

Prime Focus Antenna
A parabolic dish having the feed/LNA assembly at the focal point directly in the front of the antenna.

Q Signal
One of two color video signal components used to modulate the NTSC color subcarrier. It represents the color range from yellowish to green to magenta.

Radio Frequency
The approximately 10 kHz to 100 GHz electromagnetic band of frequencies used for man-made communication.

Raster
The random pattern of illumination seen on a television screen when no video signal is present.

Reed Switch
A mechanical switch which uses two thin slivers of metal in a glass tube to make and break electrical contact and thus to count pulses which are sent to the antenna actuator controller. The position of the slivers of metal is governed by a magnetic field applied by a bar or other type of magnet.

Reference Signal
A highly stable signal used as a standard against which other variable signals may be compared and adjusted.

Repeater
A device that receives and re-broadcasts signals to extend the range of an original broadcast facility.

Return Loss
A ratio of the amount of reflected signal to the total available signal entering a device expressed in decibels.

Retrace
The blanked-out line traced by the scanning beam of a picture tube as it travels from the end of any horizontal line to the beginning of either the next horizontal line or field.

SAW (Surface Acoustic Wave) Filter
A solid state filter that yields a sharp transition between regions of transmitted and attenuated frequencies.

SLM
Abbreviation for signal level meter.

Satellite Receiver
The indoors electronic component of an earth station which downconverts, processes and prepares satellite signals for viewing or listening.

Scanning
The organized process of moving the electron beam in a television picture tube so an entire scene is drawn as a sequential series of horizontal lines connected by horizontal and vertical retraces.

Scrambling
A method of altering the identity of a video or audio signal in order to prevent its reception by persons not having authorized decoders.

Screening
A metal, concrete or natural material that screens out unwanted TI from entering an antenna or a metal shield that prevents the ingress of unwanted RF signals in an electronic circuit.

Serrated Vertical Pulse
The television vertical sync pulse which is subdivided into six serrations. These sub-pulses occur at twice the horizontal scanning frequency.

Servo Hunting
An oscillatory searching of the feedhorn probe when use of inadequate gauge control cables results in insufficient voltage at the feedhorn.

Side Lobe
A parameter used to describe an antenna's ability to detect off-axis signals. The larger the side lobes, the more noise and interference an antenna can detect.

Sidebands
The signals falling in the band of frequencies on either side of a carrier resulting from the process of modulation.

Single Channel Per Carrier (SCPC)
A satellite transmission system that employs a separate carrier for each channel, as opposed to frequency division multiplexing that combines many channels on a single carrier.

Signal-to-Noise Ratio (S/N)
The ratio of signal power to noise power in a specified bandwidth, usually expressed in decibels.

Skew
A term used to describe the adjustment necessary to fine tune the feedhorn polarity detector when scanning between satellites.

Slant Range
The distance that a signal travels from a satellite to a TVRO.

Snow
Video noise or sparklies caused by an insufficient signal-to-noise input ratio to a television set or monitor.

Solar Outage
The loss of reception that occurs when the sun is positioned directly behind a target satellite. When this occurs, solar noise drowns out the satellite signal and reception is lost.

Sparklies
Small black and/or white dashes in a television picture indicating an insufficient signal-to-noise ratio. Also known as "snow."

Spherical Antenna
An antenna system using a section of a spherical reflector to focus one or more satellite signals to one or a series of focal areas.

Splitter
A device that takes a signal and splits it into two or more identical but lower power signals.

Subcarrier
A signal that is transmitted within the bandwidth of a stronger signal. In satellite transmissions a 6.8 MHz audio subcarrier is often used to modulate the C-band carrier. In television, a 3.58 MHz subcarrier modulates the video carrier on each channel.

Surface Acoustic Wave
A sound or acoustic wave traveling on the surface of the optically polished surface of a piezoelectric material. This wave travels at the speed of sound but can pass frequencies as high as several gigahertz. See SAW Filter.

Synchronizing Pulse
Pulses imposed on the composite baseband video signal used to keep the television picture scanning in perfect step with the scanning at the television camera.

TVRO
A television receive-only earth station designed only to receive but not to transmit satellite communications.

Tap
A device that channels a specific amount of energy out of the main distribution system to a secondary outlet.

Television Receive-Only (TVRO)
A satellite system that can only receive but not transmit signals.

Terrestrial Interference (TI)
Interference of earth-based microwave communications with reception of satellite broadcasts.

Tilt
The uneven attenuation of a broadband signal as it travels through a coaxial cable. In general, attenuation increases as signal frequency increases.

Thermal Noise
Random, undesired electrical signals caused by molecular motion, known more familiarly as noise.

Trace
The movement of the electron beam from left to right on a television or oscillosope screen.

Threshold
A minimal signal to noise input required to allow a video receiver to deliver an acceptable picture.

Transponder
A microwave repeater, which receives, amplifies, downconverts and re-transmits signals from a communication satellite.

Trap
An electronic device that attenuates a selected band of frequencies in a signal. Also known as a notch filter.

UHF
Ultrahigh frequencies ranging from 300 to 3,000 MHz. North American TV channels 14 through 83. European TV channels 21 to 69.

Upconverter
A device that increases the frequency of a transmitted signal.

Uplink
The earth station electronics and antenna which transmits information to a communication satellite.

VHF
Very high frequencies in the range from 54 MHz to 216 MHz, NTSC TV channels 2 through 13.

VSWR (Voltage Standing Wave Ratio)
The ratio between the minimum and maximum voltage on a transmission line. An ideal VSWR is 1.0. Ghosting can result as the VSWR increases. It is also a measure of the percentage of reflected power to the total power impinging upon a device.

Vertical Blanking Pulse
A pulse used during the vertical retrace period at the end of each scanning field to extinguish illumination from the electron beam.

Vertical Sync Pulse
A series of pulses which occur during the vertical blanking interval to synchronize the scanning process at the television with that created at the studio. See also Serrated Vertical Pulse.

VHF
Very high frequency range from 30 to 300 MHz

APPENDIX F

Video Signal
That portion of the transmitted television signal containing the picture information.

Voltage Tuned Oscillator (VTO)
An electronic circuit whose output oscillator frequency is adjusted by voltage. Used in downconverters and satellite receivers to select from among transponders.

Video Monitor
A television that accepts unmodulated baseband signals to reproduce a broadcast.

Yagi Antenna
An antenna consisting of a driven element that is typically a folded dipole, a parasitic reflector and one or more directors.

APPENDIX G. MANUFACTURERS

MANUFACTURERS of SMATV COMPONENTS

North America

AML SPECIALTY
5482 Complex Street, Suite 108
San Diego, CA 92123
Telephone: 619-569-7425
Microwave transmitters

ANTENNA TECHNOLOGY CORP.
1140 East Greenway
Mesa, AZ 85203
Telephone: 602-264-7275
FAX: 602-898-7667

AUGUT BROADBAND
1311 Commerce Lane
Jupiter, FL 33458
Telephone: 800-327-6690
Distribution amplifiers

AVCOM of VIRGINIA
500 Southlake Boulevard
Richmond, VA 23236
Telephone: 804-794-2500
FAX: 804-794-8284
Spectrum analyzers, other test equipment

BLONDER-TONGUE LABORATORIES
One Jake Brown Road
Old Bridge, NJ 08857
Telephone: 908-679-4000
FAX: 908-679-4353
Headend elec. / line distribution equipment

CALIFORNIA AMPLIFIER
460 Calle San Pablo
Camarillo, CA 93012
Telephone: 805-987-9000
FAX: 805-987-8359
Downconverters, LNBs

CHANNEL MASTER
P.O. Box 1416
Industrial Park Drive
Smithfield, NC 27577
Telephone: 919-934-9711
FAX: 919-934-9711
Satellite/off-air antennas, microwave links

CHANNEL PLUS
MULTIPLEX TECHNOLOGY
3200 East Birch
Brea, CA 92621
Telephone: 800-999-5225
              714-996-4100
Modulators, amps, distrib./control equip.

COMTECH ANTENNA CORPORATION
3100 Communications Road
St. Cloud, FL 32769
Telephone: 892-6111
Satellite antennas

DH SATELLITE
P.O. Box 239
Prairie du Chien, WI 53821
Telephone: 608-326-6041
Satellite antennas

DX COMMUNICATIONS
10 Skyline Drive
Hawthorne, NY 10532
Telephone: 914-347-4040
Headend equipment

ELECTROLINE
8750 8th Avenue
Montreal, PQ H1Z 2W4, Canada
Telephone: 514-374-6335
              800-461-3344
FAX: 514-374-9370
Addressable subscriber control equip.

GENERAL INSTRUMENTS
2200 Byberry Road
Hatboro, PA 19040
Telephone: 215-674-4800
Receiver, modulators & line distribution eqp.

HOLLAND ELECTRONICS CORP.
5308 Derry Avenue, Suite W
Agoura Hills, CA 90301
Telephone: 818-597-0015
Headend and distribution equipment

HUGHES AIRCRAFT COMPANY
P.O. Box 2940
Mail Stop 243/1218
Torrance, CA 90509
Telephone: 213-517-6233
18 GHz transmission equipment

INTERNATIONAL SATELLITE SYSTEMS
104 Constitution, #4
Menlo Park, CA 94026
Telephone: 800-227-6288
Receiver, modulators and stereo processors

LASER VISION
418-A Stump Road
Montgomery, PA 18936
Telephone: 215-669-6882
Laser and microwave comm. equipment

MAGNAVOX CATV SYSTEMS
100 Fairgrounds Drive
Mamiluius, NY 13104
Telephone: 800-448-5171
Headend and distribution products

APPENDIX G

MICROWAVE FILTER COMPANY
6743 Kinne Street East
East Syracuse, NY 13057
Telephone: 800-448-1666
             315-437-3953
FAX: 315-463-1467
Filters for satellite and wireless reception

MICROWAVE RADIO CORPORATION
20 Alpha Road
Chelmsford, MA 01824
Telephone: 508-250-1110
Microwave transmitters, receivers
and antennas

NEXUS ENGINEERING CORPORATION
7725 Lougheed Highway
Burnaby, BC V5A 1W8
Canada
Telephone: 604-420-5322
FAX: 604-420-5941
Headend electronics

NORSAT INTERNATIONAL
302 - 12886 78th Avenue
Surrey, BC V3W 8E7
Canada
Telephone: 604-597-6200
FAX: 604-597-6214
Sat. receivers, modulators, downconverters

ORBITRON
351 South Peterson Street
Spring Green, I 53588
Telephone: 608-588-2933
Satellite antennas

PICO MACOM
12500 Foothill Blvd.
Lakeview Terrace, CA 91342
Telephone: 800-421-6511
             818-897-0028

QUINTECH ELECTRONICS & COMM.
650 South 13th Street
Indiana, PA 15701
Telephone: 412-357-6294
Amplifiers, converters and subscriber control
equipment

RESEARCH CONCEPTS
10679 Widmer
Lenexa, KS 66215
Telephone: 913-469-4125
Antenna controllers

R.L. DRAKE COMPANY
P.O. Box 3006
Miamisburg, OH 45343
Telephone: 513-866-2421
FAX: 513-866-0806
Satellite receivers/decoders

ROHN
P.O. Box 2000
Peoria, IL 61656
Telephone: 309-697-4400
Towers and shelters

SCIENTIFIC ATLANTA
P.O. Box 105027
Atlanta, GA 30348
Telephone: 404-903-5099
Receivers, modulators, satellite
antennas and distribution

SENCORE
3200 Sencore Drive
Sioux Falls, SD 57107
Telephone: 800-736-2673
             605-339-0100
FAX: 605-335-6379

STANDARD COMMUNICATIONS
P.O. Box 92151
Los Angeles, CA 90009-2151
Telephone: 800-243-1357

SUPERIOR SATELLITE ENGINEERS
P.O. Box 2549
Roseville, CA 95746
Telephone: 916-791-3315
Satellite antennas

TELESCRIPT INDUSTRIES
6735 Odessa Avenue
Van Nuys, CA 91406
Telephone: 818-989-7979
Infrared transmitters

TIMES FIBER COMMUNICATIONS
358 Hall Avenue
Wallingford, CT 66492
Telephone: 800-243-6904
Cable, fiber optics, off-premises
addressable converters

TONER CABLE EQUIPMENT
969 Horsham Road
Horsham, PA 19044
Telephone: 215-675-2053
FAX: 215-675-7543
Wide variety of SMATV equipment

WESTEC COMMUNICATIONS
14405 North Scottsdale Road
Scottsdale, AZ 85254
Telephone: 602-948-4484
18 GHz communication system

APPENDIX G

MANUFACTURERS of WIRELESS COMPONENTS

ALUMA TOWER COMPANY
1639 Old Dixie Highway, Box 2806
Vero Beach, FL 32961-2806
Telephone: 407-567-3423
FAX: 407-567-3432

ANDREW CORPORATION
10500 West 153rd Street
Orland Park, IL 60462
Telephone: 708-349-3300
FAX: 708-349-5442

BOGNER BROADCAST COMPANY
603 Cantiague Rock Road
Westbury, NY 11590
Telephone 516-997-7800
FAX: 516-997-7721

CALIFORNIA AMPLIFIER
460 Calle San Pablo
Camarillo, CA 93012
Telephone: 805-987-9000
FAX: 805-987-8359

CHANNEL MASTER
P.O. Box 1416
Industrial Park Drive
Smithfield, NC 27577
Telephone: 919-934-9711
FAX: 919-934-0380

COMBAND TECHNOLOGIES
1122 Executive Blvd.
Chesapeake, VA 23320-3636
Telephone: 804-436-0333

COMWAVE
Crestwood Industrial Park, Box 69
Mountaintop, PA 18707
Telephone: 717-474-6751
FAX: 717-474-5469

CED
2500 N.W. 39th Street
Miami, FL 33142
Telephone: 305-633-8020
FAX: 305-635-5445

CONFIER CORPORATION
P.O. Box 1025
1400 Roosevelt
Burlington, IA 52601
Telephone: 319-752-3607 or 800-843-5419
FAX: 319-753-5508

EMCEE BROADCAST PRODUCTS
P.O. Box 68, Susquehanna Street Ext.
White Haven, PA 18661
Telephone: 717-443-9575 or 800-233-6193
FAX: 717-443-9257

FUTURE VISION
3015 Compton Court
Alpharetta, GA 30203
Telephone: 404-263-6677
FAX: 404-263-6647

JERROLD COMMUNICATIONS - TOCOM
P.O. Box 569240
Dallas, TX 75356-9240
Telephone: 214-438-7691

ITS CORPORATION
375 Valley Brook Road
McMurray, PA 15317
Telephone: 412-941-1500
FAX: 412-941-4603

LANCE INDUSTRIES
13001 Bradley Avenue
Sylmar, CA 91342
Telephone: 818-367-1811
FAX: 818-362-3594

LINDSAY SPECIALTY PRODUCTS
50 Mary Street West
Lindsay, Ontario K9V 4S7
Canada
Telephone: 800-465-7046

MICROWAVE FILTER COMPANY
6743 Kinne Street
East Syracuse, NY 13057
Telephone: 800-448-1666 or 315-437-3953
FAX: 315-463-1467

NEXUS ENGINEERING CORPORATION
7725 Lougheed Highway
Burnaby, BC V5A 4K4
Canada
Telephone: 604-420-5322
FAX: 604-939-8720

NORSAT INTERNATIONAL
302 - 12886 78th Avenue
Surrey, BC V3W 8ET
Canada
Telephone: 604-597-6200

PACIFIC MONOLITHICS
245 Santa Ana Court
Sunnyvale, CA 94086
Telephone: 408-732-8000

PICO MACOM
12500 Foothill Blvd.
Lakeview Terrace, CA 91342
Telephone: 818-897-0028
800-421-6511
FAX: 818-899-1165

RONARD INDUSTRIES
P.O. Box 708
Michigan City, IN 46360
Telephone: 219-383-2700
FAX: 219-872-6681

TEXSCAN CORPORATION
10841 Pellicano Drive
El Paso, TX 79935
Telephone: 800-351-2345
915-594-3555

VISTA VISION
296 Wyoming Avenue
Wyoming, PA 18644
Telephone: 717-693-4221

ZENITH CABLE PRODUCTS
1000 North Milwaukee Avenue
Glenview, IL 60025
Telephone: 708-391-7702
FAX: 708-391-8569

THE AFFORDABLE STANDARD REFERENCE BOOK
1996/97 WORLD SATELLITE YEARLY
Footprints, Programming and Technical Use

This affordable book, written in the concise style that has become the trademark of Dr. Frank Baylin, provides the information required to easily determine the satellite programming available and the equipment necessary to receive these satellite signals from any point on our globe. It is organized into five easy-to-use sections that are separated by tabs.

SECTION I outlines required:
- METHODS to receive satellite audio and video signals
- to interpret satellite footprints
- to size antennas and select equipment
- to aim an receive dish
- AS WELL AS an overview of worldwide broadcast standards

SECTION II presents:
- the latest in audio and video compression
- an overview of encryption technology as well as a study of security methods and where we are heading

SECTION III presents:
- a complete listing of footprint maps and other information about over 350 worldwide satellites – past, present and planned
- easy-to-read and accurate footprints of all active satellites

SECTION IV lists:
- programming available on each active satellite
- world video standards and scrambling systems

SECTION V lists:
- Addresses and telephone numbers of satellite manufacturers, service companies, programmers and major satellite system operators

Over 800 pages / 8-1/2 x 11" / illustrations / glossary / index / ISBN 0-917893-26-3

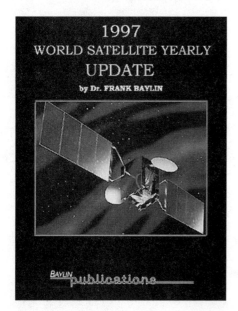

WORLD SATELLITE YEARLY 1997 UPDATE
by Dr. Frank Baylin

All the latest information including footprints and characteristics of over 25 newly launched satellites as well as the latest in worldwide video transponder loading. This book is organized into two easy-to-use sections that are separated by tabs.

SECTION I includes:
- A complete listing of footprint maps and other information describing over 25 newly launched satellites
- Footprints of satellites, such as EchoStar, not previously available in our *1996/97 World Satellite Yearly*

SECTION II includes:
- A complete listing of video programming available on every active satellite in the world.

288 pages / 8-1/2 by 11" / illustrations / glossary / ISBN 0-917893-30-1

DIGITAL SATELLITE TV
by Dr. Frank Baylin

A clear explanation of all aspects of worldwide digital satellite television. This completely rewritten 5th edition of the classic book Ku-Band Satellite TV is expanded with additional sections on the Internet and data transmissions, digital modulation methods, video and audio compression, MPEG-2 and DVB standards, design of digital satellite receivers, worldwide scrambling methods, digital link analysis, IF distribution systems for the new digital broadcasts, mobile applications and much more...

This comprehensive book is organized into 16 complete chapters that, among other topics, include:

- ✓ The history and background of satellite communications
- ✓ Communication fundamentals, analog and digital modulation, and frequency allocations
- ✓ Analog television operation, NTSC/PAL/SECAM broadcast standards, and mono and stereo audio processing methods
- ✓ Digital compression, MPEG-2 systems/transport layers, the DVB standard, modulation, multiplexing and error correction, the future of MPEG-2 DVB
- ✓ Program sources and uplink operation
- ✓ Satellite design, operation, launching and stationkeeping, and future trends
- ✓ Operation of dishes, mounts, actuators, feeds and polarization selection, LNBs, downconversion methods, LNB electronic design and operation, judging receiver performance, customer interfaces and modulators
- ✓ The design of both analog and digital satellite receivers
- ✓ History, structure and operation of the INTERNET, uncorking the bottleneck and the satellite INTERNET providers
- ✓ Evaluating receive components and an overview of encryption and scrambling
- ✓ Detailed installation methods for fixed and tracking dish systems
- ✓ Retrofitting older systems to upgrade from C- to Ku-band, from analog to digital, and from single to dual-band reception
- ✓ Analog and digital link analysis, adjacent satellite interference and dish sizing
- ✓ Components of TV distribution systems, the headend, conventional and IF methods
- ✓ Marine and terrestrial, mobile applications and installation methods
- ✓ Satellite reception problems, troubleshooting and repair methods

KNOWLEDGE is the KEY to your FUTURE!

480 pages / 8.5 by 11 inches / over 400 photos, illustrations, table and charts / appendices / 0-917893-22-0

The GPS MANUAL
Principles and Applications
by Steve Dye with Dr. Frank Baylin

The Global Positioning System (GPS) allows anyone equipped with a suitable receiver and equipment to locate his or her position anywhere on the globe. Originally intended by the U.S. Department of Defense as a means of accurately guiding their missiles to their targets, GPS has since then assumed many more roles. By releasing a frequency and an access code for civilian use, a plethora of applications have appeared, thus creating a billion dollar industry. GPS is in widespread use and continues to find new applications every day. Some applications are as diverse as they are ingenious, some save money, some save lives.

This book, written in a style understandable by any technically interested person, explores GPS technology and its applications. It serves as a useful guide for people wishing to familiarize themselves with modern technology as well as for those considering a receiver for personal use. This clear book also serves as a useful tool for corporate business managers considering adopting GPS as a business strategy but needing a background knowledge of GPS.

This book examines the principles behind GPS and points the reader to the variety of applications GPS creates. By using simple, clearly written text accompanied by many concise diagrams the basic principles are easily understood, educating the reader thoroughly.

BE on the CUTTING EDGE

- Chapter 1: Brief history of GPS and a look at other navigational systems including terrestrial networks as well as an introduction to satellite communications basics
- Chapter 2: Navigational methods used prior to GPS with an explanationof earlier methods.
- Chapter 3: GPS Principles. The basic principles behind GPS, the use of time of arrival for pseudo ranging, pseudo random codes, signal correlation, velocity measurement, factors affecting accuracy, receiver basics, sequence of operation and more.
- Chapter 4: Improving the Accuracy : Differential GPS, carrier phase techniques precision orbit predictions, Wide Area Augmentation System.
- Chapter 5: Industrial, professional and personal applications of GPS including: Avionics, marine, agricultural, GIS, time referencing, surveying and more.
- Chapter 6: GPS products. An overview of the market and an explanation of the various types of GPS receiver available ranging from lower end handheld receivers to expensive receivers used for surveying and guiding aircraft. Photographs and specifications of these products featured.
- Chapter 7: Choosing a Personal GPS receiver. What to look for and what you are getting.
- Chapter 8: GPS manufacturer and vendor addresses and contacts, Internet addresses and web sites.
- Chapter 9: GPS Dictionary – an A-Z listing of GPS terms and meanings.

NEW BUSINESS OPPORTUNITIES!

248 pages / 8-1/2 by 11" / figures, photos, tables / ISBN 0-917893-29-8

EUROPEAN SCRAMBLING SYSTEMS
THE BLACK BOOK - New 5th Edition!

by John McCormac

The "bible" of the black arts of signal security analyzes all of the latest hacks as well as satellite broadcasting scrambling and encryption systems. While emphasizing European systems, this 576-page book features schematics and some circuit diagrams applicable to systems worldwide.
Topics include:

- The principles of security and descrambler building blocks
- Audio and video scrambling techniques
- Dirty tricks and information acquisition
- The information wars
- Smart cards and stupid mistakes
- Cracking the code (DES and RSA)
- Video manipulative scrambling systems
- Cable and MMDS distribution
- Descrambler hook-ups
- The latest on the **DirecTv** DDS system hack INCLUDING source code

THE INTERNATIONAL BEST SELLING 'HACKER'S BIBLE'

576-pages / 5-3/4" x 8-1/4" / photos, illustrations, tables / glossary / ISBN 1-873556-22-5

1997 PRIVATE TELECOMMUNICATIONS DIRECTORY

The 1997 Private Telecommunications Directory is the source for information on cable and telephone operators serving multiple dwelling unit properties such as apartments, condominiums, retirement communities and mobile home parks as well as colleges, universities, hotels, motels, hospitals and prisons. With multiple sorts, this easy-to-use directory includes specialized industry data on operators that will be useful as a year-round resource. Listings are also included for suppliers, programmers and consultants to the private cable and telephone industries.

148 pages / 8-1/2 by 11" / photos, illustrations

ALL THE DETAILS of MMDS INSTALLATION

THE MMDS
(Wireless Cable)
INSTALLATION MANUAL

by Baylin, Berkoff and Hatfield

The most thorough installation manual on MMDS available, this book has all the details. Written by three seasoned and experienced experts it includes:

- An overview of MMDS history and technology
- Company policies and operational practices that lead to success
- Operational practices of an efficient installation dispatch and control system
- The details of technical basics such as cutting fittings, weatherproofing and grounding
- The lastest OSHA safety regulations
- Customer relations and creating goodwill
- Detailed receive-site installation practices - interior and exterior
- Quality control procedures
- Service and troubleshooting methods
- Safety issues and practices
- A clear overview of headend operation

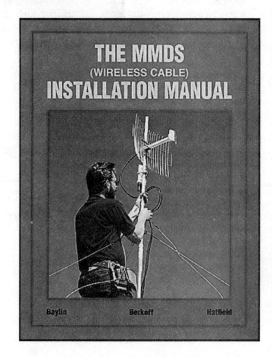

230 pages / 8.5 x 11 / photos, illustrations, diagrams, tables/appendices, glossary / ISBN 0-917893-28-X

The Perfect Mate for "Wireless Cable and SMATV"

A MUST for EVERY MMDS INSTALLER in the WORLD!

A copy of this manual should be in the hands of each and every installer in every MMDS system throughout the world. Every detail is spelled out and excellently illustrated with photos, tables and charts.

SATfinder CD-ROM

COMPANY INFORMATION lists over 10,000 companies involved in the satellite communications industry. Search from among 31 categories of companies or by company name for a particular company. Search by product or service and then get full details including contact names and numbers.

SATELLITE INFORMATION lists over 450 planned and operational satellites. Details on owner/operator, manufacturer, launch information, bandwidth & operating frequencies, power levels and many other important details.

MAP INFORMATION gives the EIRP, G/T and SFD of most operational satellites. There are over 500 maps - decide on which satellite or country for which you wish to see coverage and then get the full details.

WHAT'S on SATELLITE list the video activity on over 100 satellites. Frequencies, bandwidth, EIRP, video format and many other details provided.

STANDARDS & STATISTICS gives the broadcast standards and basic statistical information on most countries in the world. Uplink Information gives the equipment specifications and contact information necessary to do a satellite uplink from anywhere in the world.

PLUS SATMASTER! (see more complete description under SATMASTER listing). All necessary reference, system design and analysis tools into a single, easy-to-use package. Satmaster is designed to be very simple to use and works just like any other Windows application with pull-down menus, toolbar for shortcuts, dialog boxes, text and graphics windows, a status bar and a comprehensive help system.

The World's Most Complete Communication Satellite Reference Source

The International Satellite Directory on CD-ROM! and much more...

COST: Year 1
$975 - includes the original CD-ROM plus a new update 6 month later and a copy of the current *International Satellite Directory*. Pay an additional $200 for a total of $1,150 and receive 4 CD-ROMs, the original plus an update every quarter.

COST: Year 2
Only $775 or $975 if you wish the 3 updates during the year.

SHIPPING: $9.50 in North America; Air mail $55 elsewhere

NEW, POWERFUL, EASY-TO-USE & COMPREHENSIVE

SATELLITE TOOLBOX SOFTWARE
Designed by Dr. Frank Baylin

Windows Version 1.0

PROFESSIONAL, EASY-to-USE, MENU-DRIVEN SOFTWARE
Worldwide Satellite Programming
Antenna Aiming & Sizing
Receive-Site Analysis
Graphs, Tables & Reports

SATELLITE DATABASE: Lists of all the world's communication spacecraft and the programming they relay. Select any of the world's satellites to view all the video and audio programming available as well as transponder bandwidths, center frequencies or polarization formats. Or enter site latitude and longitude and print a list of those satellites viewable from your receive site.

AIMING SECTION: To activate simply enter your site coordinates. The software **automatically calculates magnetic variation**, compass heading, azimuth bearing as well as all the aiming angles for a polar or modified polar mount to any satellite within view or any group of satellites that you select. A datafile of major cities worldwide with latitudes and longitudes is provided. In addition, a similar database on every city, town and village in ITU regions I, II and III is available at a small additional cost.

ANTENNA SIZING – Analog or Digital: This component makes dish sizing, choice of LNB noise temperature and other system parameters simple. Choose between digital, NTSC, PAL or other broadcast formats to calculate G/T, S/N, C/N, etc... Print detailed analyis reports with your business name as the source. Save an unlimited number of files for each customer site anywhere on the globe.

ANALYSIS – Analog or Digital: This component allows you to easily calculate virtually every parameter relating to satellite reception including antenna gain, side lobe gain, focal distance as well as path loss, slant distance, G/T, C/N, video and audio S/N and much more. Create graphs or tables of how any one variable affects the outcome of any other. How would changing antenna size or LNB noise temperature effect audio signal-to-noise ratio? Answer a multidude of other pertinent questions.

SYSTEM REQUIREMENTS
BASIC SYSTEM
Windows 3.0 or higher version
A minimum of 640 kb of RAM
and
3 MB of harddrive space

ADDITIONAL DATABASES
ITU1 – 1.96 MB
ITU2 – 2.13 MB
ITU3 – 3.07 MB

MAIN PROGRAM........$95

CITY DATABASES:
   ITU1, 2 or 3...............$70
   Any 2 databases....$105
   All 3 databases......$120

The program runs without a hitch and the documentation is excellent. In light of the wide range of calculations and data, the program ... would be a bargain at twice Baylin Publication's price –
Satellite Retailer Magazine

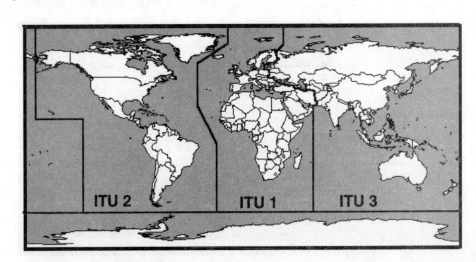

SATMASTER
THE ADVANCED SOFTWARE DESIGN TOOL

Used by broadcasters, SNG operators, dish installation companies and dealers serving the satellite TV industry. Written in C++ for speedy operation, dish aiming, link budget analysis and solar outage prediction provide the backbone of the package. The link budget facilities employ rain attenuation and atmospheric absorption prediction modeling even for low elevations. These and other useful features, such as an integral magnetic variation calculation and tens of thousands of stored town/city co-ordinates, combine to enable a system to be either designed for any global location. Thus Satmaster for Windows integrates all necessary reference, system design and analysis tools into a single easy-to-use package.

Designed to be **very simple to use**, it works just like any other Windows application with pull-down menus, toolbar for shortcuts, dialog boxes, text and graphics windows, a status bar and a comprehensive help system. **Satmaster also:**

- ✓ Prediction of accurate global solar outage, presenting annual lists of data and time windows. All international time zones are supported with convenient reference data.
- ✓ Accurately locates north/south bearing anywhere in the world for polar mounts by producing selected monthly tabulation of solar transit time.
- ✓ Creates monthly tables of time when the sun and a selected satellite have identical azimuth. This enables a fixed dish to be installed where local magnetic anomalies may affect compass readings.
- ✓ Plot graphs of parabolic surfaces clearly showing focal distance and calculates focal length.
- ✓ Calculate elevation, true azimuth, compass bearing and skew angles for any geostationary satellite for any specific location. Automatically calculations magnetic variation. Throw all those out of date maps away!
- ✓ Generation of tables of essential 'look angles' for fixed and motorized systems for multiple locations.
- ✓ Generates graphs and tables relating to digital modulation, rain attenuation, atmospheric absorption and more.
- ✓ Handles all common modulation schemes including BPSK, QPSK, 8-PSK or 16QAM with any FEC code rate. A useful option is provided such that for a given input back-off total the corresponding output back-off total and intermodulation interference may be estimated from the stored parameters of a 'typical' TWTA or an SSPA transponder. Ideal software for SMG, Ku trucks, data transmission and radio station feeds.
- ✓ Calculation of analog and digital downlink budgets as well as detailed up-down SCPC digital calculations. Takes co-channel, adjacent satellite and transponder intermodulation interference into account. Finds uplink power requirements, transponder bandwidth and power usage per carrier. Atmospheric losses and appropriate rain fade margins are automatically calculated, anywhere in the world, according to desired link availability. Use information rates from 56 kbps to 40 Mbps.
- ✓ Calculates short form downlink budgets for both analog and digital modulated TV broadcast signals. Ideal for TVRO dish sizing! Fast minimum dish size optimize for any selected target value of C/N, C/N_o, S/N CCIR Grade, E_b/N_o or LNB output level. Can home in on either 'clear sky' or 'degraded sky' target values. Alters any link parameter and instantly to see the result!. Find LNB output carrier levels and the corresponding antenna size needed to design small IF distribution systems.
- ✓ Supplied with a listing of tens of thousands of town and city records that covers virtually every country in the world. Full editable database for extensions if needed.
- ✓ Display world and regional maps of CCIR rain-climatic zones, seasonal water vapor density and mean surface temperatures (necessary for detailed link analysis). Partially completed forms are preserved if other reference data must be consulted.
- ✓ Printing of professional looking reports to any Windows compatible printer, with your company name at the header.
- ✓ Multiple windows that may be opened to mixing graphs, tables, link budgets and maps.
- ✓ Generation of beamwidth and lobe patterns for uniformly illuminated antennas for judging the likelihood of potential interference from adjacent satellites.
- ✓ Complete with a wide selection of popular footprint maps. However, you may use another application and a scanner to generate more. The program displays *.bmp and *.wmf files.
- ✓ All link budget modules calculate accurate rain depolarization, atmospheric absorption, tropospheric scintillations and rain fade margins with corresponding noise increase on the downlink. Calculations are performed for any desired signal availability and earth-space path and apply even at low elevations.
- ✓ An in-built conventional text editor for project notes.
- ✓ Input data from previous activated parameter file is automatically transferred to all supplementary dialogs to reduce from filled requirements. A 'Clear' button is provide if previous data is not needed.
- ✓ Performance of useful calculations and conversions, and includes an expression evaluator for the more complex calculations.
- ✓ All output can be directly pasted into common spreadsheet packages, as tab delimited text, for editing, customization or to be included in other documents.
- ✓ 40,000+ word context sensitive help file packed with basic information and theory on all input data fields, accessed by pressing F1. This gives advice on on typical values to enter such as those for co-channel and adjacent satellite interference.

Baylin Publications
1905 Mariposa
Boulder, CO 80302, USA
Telephone: 303-449-4551
FAX: 303-939-8720
E-Mail: fbaylin@baylin.com

Distributed from England by:
Swift Television Publications
17 Pittsfield
Cricklade, Wiltshire SN6 6AN, U.K.
Telephone: 44 (0) 1793-750620
FAX: 44 (0) 1793-752399
U.K. Master Distributor: 44 (0) 118-9414468

ITEM	PRICE $	PRICE £	NUMBER	TOTAL
World Satellite TV & Scrambling Methods – 3rd edition	$40	£29		
The GPS Manual - Principles and Applications	40	30		
Digital Satellite Television (5th edition Ku-Band Satellite TV)	50	36		
The Digital Revolution - Satellite Television Manual	40	27		
Satmaster Pro for Windows	195	99		
Satellite Toolbox Software ❑ DOS ❑ Windows	95	59		
– ITU Region 1, 2 and 3 Databases	See page 384 for prices			
Update to Satellite Toolbox Databases	20	NA		
Wireless Cable and SMATV	50	35		
The MMDS Installation Manual	40	27		
Satellite & Cable TV Scrambling – 2nd edition	20	21		
Home Satellite TV Manual, English – 4th revised edition	30	25		
Install, Aim & Repair Your Satellite TV System – 2nd edition	10	12		
Satellite TV Installation Video - English	40	27		
Install Your Digital Satellie TV System – Video	20	NA		
Home Satellite TV Manual ❑ Spanish ❑ Portuguese	30	25		
The "How To" Book of Satellite Communications – 2nd edition	25	22		
1996/97 World Satellite Yearly	90	59		
1997 World Satellite Yearly Update	60	39		
The Private Telecommunications Directory	190	NA		
What's On Satellite? (3 issues) Call or E-mail for shipping cost	95	NA		
1997 International Satellite Directory	260	173		
European Scrambling Systems - 5th edition	60	34		
Miniature Satellite Dishes – The New Digital TV – 2nd edition	20	14		
The Satellite Book – 4th edition	50	32		
SATfinder CD-ROM	see page 383 for prices			
Method of Payment / Shipping Method	Total for this page			
	Merchandise Order Total – from previous page			
	Sales Tax – Colorado Residents only (3.7%)			
	Shipping (see below for detailed costs)			
	Grand Total			

Method of Payment:
Check Enclosed ❑
Cash ❑
Credit Card ❑
COD ❑

Shipping Method:
Air Mail ❑
Surface ❑
UPS Blue ❑
UPS Red ❑

Visa, Mastercard & Amex – U.S. Office
Visa, Mastercard, Eurocard & Access – U.K. Office

SHIPPING INSTRUCTIONS

to UNITED STATES: Add $4.00 for each item ordered. UPS regular, 2nd day or overnight may be requested at additional charge. UPS COD orders are an additional $5.00. MasterCard, Visa and Amex cards accepted.

to CANADA: Add $5.00 per item ordered. Please remit funds in U.S.$ or by credit card. Orders are shipped via book rate mail. UPS shipping can be requested. No COD is available to Canada.

WORLDWIDE (from our US office): Remit payment in U.S. funds as checks drawn on U.S. banks, money orders, Master, Visa or Amex credit cards, or cash. Shipments are by surface mail unless additional air mail charges can be added onto credit card billing.

from UK OFFICE: Add £3 postage per book within UK. Add £5 per book to Europe; beyond Europe add 30%. Optional insurance £4 extra. Remit payment in UK £ Sterling drawn on UK bank, Eurocheque, Postal Order,

ORDER TODAY
(Quantity Discounts Available)

Name:_____

Company:_____

Address:_____

_____

_____

Country:_____

Telephone:_____

FAX:_____

Credit Card No.:_____

Expiration Date:_____

Signature:_____